Lecture Notes in Computer Science 1645

Edited by G. Goos, J. Hartmanis and J. van Leeuwen

Springer

Berlin
Heidelberg
New York
Barcelona
Hong Kong
London
Milan
Paris
Singapore
Tokyo

Maxime Crochemore   Mike Paterson (Eds.)

Combinatorial Pattern Matching

10th Annual Symposium, CPM 99
Warwick University, UK, July 22-24, 1999
Proceedings

 Springer

Series Editors

Gerhard Goos, Karlsruhe University, Germany
Juris Hartmanis, Cornell University, NY, USA
Jan van Leeuwen, Utrecht University, The Netherlands

Volume Editors

Maxime Crochemore
Institute Gaspard-Monge, University of Marne-la-Vallée
F-77454 Marne-la-Vallée Cedex 2, France
E-mail: Maxime.Crochemore@univ-mlv.fr

Mike Paterson
Department of Computer Science, University of Warwick
Coventry, CV4 7AL, England
E-mail: Mike.Paterson@dcs.warwick.ac.uk

Cataloging-in-Publication data applied for

Die Deutsche Bibliothek - CIP-Einheitsaufnahme

Combinatorial pattern matching : 10th annual symposium ;
proceedings / CPM 99, Warwick, UK, July 22 - 24, 1999. Maxime
Crochemore ; Mike Paterson (ed.). - Berlin ; Heidelberg ; New York ;
Barcelona ; Hong Kong ; London ; Milan ; Paris ; Singapore ; Tokyo
: Springer, 1999
  (Lecture notes in computer science ; Vol. 1645)
  ISBN 3-540-66278-2

CR Subject Classification (1998): F.2.2, I.5.4, I.5.0, I.7.3, H.3.3, E.4, G.2.1

ISSN 0302-9743
ISBN 3-540-66278-2 Springer-Verlag Berlin Heidelberg New York

© Springer-Verlag Berlin Heidelberg 1999
Printed in Germany

Typesetting: Camera-ready by author
SPIN: 10703545   06/3142 – 5 4 3 2 1 0   Printed on acid-free paper

Foreword

The papers contained in this volume were presented at the Tenth Annual Symposium on Combinatorial Pattern Matching, held July 22 – 24, 1999 at the University of Warwick, England. They were selected from 26 abstracts submitted in response to the call for papers. In addition, invited lectures were given by Joan Feigenbaum from AT&T Labs Research (*Massive graphs: algorithms, applications, and open problems*) and David Jones from the Department of Biology, University of Warwick (*Optimizing biological sequences and protein structures using simulated annealing and genetic algorithms*).

The symposium was preceded by a two-day summer school set up to attract and train young researchers. The lecturers of the school were Alberto Apostolico (*Computational Theories of Surprise*), Joan Feigenbaum (*Algorithmics of network-generated massive data sets*), Leszek Gasieniec and Paul Goldberg (*The complexity of gene placement*), David Jones (*An introduction to computational molecular biology*), Arthur Lesk (*Structural alignment and maximal substructure extraction*), Cenk Sahinalp (*Quest for measuring distance between strings: exact, approximate, and probabilistic algorithms*), and Jim Storer.

Combinatorial Pattern Matching (CPM) addresses issues of searching and matching strings and more complicated patterns such as trees, regular expressions, graphs, point sets, and arrays. The goal is to derive non-trivial combinatorial properties of such structures and to exploit these properties in order to achieve superior performance for the corresponding computational problems.

Over recent years, a steady flow of high-quality research on this subject has changed a sparse set of isolated results into a fully-fledged area of algorithmics. This area is continuing to grow even further due to the increasing demand for speed and efficiency that comes from important and rapidly expanding applications such as the World Wide Web, computational biology, and multimedia systems, involving requirements for information retrieval, data compression, and pattern recognition. The objective of the annual CPM gatherings is to provide an international forum for research in combinatorial pattern matching and related applications.

The general organisation and orientation of CPM conferences is coordinated by a steering committee composed of A. Apostolico, M. Crochemore, Z. Galil, and U. Manber.

The first nine meetings were held in Paris (1990), London (1991), Tucson (1992), Padova (1993), Asilomar (1994), Helsinki (1995), Laguna Beach (1996), Aahrus (1997), and Piscataway (1998). After the first meeting, a selection of papers appeared as a special issue of *Theoretical Computer Science* in volume 92. The proceedings of the third to ninth meetings appeared as volumes 644, 684, 807, 937, 1075, 1264, and 1448 of the present LNCS series at Springer.

CPM'99 was organised by Cenk Sahinalp of the Department of Computer Science at Warwick University. The conference was supported in part by MATHFIT (a joint programme of EPSRC and the London Mathematical Society).

May 1999   M. Crochemore, M. Paterson

Programme Committee

Maxime Crochemore, *co-chair*
Leszek Gasieniec
Roberto Grossi
Tao Jiang
Heikki Mannila
Rajeev Motwani
Gene Myers
Chris Overton

Mike Paterson, *co-chair*
Cenk Sahinalp
Dan Spielman
Jim Storer
Kiem-Phong Vo
Moti Yung
Jacob Ziv

Additional Referees

The following external referees who helped with the selection of papers for CPM'99 are gratefully acknowledged:

James Abello, Cyril Allauzen, Amihood Amir, Sabria Benhamida, Adam Buchsbaum, R. DePrisco, Ramesh Hariharan, Piotr Indyk, Gad Landau, A. Moffat, S. Muthukrishnan, Kunsoo Park, R. Ravi, Giuseppina Rindone, Mikkel Thorup, Zdenek Tronicek, Suresh Venkatasubramanian, Marc Zipstein

Organising Committee

S. Cenk Sahinalp, *chair*
Leslie Goldberg
Paul Goldberg
Graham Cormode

Jonathan Sharp
Nasir Rajpoot
Mary Cryan
Hesham Al-Ammal

Table of Contents

Shift-And Approach to Pattern Matching in LZW Compressed Text

Takuya Kida, Masayuki Takeda, Ayumi Shinohara, and Setsuo Arikawa

Department of Informatics, Kyushu University 33
Fukuoka 812-8581, Japan
{kida, takeda, ayumi, arikawa}@i.kyushu-u.ac.jp

Abstract. This paper considers the Shift-And approach to the problem of pattern matching in LZW compressed text, and gives a new algorithm that solves it. The algorithm is indeed fast when a pattern length is at most 32, or the word length. After an $O(m + |\Sigma|)$ time and $O(|\Sigma|)$ space preprocessing of a pattern, it scans an LZW compressed text in $O(n + r)$ time and reports all occurrences of the pattern, where n is the compressed text length, m is the pattern length, and r is the number of the pattern occurrences. Experimental results show that it runs approximately 1.5 times faster than a decompression followed by a simple search using the Shift-And algorithm. Moreover, the algorithm can be extended to the generalized pattern matching, to the pattern matching with k mismatches, and to the multiple pattern matching, like the Shift-And algorithm.

1 Introduction

Pattern matching in compressed text is one of the most interesting topics in the combinatorial pattern matching. Several researchers tackled this problem. Eilam-Tzoreff and Vishkin [8] addressed the run-length compression, and Amir, Landau, and Vishikin [6], and Amir and Benson [2, 3] and Amir, Benson, and Farach [4] addressed its two-dimensional version. Farach and Thorup [9] and Gąsieniec, et al. [11] addressed the LZ77 compression [18]. Amir, Benson, and Farach [5] addressed the LZW compression [16]. Karpinski, et al. [12] and Miyazaki, et al. [15] addressed the straight-line programs. However, it seems that most of these studies were undertaken mainly from the theoretical viewpoint. Concerning the practical aspect, Manber [14] pointed out at CPM'94 as follows.

> It is not clear, for example, whether in practice the compressed search in [5] will indeed be faster than a regular decompression followed by a fast search.

In 1998 we gave in [13] an affirmative answer to the above question: We presented an algorithm for finding multiple patterns in LZW compressed text, which is a variant of the Amir-Benson-Farach algorithm [5], and showed that in practice the algorithm is faster than a decompression followed by a simple search.

M. Crochemore, M. Paterson (Eds.): CPM'99, LNCS 1645, pp. 1–13, 1999.

Namely, it was proved that pattern matching in compressed text is not only of theoretical interest but also of practical interest. We believe that fast pattern matching in compressed text is of great importance since there is a remarkable explosion of machine readable text files, which are often stored in compressed forms.

On the other hand, the Shift-And approach [1, 7, 17] to the classical pattern matching is widely known to be efficient in many practical applications. This method is simple, but very fast when a pattern length is not greater than the word length of typical computers, say 32. In this paper, we apply this method to the problem of pattern matching in LZW compressed text and then give a new algorithm that solves it. Let m, n, r be the pattern length, the length of compressed text, and the number of occurrences of the pattern in the original text, respectively. The algorithm, after an $O(m + |\Sigma|)$ time and $O(|\Sigma|)$ space preprocessing of a pattern, scans a compressed text in $O(n + r)$ time using $O(n+m)$ space and reports all occurrences of the pattern in the original text. The $O(r)$ time is devoted only to reporting the pattern occurrences. Experimental results on the Brown corpus show that the proposed algorithm is approximately 1.5 times faster than a decompression followed by a search using the Shift-And method. Moreover, the algorithm can be extended to (1) the generalized pattern matching, to (2) the pattern matching with k mismatches, and to (3) the multiple pattern matching.

We assume, throughout this paper, that $m \leq 32$ and that the arithmetic operations, the bitwise logical operations, and the logarithm operation on integers can be performed in constant time.

The organization of this paper is as follows: We briefly sketch the LZW compression method, and the Shift-And pattern matching algorithm. We present our algorithm and discuss the complexity in Section 3. In Section 4, we show the experimental results in comparison with both an LZW decompression followed by a search using the Shift-And method and the previous algorithm presented in [13]. In Section 5 we shall discuss the extensions of the algorithm to the generalized pattern matching, to the pattern matching with k mismatches, and to the multiple pattern matching.

2   Preliminaries

We first define some notation. Let Σ, usually called an *alphabet*, be a finite set of characters, and Σ^* be a set of strings over Σ. We denote the length of $u \in \Sigma^*$ by $|u|$. We call especially the string whose length is 0 *null string*, and denote it by ε. We denote by $u[i]$ the ith character of a string u, and by $u[i:j]$ the string $u[i]u[i+1]...u[j]$, $1 \leq i \leq j \leq |u|$. For a set A of integers and an integer k, let $A \oplus k = \{i + k \mid i \in A\}$ and $k \ominus A = \{k - i \mid i \in A\}$.

In the following subsections we briefly sketch the LZW compression method and the Shift-And pattern matching algorithm.

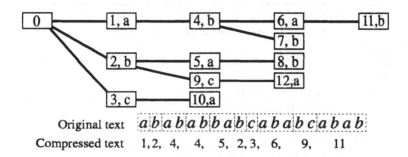

Original text $a\,b\,a\,b\,a\,b\,b\,a\,b\,c\,a\,b\,a\,b\,c\,a\,b\,a\,b$
Compressed text 1, 2, 4,   4,   5, 2, 3, 6,   9,   11

Fig. 1. Dictionary trie.

2.1 LZW Compression

The LZW compression is a very popular compression method. It is adopted as the **compress** command of UNIX, for instance. It parses a text into *phrases* and replaces them with pointers to the *dictionary*. The dictionary initially consists of the characters in Σ. The compression procedure repeatedly finds the longest match in the current position and updates the dictionary by adding the concatenation of the match and the next character. The dictionary is implemented as a trie structure, in which each node represents a phrase in it. The matches are encoded as integers associated with the corresponding nodes of the dictionary trie. The update of the dictionary is executed in $O(1)$ time by creating a new node labeled by the next character as a child of the node corresponding to the current match.

Figure 1 shows the dictionary trie for the text *ababbabcababcabab*, assuming the alphabet $\Sigma = \{a, b, c\}$. Hereafter, we identify the string u with the integer representing it, if no confusion occurs.

The dictionary trie is removed after the compression is completed. It can be reconstructed from the compressed text. In the decompression, the original text is obtained with the aid of the recovered dictionary trie. This decompression takes linear time proportional to the length of the original text. However, if the original text is not required, the dictionary trie can be built only in $O(n)$ time, where n is the length of the compressed text. The algorithm for constructing the dictionary trie from a compressed text is summarized in Figure 2.

2.2 The Shift-And Pattern Matching Algorithm

The Shift-And pattern matching algorithm was proposed by Abrahamson [1], Baeza-Yates and Gonnet [7], and Wu and Manber [17]. In the following, we present the algorithm according to the notation in [1].

Let $\mathcal{P} = \mathcal{P}[1 : m]$ be a pattern of length m, and $\mathcal{T} = \mathcal{T}[1 : N]$ be a text of length N. For $k = 0, 1, \ldots, N$, let

$$R_k = \{1 \leq i \leq m \mid i \leq k \text{ and } \mathcal{P}[1 : i] = \mathcal{T}[k - i + 1 : k]\}, \qquad (1)$$

Input.   An LZW compressed text $u_1 u_2 \ldots u_n$.
Output.   Dictionary D represented in the form of trie.
Method.
begin
    $D := \Sigma$;
    for $i := 1$ **to** $n - 1$ **do begin**
        if $u_{i+1} \le |D|$ **then**
            let a be the first character of u_{i+1}
        else
            let a be the first character of u_i;
        $D := D \cup \{u_i \cdot a\}$
    end
end.

Fig. 2. Reconstruction of dictionary trie.

and for any $a \in \Sigma$, let

$$M(a) = \{1 \le i \le m \mid \mathcal{P}[i] = a\}. \tag{2}$$

Definition 1. *Define the function* $f : 2^{\{1,2,\cdots,m\}} \times \Sigma \to 2^{\{1,2,\cdots,m\}}$ *by*

$$f(S, a) = ((S \oplus 1) \cup \{1\}) \cap M(a),$$

where $S \subseteq \{1, \cdots, m\}$ *and* $a \in \Sigma$.

Using this function we can compute the values of R_k for $k = 1, 2, \ldots, N$ by

1. $R_0 = \emptyset$,
2. $R_{k+1} = f(R_k, T[k+1])$ $(k \ge 0)$.

For $k = 1, 2, \ldots, N$, the algorithm reads the k-th character of the text, computes the value of R_k, and then examine whether m is in R_k. If $m \in R_k$, then $T[k - m + 1 : k] = \mathcal{P}$, that is, there is a pattern occurrence at position $k - m + 1$ of the text. Note that we can regard R_k as states of the KMP automaton, and f acts as the state transition function.

    When $m \le 32$, we can represent the sets R_k and $M(a)$ as m-bit integers. Then, we can calculate the integers R_k by

1. $R_0 = 0$,
2. $R_{k+1} = ((R_k \ll 1) + 1) \; \& \; M(T[k+1])$ $(k \ge 0)$,

where '$\ll$' and '$\&$' denote the bit-shift operation and the bitwise logical product, respectively. We can get a pattern occurrence if $R_k \& 2^{m-1} \ne 0$. For example, the values of R_k for $k = 0, 1, \ldots$ are shown in Figure 3, where $T = ababababbabcababc$ and $\mathcal{P} = ababc$.

The time complexity of this algorithm is $O(mN)$. However, the bitwise logical product, the bit-shift, and the arithmetic operations on 32 bit integers can be performed at high speed, and thus be considered to be done in $O(1)$ time. Then we can regard the time complexity as $O(N)$ if m is at most 32 (in fact such a case occurs very often).

```
original text:   a  b  a  b  a  b  b  a  b  c  a  b  a  b  c
              a  0  1  0  1  0  1  0  0  1  0  0  1  0  1  0  0
              b  0  0  1  0  1  0  1  0  0  1  0  0  1  0  1  0
       R_k:   a  0→ 0→ 0→ 1→ 0→ 1→ 0→ 0→ 0→ 0→ 0→ 0→ 0→ 0→ 1→ 0→ 0
              b  0  0  0  0  1  0  1  0  0  0  0  0  0  0  0  1  0
              c  0  0  0  0  0  0  0  0  0  0  0  0  0  0  0  0  1
                                                              △
```

Fig. 3. Behavior of the Shift-And algorithm.

The symbol $\triangle$ indicates that a pattern occurrence is found at that position.

3   Proposed Algorithm

We want to design a new pattern matching algorithm that runs on an LZW compressed text and simulates the behavior of the Shift-And algorithm on the original text. Assume that the text is parsed as $u_1 u_2 \ldots u_n$. Let $k_i = |u_1 u_2 \ldots u_i|$ for $i = 0, 1, \ldots, n$. Our idea is to compute only the values of R_{k_i} for $i = 1, 2, \ldots, n$, to achieve a linear time complexity which is proportional not to the original text length N but to the compressed text length n.

Definition 2. *Let $\widehat{f}$ be the function f extended to $2^{\{1,\ldots,m\}} \times \Sigma^*$ by*

$$\widehat{f}(S, \varepsilon) = S \quad \text{and} \quad \widehat{f}(S, ua) = f(\widehat{f}(S, u), a),$$

where $S \subseteq \{1, \cdots, m\}$, $u \in \Sigma^*$ and $a \in \Sigma$.

Lemma 1. *Suppose that the text is $T = xuy$ with $x, u, y \in \Sigma^*$ and $u \neq \varepsilon$. Then,*

$$R_{|xu|} = \widehat{f}(R_{|x|}, u).$$

Proof. It follows directly from the definition of $\widehat{f}$. ∎

Let D be the set of phrases in the dictionary. If we have the values of $\widehat{f}$ for the domain $2^{\{1,\ldots,m\}} \times D$, we can compute the value $R_{k_{i+1}} = \widehat{f}(R_{k_i}, u_{i+1})$ from R_{k_i} and u_{i+1} for each $i = 0, 1, \ldots, n-1$. As shown later, we can perform the computation only in $O(1)$ time by executing the bit-shift and the bitwise logical operations, using the function $\widetilde{M}$ defined as follows.

Definition 3. *For any $u \in \Sigma^*$, let $\widehat{M}(u) = \widehat{f}(\{1, \ldots, m\}, u)$.*

Lemma 2. *For any $S \subseteq \{1, \ldots, m\}$ and any $u \in \Sigma^*$,*

$$\widehat{f}(S, u) = ((S \oplus |u|) \cup \{1, 2, \ldots, |u|\}) \cap \widehat{M}(u).$$

Proof. By induction on $|u|$. It is easy for $u = \varepsilon$. Suppose $u = u'a$ with $u' \in \Sigma^*$ and $a \in \Sigma$. We have, from the induction hypothesis,

$$\widehat{f}(S, u') = ((S \oplus |u'|) \cup \{1, 2, \ldots, |u'|\}) \cap \widehat{M}(u').$$

It follows from the definition of f that, for any $S_1, S_2 \subseteq \{1, 2, \ldots, m\}$ and for any $a \in \Sigma$, $f(S_1 \cap S_2, a) = f(S_1, a) \cap f(S_2, a)$ and $f(S_1 \cup S_2, a) = f(S_1, a) \cup f(S_2, a)$. Then,

$$\widehat{f}(S, u) = (f(S \oplus |u'|, a) \cup f(\{1, 2, \ldots, |u'|\}, a)) \cap f(\widehat{M}(u'), a)$$
$$= ((S \oplus |u|) \cup \{1, 2, \ldots, |u|\}) \cap \widehat{M}(u).$$

∎

Lemma 3. *The function which takes as input $u \in D$ and returns in $O(1)$ time the m-bit representation of the set $\widehat{M}(u)$, can be realized in $O(|D| + m)$ time using $O(|D|)$ space.*

Proof. Since $\widehat{M}(u) \subseteq \{1, \ldots, m\}$, we can store $\widehat{M}(u)$ as an m-bit integer in the node u of the dictionary trie D. Suppose $u = u'a$ with $u' \in D$ and $a \in \Sigma$. $\widehat{M}(u)$ can be computed in $O(1)$ time from $\widehat{M}(u')$ and $M(a)$ when the node u is added to the dictionary trie since $\widehat{M}(u) = f(\widehat{M}(u'), a) = ((\widehat{M}(u') \oplus 1) \cup \{1\}) \cap M(a)$. Since the table $M(a)$ is computed in $O(|\Sigma| + m)$ time using $O(|\Sigma|)$ space and $\Sigma \subseteq D$, the total time and space complexities are $O(|D| + m)$ and $O(|D|)$, respectively. ∎

Now we have the following theorem from Lemmas 1, 2, and 3.

Theorem 1. *The function which takes as input $(S, u) \in 2^{\{1, \ldots, m\}} \times D$ and returns in $O(1)$ time the m-bit representation of the set $\widehat{f}(S, u)$, can be realized in $O(|D| + m)$ time using $O(|D|)$ space.*

Since $|D| = O(n)$, we can perform in $O(n + m)$ time the computation of R_{k_i} for $i = 1, \ldots, n$ by executing the bit-shift and the bitwise logical operations. However, we have to examine whether $m \in R_j$ for every $j = 1, 2, \ldots, N$. For a complete simulation of the move of the Shift-And algorithm, we need a mechanism for enumerating the set $Output(R_{k_i}, u_{i+1})$ defined as follows.

Definition 4. *For $S \subseteq \{1, \ldots, m\}$ and $u \in D$, let*

$$Output(S, u) = \{1 \le i \le |u| \mid m \in \widehat{f}(S, u[1:i])\}.$$

To realize the procedure enumerating the set *Output*, we define the following sets.

Definition 5. *For any $u \in D$, let*

$$U(u) = \{1 \leq i \leq |u| \mid i < m \text{ and } m \in \widehat{M}(u[1:i])\}, \text{ and}$$
$$V(u) = \{1 \leq i \leq |u| \mid i \geq m \text{ and } m \in \widehat{M}(u[1:i])\}.$$

Then, we have the following lemma.

Lemma 4. *For any $S \subseteq \{1, \ldots, m\}$ and any $u \in \Sigma^*$,*

$$Output(S, u) = ((m \ominus S) \cap U(u)) \cup V(u).$$

Proof. By Lemma 2 and Definitions 4 and 5, we obtain:

$$Output(S, u) = \{1 \leq i \leq |u| \mid i < m \text{ and } m \in (S \oplus i) \cap \widehat{M}(u[1:i])\}$$
$$\cup \{1 \leq i \leq |u| \mid m \leq i \text{ and } m \in \widehat{M}(u[1:i])\}$$
$$= ((m \ominus S) \cap U(u)) \cup V(u).$$

■

Since $U(u) \subseteq \{1, \ldots, m\}$, we can store the set $U(u)$ as an m-bit integer in the node u of the dictionary trie D.

Lemma 5. *The function which takes as input $u \in D$ and returns in $O(1)$ time the m-bit representation of $U(u)$, can be realized in $O(|D|+m)$ time using $O(|D|)$ space.*

Proof. By the definition of U, for any $u = u'a$ with $u' \in \Sigma^*$ and $a \in \Sigma$,

$$U(u) = U(u') \cup \{|u| \mid |u| < m \text{ and } m \in \widehat{M}(u)\}.$$

Then, we can prove the lemma in a similar way to the proof of Lemma 3.   ■

To eliminate the cost of performing the operation $\ominus$ in $(m \ominus S) \cap U(u)$, we store the set $U'(u) = m \ominus U(u)$ instead of $U(u)$. Then, we can obtain the integer representing the set $S \cap U'(u)$ by one execution of the bitwise logical product operation. For an enumeration of the set, we repeatedly use the logarithm operation to find the leftmost bit of the integer that is one. Assuming that the logarithm operation can be performed in constant time, this enumeration takes only linear time proportional to the set size.

Next, we consider $V(u)$. Since the set $V(u)$ cannot be represented as an m-bit integer, we shall represent it as a linked list as shown in the proof of the next lemma.

Lemma 6. *The procedure which takes as input $u \in D$ and enumerates the set $V(u)$, can be realized in $O(|D| + m)$ time using $O(|D|)$ space, so that it runs in linear time with respect to $|V(u)|$.*

Proof. By the definition of V, for any $u = u'a$ with $u' \in \Sigma^*$ and $a \in \Sigma$,

$$V(u) = V(u') \cup \{|u| \mid m \le |u| \text{ and } m \in \widehat{M}(u)\}.$$

We use the function $Prev(u)$ that returns the node of the dictionary trie D that represents the longest proper prefix v of u such that $|v| \in V(u)$. Then, we have

$$V(u) = V(Prev(u)) \cup \{|u| \mid m \le |u| \text{ and } m \in \widehat{M}(u)\}.$$

The function $Prev(u)$ can be realized to answer in $O(1)$ time, using $O(|D|)$ time and space. Therefore it is sufficient to store in every node u of the dictionary trie D the value $Prev(u)$ and the boolean value $in\_V(u)$ indicating whether $|u| \in V(u)$. The proof is now complete. ∎

From Lemmas 4, 5, and 6, we have the following theorem.

Theorem 2. *The procedure which takes as input $(S, u) \in 2^{\{1,\dots,m\}} \times D$ and enumerates the set Output(S, u), can be realized in $O(|D| + m)$ time using $O(|D|)$ space, so that it runs in linear time with respect to $|$Output$(S, u)|$.*

Now we can simulate the behavior of the Shift-And algorithm on an uncompressed text completely. The algorithm is summarized as in Figure 4. The behavior of the new algorithm is illustrated in Figure 5.

Theorem 3. *The algorithm of Figure 4 runs in $O(|\Sigma| + m + n + r)$ time using $O(|\Sigma| + m + n)$ space, where r is the number of pattern occurrences.*

4   Experimental Results

In order to estimate the performance of the proposed algorithm, we carried out some experiments on the following four methods.

Method 1. A decompression followed by the Shift-And algorithm.
Method 2. Our previous algorithm presented in [13].
Method 3. The new algorithm proposed in this paper.
Method 4. Searching the uncompressed text, using the Shift-And algorithm.

In our experiments we used the Brown corpus as the text to be searched. The uncompressed size is about 6.8Mb and the compressed size is about 3.4Mb. The experiments were performed in the following two different situations.

Situation 1. Workstation (SPARCstation 20) with remote disk storage. The file transfer ratio is 0.96 Mbyte/sec.
Situation 2. Workstation (SPARCstation 20) with local disk storage. The file transfer ratio is 3.27 Mbyte/sec.

Input.   An LZW compressed text $u_1 u_2...u_n$ and a pattern $\mathcal{P}$.
Output.   All positions at which $\mathcal{P}$ occurs.
begin

    /* We represent the set $V(u)$ by the functions $Prev(u)$ and $in\_V(u)$.
      See the proof of Lemma 6. */

    /* Preprocessing */
    Construct the table M from $\mathcal{P}$;
    $D := \emptyset;$ $U'(\varepsilon) := \emptyset;$ $in\_V(\varepsilon) := false;$ $Prev(\varepsilon) := \varepsilon;$
    for each $a \in \Sigma$ **do call** $Update(\varepsilon, a);$

    /* Text scanning */
    $k := 0;$ $R := \emptyset;$
    for $\ell := 1$ **to** n **do begin**
        call $Update(u_{\ell-1}, u_\ell);$ /* We assume $u_0 = \varepsilon$./
        for each $p \in \big(R \cap U'(u_\ell)\big) \cup V(u_\ell)$ **do**
            report a pattern occurrence at position $k + p - m + 1;$
        $R := \big((R \oplus |u_\ell|) \cup \{1, 2, \ldots , |u_\ell|\}\big) \cap \widehat{M}(u_\ell);$
        $k := k + |u_\ell|$
    end
end.

procedure $Update(u, v)$
begin
    if $v \le |D|$ **then**
        let a be the first character of v
    else
        let a be the first character of u;
    $D := D \cup \{u \cdot a\};$
    $\widehat{M}(u \cdot a) := \big((\widehat{M}(u) \oplus 1) \cup \{1\}\big) \cap M(a);$
    if $|u \cdot a| < m$ **then**
        if $m \in \widehat{M}(u \cdot a)$ **then**
            $U'(u \cdot a) := U'(u) \cup \{m - |u \cdot a|\}$
        else
            $U'(u \cdot a) := U'(u)$
    else begin
        $U'(u \cdot a) := \emptyset;$
        if $m \in \widehat{M}(u \cdot a)$ **then**
            $in\_V(u \cdot a) := true$
        else
            $in\_V(u \cdot a) := false;$
        if $in\_V(u) = true$ **then**
            $Prev(u \cdot a) := u$
        else
            $Prev(u \cdot a) := Prev(u)$
    end
end;

Fig. 4. Pattern matching algorithm in LZW compressed text

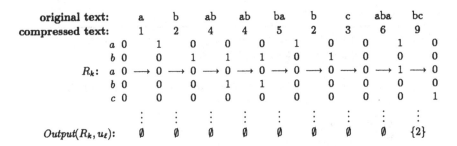

Fig. 5. Behavior of the algorithm.

Table 1. CPU time and elapsed time.

method	CPU time (sec)	elapsed time (sec) Situation 1	Situation 2
Method 1	7.52	8.16	7.62
Method 2	6.57	7.31	6.83
Method 3	5.15	6.05	5.41
Method 4	3.09	9.36	3.25

The searching times, measured in both the CPU time and the elapsed time, are shown in Table 1, where we included the preprocessing time.

Although the time complexities of our algorithms are linear with respect to the compressed text size n not to the original size N, the LZW compression of typical English texts normally gives $n = N/2$ and thus the constant factor is crucial. It is observed from Table 1 that, in the CPU time comparison, our algorithms (Methods 2 and 3) are slower than the uncompressed case (Method 4) whereas they are faster than a decompression followed by a search (Method 1). It is also observed that the new algorithm (Method 3) is about 1.3 times faster than the previous one (Method 2).

In general, the searching time is the sum of (1) the file I/O time and (2) the CPU time consumed for compressed pattern matching. Text compression reduces the file I/O time at the same ratio as the compression ratio while it may increase the CPU time. When the data transfer is slow, we have to give a weight to the reduction of the file I/O time, and a good compression ratio leads to a fast search. In fact, even a decompression followed by a simple search (Method 1) was faster than the uncompressed search (Method 4) in Situation 1. It should be noted that, in this situation, the previous algorithm (Method 2) and the new algorithm (Method 3) are faster than the uncompressed case (Method 4), and especially the latter is approximately 1.5 times faster than the uncompressed case.

On the contrary, in the situations that the data transfer is ralatively fast, the CPU time becomes a dominant factor. It is observed that, like in the CPU time comparison, Methods 2 and 3 are slower than Method 4 while they are faster than Method 1 in the elapsed time comparison in Situation 2.

Thus we conclude that, for the LZW compression, the compressed search is indeed faster than a decompression followed by a fast search, and that the Shift-And approach is effective in the LZW compressed pattern matching. When the data transfer is slow, e.g. network environments, the compressed search can be faster than the uncompressed search.

5   Extensions

In this section, we mention how to extend our algorithm.

5.1   Generalized Pattern Matching

The generalized pattern matching problem [1] is a pattern matching problem in which a pattern element is a set of characters. For instance, $(b + c + h + l)ook$ is a pattern that matches the strings book, cook, hook, and look. Formally, let $\Delta = \{X \subseteq \Sigma \mid X \neq \emptyset\}$ and $\mathcal{P} = X_1...X_m$ $(X_i \in \Delta)$. Then we want to find all integers i such that $T[i : i + m - 1] \in \mathcal{P}$.

It is not difficult to extend our algorithm to the problem. We have only to modify some equations: For example, we modify Equations (1) and (2) in Section 2.2 as follows.

$$R_k = \{1 \leq i \leq m \mid \mathcal{P}[1 : i] \ni T[k - i + 1 : k]\}, \tag{1'}$$

$$M(a) = \{1 \leq i \leq m \mid \mathcal{P}[i] \ni a\}. \tag{2'}$$

5.2   Pattern Matching with k Mismatches

This problem is a pattern matching problem in which we allow up to k characters of the pattern to mismatch with the corresponding text [10]. For example, if $k = 2$, the pattern pattern matches the strings postern and cittern, but does not match eastern. The idea stated in [7] to solve this problem is to count up the number of mismatches using $\lceil m \log_2 m \rceil$ bits instead of using one bit to see whether $\mathcal{P}[i] = T[k]$. This technique can be used to adapt our algorithm for the problem.

5.3   Multiple Pattern Matching

Suppose we are looking for multiple patterns in a text. One solution is to keep one bit vector R per pattern and perform the Shift-And algorithm in parallel, but the time complexity is linearly proportional to the number of patterns. The solutions in [7] and in [17] are to coalesce all vectors, keeping all the information in only one vector. Such technique can be used to adapt our algorithm for the multiple pattern matching problem in LZW compressed text.

6   Conclusion

In this paper we addressed the problem of searching in LZW compressed text directly, and presented a new algorithm. We implemented the algorithm, and showed that it is approximately 1.5 times faster than a decompression followed by a search using the Shift-And algorithm. Moreover we showed that our algorithm has several extensions, and is therefore useful in many practical applications. Some future directions of this study will be extensions to the pattern matching with k differences, and to the regular expression matching, and will be to develop a compression method which enables us to scan compressed texts faster.

References

[1] K. Abrahamson. Generalized string matching. *SIAM J. Comput.*, 16(6):1039–1051, December 1987.

[2] A. Amir and G. Benson. Efficient two-dimensional compressed matching. In *Proc. Data Compression Conference*, page 279, 1992.

[3] A. Amir and G. Benson. Two-dimensional periodicity and its application. In *Proc. 3rd Symposium on Discrete Algorithms*, page 440, 1992.

[4] A. Amir, G. Benson, and M. Farach. Optimal two-dimensional compressed matching. In *Proc. 21st International Colloquium on Automata, Languages and Programming*, 1994.

[5] A. Amir, G. Benson, and M. Farach. Let sleeping files lie: Pattern matching in Z-compressed files. *Journal of Computer and System Sciences*, 52:299–307, 1996.

[6] A. Amir, G. M. Landau, and U. Vishkin. Efficient pattern matching with scaling. *Journal of Algorithms*, 13(1):2–32, 1992.

[7] R. Baeza-Yaltes and G. H. Gonnet. A new approach to text searching. *Comm. ACM*, 35(10):74–82, October 1992.

[8] T. Eilam-Tzoreff and U. Vishkin. Matching patterns in a string subject to multilinear transformations. In *Proc. International Workshop on Sequences, Combinatorics, Compression, Security and Transmission*, 1988.

[9] M. Farach and M. Thorup. String-matching in Lempel-Ziv compressed strings. In *27th ACM STOC*, pages 703–713, 1995.

[10] Z. Galil and R. Giancarlo. Data structures and algorithms for approximate string matching. *Journal of Complexity*, 4:33–72, 1988.

[11] L. Gąsieniec, M. Karpinski, W. Plandowski, and W. Rytter. Efficient algorithms for Lempel-Ziv encoding. In *Proc. 4th Scandinavian Workshop on Algorithm Theory*, volume 1097 of *Lecture Notes in Computer Science*, pages 392–403. Springer-Verlag, 1996.

[12] M. Karpinski, W. Rytter, and A. Shinohara. An efficient pattern-matching algorithm for strings with short descriptions. *Nordic Journal of Computing*, 4:172–186, 1997.

[13] T. Kida, M. Takeda, A. Shinohara, M. Miyazaki, and S. Arikawa. Multiple pattern matching in LZW compressed text. In J. A. Atorer and M. Cohn, editors, *Proc. of Data Compression Conference '98*, pages 103–112. IEEE Computer Society, March 1998.

[14] U. Manber. A text compression scheme that allows fast searching directly in the compressed file. In *Proc. 5th Annu. Symp. Combinatorial Pattern Matching*, volume 807 of *Lecture Notes in Computer Science*, pages 113–124. Springer-Verlag, 1994.

[15] M. Miyazaki, A. Shinohara, and M. Takeda. An improved pattern matching algorithm for strings in terms of straight-line programs. In *Proc. 8th Annu. Symp. Combinatorial Pattern Matching*, volume 1264 of *Lecture Notes in Computer Science*, pages 1–11. Springer-Verlag, 1997.

[16] T. A. Welch. A technique for high performance data compression. *IEEE Comput.*, 17:8–19, June 1984.

[17] S. Wu and U. Manber. Fast text searching allowing errors. *Comm. ACM*, 35(10):83–91, October 1992.

[18] J. Ziv and A. Lempel. A universal algorithm for sequential data compression. *IEEE Trans. Inform. Theory*, IT-23(3):337–349, May 1977.

A General Practical Approach to Pattern Matching over Ziv-Lempel Compressed Text

Gonzalo Navarro[1] and Mathieu Raffinot[2]

[1] Dept. of Computer Science, University of Chile.
Blanco Encalada 2120, Santiago, Chile.
gnavarro@dcc.uchile.cl.
Partially supported by Fondecyt grant 1-990627.
[2] Institut Gaspard Monge, Cité Descartes,
Champs-sur-Marne, 77454 Marne-la-Vallée Cedex 2, France.
raffinot@monge.univ-mlv.fr

Abstract. We address the problem of string matching on Ziv-Lempel
compressed text. The goal is to search a pattern in a text without uncompressing it. This is a highly relevant issue to keep compressed text
databases where efficient searching is still possible. We develop a general technique for string matching when the text comes as a sequence of
blocks. This abstracts the essential features of Ziv-Lempel compression.
We then apply the scheme to each particular type of compression. We
present the *first* algorithm to find all the matches of a pattern in a text
compressed using LZ77. When we apply our scheme to LZ78, we obtain
a much more efficient search algorithm, which is faster than uncompressing the text and then searching on it. Finally, we propose a new hybrid
compression scheme which is between LZ77 and LZ78, being in practice
as good to compress as LZ77 and as fast to search in as LZ78.

1 Introduction

String matching is one of the most pervasive problems in computer science, with
applications in virtually every area. It is also one of the oldest and richest area of
development. The *string matching problem* is: given a pattern $P = p_1...p_m$ and a
text $T = t_1...t_u$, both sequences of symbols over a finite alphabet Σ of size σ, find
all the occurrences of P in T. There are many algorithms to solve this problem,
from classical to very recent [19, 8, 4, 14, 27, 9, 25]. The complexity of this
problem is $O(u)$ in the worst case and $O(u \log(m)/m)$ on average, where $u = |T|$
and $m = |P|$, and there exist variants of [8, 9] which achieve this complexity. In
practice, however, [27, 25] are the fastest algorithms in most cases.

Another old and rich area in computer science is text compression. Its aim is
to exploit the redundancies of the text to reduce its space usage. There are many
different compression schemes [5], among which the Ziv-Lempel family [31, 32]
is one of the best in practice because of their good compression ratios combined
with efficient compression and decompression times. Other compression schemes
are Huffman coding [15] and arithmetic coding [29], among others.

M. Crochemore, M. Paterson (Eds.): CPM'99, LNCS 1645, pp. 14–36, 1999.

Today's textual databases are an excellent example of applications where both problems are crucial: the texts should be kept compressed to save space and I/O time, and they should be efficiently searched. Surprisingly, these two combined requirements are not easy to achieve together, as the only solution before the 90's was to process queries by uncompressing the texts and then searching into them.

The *compressed matching problem* was first defined by Amir and Benson [1] as the task of performing string matching in a compressed text without decompressing it. Given a text T, a corresponding compressed string $Z = z_1 \ldots z_n$, and a pattern P, the compressed matching problem consists in finding all occurrences of P in T, using only P and Z. A naive algorithm, which first decompresses the string Z and then performs standard string matching, takes time $O(u + m)$. An optimal algorithm takes worst-case time $O(n + m)$, where $n = |Z|$. In [2], a new criterion, called *extra space*, for evaluating compressed matching algorithms, was introduced. According to the extra space criterion, algorithms should use at most $O(n)$ extra space, optimally $O(m)$ in addition to the n-length compressed file.

We define now a variation where we are required to report all the matching positions. That is, given P and Z, report all the $|x|$ such that $T = xPy$. The optimal algorithm for this problem takes $O(m + n + R)$ time, where R is the number of matches.

Two different approaches have emerged in the last years to combine compression and searching in textual databases. A first one is strongly oriented to natural language texts, which are assumed to be composed of words which follow some statistical rules. The basic idea is to compress the text using Huffman, where the words instead of the characters are taken as the symbols [7, 22]. As Huffman assigns a fixed code to each symbol, searching a given string is a matter of compressing it and searching it in the compressed text using a classical string matching algorithm with minor modifications [24, 23]. Despite its simplicity, this approach is very effective on natural language text, with better compression ratios than those of the Ziv-Lempel family, and search time which is between 2 and 8 times faster than the fastest algorithms for standard string matching over the uncompressed text. They are also able to search for complex patterns (such as regular expressions) and allow errors in the matches, provided that words are matched against words. The average search time for a simple pattern is close to $O(m + n\log(u/n)/(u/n))$. The extra space is $O(\sqrt{u})$, which is the same space necessary to decompress the text. A weakness of this scheme is that it does not work well on small texts (say, less than 10 Mb), since in that case the vocabulary is almost as big as the text itself. Also, it can be applied only to natural language texts.

Another practical approach is an ad-hoc technique [20], which however is not so fast, obtains compression ratios of near 70% (against 30% to 40% of Ziv-Lempel algorithms), and relies on the ASCII encoding.

The second line of research considers Ziv-Lempel compression, which is based on finding repetitions in the text and replacing them with references to similar strings previously appeared. LZ77 is able to reference any substring of the text already processed, while LZ78 references only a single previous reference plus a new letter that is added. In both cases, the referenced text to be found is normally limited by a *window* which precedes the current text position.

String matching in Ziv-Lempel compressed texts is much more complex, since the pattern can appear in different forms across the compressed text. In [2] a compressed matching algorithm for LZ78 is presented, which works in time and space $O(m^2 + n)$. For LZ77, the only result is [11], which is a randomized algorithm to determine in time $O(m + n \log^2(u/n))$ whether a pattern is present or not in an LZ77-compressed text, but they do not find all the pattern occurrences. Other algorithms for different specific search problems have been presented in [13, 17]. This second branch is rather theoretical and, to the best of our knowledge, no actual implementations have been developed.

In this paper we aim at efficient algorithms for string matching on Ziv-Lempel compressed texts. We present new theoretical developments but also give practical implementations and experiments on our algorithms. Our main results are

- We develop a general technique for string matching on a text which is given as a sequence of blocks. This abstracts the essential features of Ziv-Lempel compressed texts and is the basis for the algorithms which run over specific members of the family.
- We apply our technique to LZ77-compressed texts. The result is the *first* algorithm to search under this compression scheme (recall that [11] cannot find all the occurrences of the pattern). The algorithm, however, is $O(u)$ time at best. In practice, the algorithm is slower than uncompressing the text and searching it with a classical algorithm.
- We apply the technique to the LZ78 compression scheme. The result is an algorithm which turns out to be a practical implementation of the theoretical proposal of [2]. This algorithm is $O(n + R)$ time in the worst and average case, and is in practice twice as fast as decompressing and searching.
- We propose a hybrid compression scheme which is between LZ77 and LZ78, which keeps some of the good features of LZ77 and which can be searched in $O(\min(u, n \log m) + R)$ time on average (and $O(\min(u, mn) + R)$ in the worst case). In practice, the compression efficiency is similar to LZ77 and the search time is similar to LZ78.

In all cases our preprocessing cost is $O(\sigma + m)$ and our extra space is $O(n + R)$, almost the same necessary to decompress the text. Our approach is practical and relies on *bit-parallelism*. Bit-parallelism is a general technique to take advantage of the fact that the computer operates in parallel over all the bits of the machine word, so that if a process is so simple that it can be expressed with bit operations we can perform many of those steps in a single operation of the processor. If we call w the length in bits of the machine word (typically 32 or 64), then the possible speedups are up to $O(w)$. The complexity results presented assume that $m = O(w)$, otherwise we have to multiply the u and n of our complexities by m/w.

2   String Matching on Blocks

We describe now a general technique for string matching when the text is presented as a sequence of atomic strings (here called "blocks") instead of a sequence of characters. This technique is the basis for all the different searching algorithms on Ziv-Lempel compressed text, which are described in the next sections.

Our general assumption is that the blocks either have just one letter (that we can access directly) or are formed by a concatenation of previously seen blocks. We describe an online algorithm where we process the text block by block. At any moment of the search we denote T' the text already processed (of $|T'|$ characters). When we finish the search, $T' = T$, i.e. the original text.

The method works as follows. We process the blocks one by one. For each new block B, we compute a *description* for B which has all the information of the block which is relevant for the search. This description is denoted $D(B) = (L, O, S, P, M)$, where

- $L = |B|$, that is, the length of B in characters;
- $O = \text{Offs}(B) =$ the length in characters of the text we had processed when B appeared;
- $S = \text{Suff}(B) =$ all the pattern positions[1] which either start a complete occurrence of B inside the pattern, or start a proper pattern suffix which matches with a prefix of B. Formally,

$$\text{Suff}(B) = \{|x|, P = xBy\} \cup \{|x|, |x| > 0 \land |z| > 0 \land P = xz \land B = zy\} ;$$

- $P = \text{Pref}(B) =$ all the pattern positions which either follow a complete occurrence of B inside the pattern, or follow a proper pattern prefix which matches with a suffix of B. Formally,

$$\begin{aligned} \text{Pref}(B) = \{|xB|, \ P = xBy \ \land \ |y| > 0\} \ \cup \\ \{|z|, \ |z| > 0 \ \land \ |y| > 0 \ \land \ P = zy \ \land \ B = x \} ; \end{aligned}$$

- $M = \text{Matches}(B) =$ all the block positions where the pattern occurs ($\emptyset$ if $|B| < |P|$). Formally,

$$\text{Matches}(B) = \{|x|, \ B = xPy\} .$$

Figure 1 illustrates these concepts.

The description $D(B)$ of a new block B is obtained in two forms: (a) the block is an explicit letter and then we obtain the description directly, or (b) the block is a concatenation of other blocks previously known, and we obtain its description by operating on the descriptions of the previous blocks.

Once the description of the new block is computed, we use that description to update the state of the search. This concludes the processing of the block and we move to the next one. The state of the search contains the matches that have already occurred and the potential matches in progress, that is,

---

[1] To simplify the notation, we number pattern positions starting at zero.

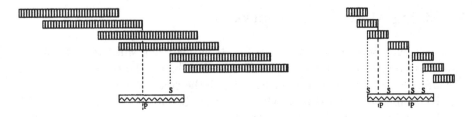

Fig. 1. Prefixes (P) and suffixes (S) for a long and a short block. The pattern has the diagonal tiling and the possible blocks have a bar tiling. The suffixes (dotted lines) and prefixes (dashed lines) are pattern positions. Prefixes are marked after the position where they finish, suffixes are marked at the position they start.

- $\text{Res}(T') =$ the text positions that matched up to now, formally

$$\text{Res}(T') \quad = \quad \{|x|, \; T' = xPy\} \; ;$$

- $\text{Active}(T') =$ the set of positions following the pattern prefixes which match a suffix of the current text. Formally,

$$\text{Active}(T') \quad = \quad \{|x|, \; |x| > 0 \; \wedge \; |y| > 0 \; \wedge \; P = xy \; \wedge \; T' = zx\} \; .$$

Hence, when we complete the text processing and T' is not a text prefix anymore but the whole text, $\text{Res}(T)$ is our answer. The initial state of the search is $T' = \epsilon$, and $\text{Res}(\epsilon) = \text{Active}(\epsilon) = \emptyset$.

We have defined already the information we keep, and consider now how to compute that information. For the formulas that follow, we define some auxiliary functions, namely

- $\text{Left}_i(X) \;=\; \{x-i, \; x \in X\} \cup \{m-i, m-i+1, \ldots, m-1\}$, which receives a set of Suff() positions not smaller than i, subtracts i to all them and then adds new pattern positions filling the hole left by the shift.
- $\text{Right}_i(X) \;=\; \{x+i, \; x \in X\} \cup \{1, 2, \ldots, i\}$, which does the same for Pref() positions, in the other direction.
- $\text{Add}_i(X) \;=\; \{i+x, \; x \in X\}$, which adds i to all the elements of the set.
- $\text{Subtr}_i(X) \;=\; \{i-x, \; x \in X\}$, which subtracts all the elements of the set from i.

2.1   Description of a Letter

The base case of our scheme is to obtain the description of a block which is a letter a. The following is obtained by direct application of the general formulas.

- $|B| \;=\; 1$
- $\text{Offs}(B) \;=\; |T'|$
- $\text{Suff}(B) \;=\; \{|x|, \; P = xay\}$
- $\text{Pref}(B) \;=\; \{|xa|, \; P = xay \; \wedge \; |y| > 0\}$
- $\text{Matches}(B) \;=\;$ if $P = a$ then $\{0\}$ else $\emptyset$

2.2   Concatenating Two Blocks

Assume that our block B is defined as the concatenation of one or more previous blocks. If only one previous block B' is referenced, we just copy its definition. We show now how to concatenate two blocks, since the case of more than two blocks is a simple iteration over this procedure. We are given two blocks B_1 and B_2, and we have to obtain the description for their concatenation $D(B) = D(B_1 B_2) = D(B_1) \cdot D(B_2)$ (where we define $\cdot$ as the concatenation of block descriptions). The formulas are as follows

- $|B| = |B_1| + |B_2|$
- $\text{Offs}(B) = |T'|$
- $\text{Suff}(B) = \text{Suff}(B_1) \cap \text{Left}_{|B_1|}(\text{Suff}(B_2))$
- $\text{Pref}(B) = \text{Pref}(B_2) \cap \text{Right}_{|B_2|}(\text{Pref}(B_1))$
- $\text{Matches}(B) = \text{Matches}(B_1) \cup \text{Add}_{|B_1|}(\text{Matches}(B_2))$
  $\cup (\text{Subtr}_{|B_1|}(\text{Pref}(B_1) \cap \text{Suff}(B_2)) \cap \{0, 1, 2, \ldots, |B| - m\})$

We explain now the rationale for the formulas (see Figure 2). The first two are immediate. For $\text{Suff}(B)$, note that $\text{Suff}(B_1 B_2)$ considers that either a prefix of B_1 may be a suffix of P or B_1 may be completely inside P followed by a prefix of B_2 matching the a suffix of P. That is, if the number i belongs to $\text{Suff}(B_1 B_2)$ then either

- $i \geq m - |B_1|$, that is, a prefix of $B_1 B_2$ is a suffix of P. Notice that in this case also a prefix of B_1 is a suffix of P. Since $\text{Left}_{|B_1|}$ will add all these positions, they will appear in the result if and only if they are present in $\text{Suff}(B_1)$, which is correct.
- $i < m - |B_1|$, that is, B_1 appears inside P and is immediately followed by an occurrence of B_2 (which can be a complete occurrence or share a prefix with the pattern suffix). If we subtract $|B_1|$ to the elements in $\text{Suff}(B_2)$, then we are interested in the positions which also appear in $\text{Suff}(B_1)$ (which since $i < m - |B_1|$ can only correspond to complete occurrences of B_1).

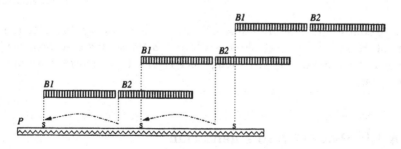

Fig. 2. Suffixes of the concatenation of two blocks. It is possible that the result involves only B_1 (rightmost pair) or that it involves both. In this case B_1 is completely inside the pattern and B_2 may or may not be totally inside (leftmost and middle pairs, respectively).

The rationale for Pref() is analogous to Suff(). For Matches(B), there are
three parts. The first one is the matches which are inside B_1, and the second one
is the same for B_2 (displaced since now B_2 comes after B_1 in B). The third one
accounts for matches that appear only when B_1 and B_2 are concatenated. If a
prefix of the pattern is at the end of B_1, and the corresponding suffix is at the
beginning of B_2, then we have the pattern in B_1B_2. The *Subtr* converts pattern
to block positions and the final set which is intersected with the results ensures
that we have really prefixes and suffixes instead of substrings of the blocks.

2.3 Updating the Search State

We want now to update the state of our search by processing a new block B
whose description has just been computed. The formulas to obtain the new
Res($T'B$) and Active($T'B$) values from the old Res(T') and Active(T') ones are

- Active($T'B$) = Right$_{|B|}$(Active(T')) $\cap$ Pref(B)
- Res($T'B$) = Res(T') $\cup$ Add$_{|T'|}$(Matches(B)) $\cup$
  Subtr$_{|T'|}$(Active(T') $\cap$ Suff(B) $\cap$ $\{m-|B|, m-|B|+1, \ldots, m-1\}$)

The new Active($T'B$) value considers that, since a new block B has been
added to T', the pattern prefixes that are suffixes of $T'B$ are those that are
already suffixes of B (i.e. Pref(B)), or those which are suffixes of T' and are
followed by B in the pattern. As before, Right does the trick of considering both
cases in a single formula.

The new value Res($T'B$) adds to Res(T') not only the matches which are completely inside B, but also those which appear when T' is concatenated to B. For
this sake, we consider pattern prefixes which are suffixes of T' (i.e. Active(T')),
and which are followed by the corresponding pattern suffix in B. The final intersection ensures that the complete pattern has appeared. Figure 3 illustrates.

Fig. 3. Updating the state of the search. In the first case we illustrate the updating of Active(T') (a short block is added). In the second case we show how
the matches are updated (when a long block is added). In general both updates
are necessary.

3   A Bit-Parallel Implementation

Until now, we have defined our algorithms in terms of sets of pattern positions.
We present now a very well-suited implementation paradigm which allows to
convert the previous algorithms into efficient implementations.

We use the technique called *bit-parallelism* [3]. This technique takes advantage of the fact that the processor works in parallel on all the bits of the computer word. We call w the number of bits of the computer word, which is 32 or 64 in current architectures. If one is able to map the elements of a set on bits, and to express the operations to perform on them by using only the operators provided by the processor (which are rather limited, i.e. bit shifts, masking, etc.), then one can effectively parallelize the work on the set, obtaining speedups of up to $O(w)$ over the original algorithm.

This paradigm was invented in 1989 by Baeza-Yates and Gonnet [4] for a text searching algorithm called Shift-Or. If we consider $m \le w$, then we keep the state of the search in a computer word D, whose i-th bit tells whether the prefix of length i of the pattern matches the current text suffix. All the bits start with value zero, and a match is reported whenever the m-th bit of D signals a match. The update formula upon reading a new text character is

$$D' \;=\; (D << 1) \mid S[a]$$

where $S[a]$ is a mask whose i-th bit tells whether $P_i = a$, we are assuming that 0 represents a match and a 1 a mismatch, "|" is the bitwise-or of the computer word, and "$<< \ell$" is a bit shift operation which assigns the i-th bit to the $(i + \ell)$-th, setting the first ℓ bits to zero. Other operations allowed in most architectures are bitwise-and (&), shift to the other direction (>>), and, which is more sophisticated, arithmetic operations such as addition and subtraction which operate on the bit mask as if it were a number.

The Shift-Or algorithm is $O(n)$ provided $m \le w$. If the computer word is too short to hold one bit per pattern position, then $\lceil m/w \rceil$ computer words are used for the simulation, and the search takes in the worst case $O(mn/w)$ time. It is not hard to show that on average it takes $O(n)$, since $O(1)$ computer words have active states on average.

Our implementation can indeed be seen as a Shift-Or algorithm working on blocks instead of letters. The sets Pref(B), Suff(B), and Active(T') are represented by bit masks. Hence, for blocks of one letter a we have Suff(B) = $S[a]$ and Pref(B) = ($S[a] << 1$). The formulas to concatenate blocks are directly translated by noticing that Left$_\ell$ and Right$_\ell$ are converted into "$>> \ell$" and "$<< \ell$", respectively (taking care of the borders which must get active bits), and union and intersection are converted into "|" and "&" respectively. Hence, all those operations on sets are performed in $O(1)$ time if $m \le w$, and $O(m/w)$ time in general. In practical text searching we can assume $m = O(w)$.

On the other hand, the sets Res(T') and Matches(B) are explicitly stored in an array. However, it is not difficult to see that the total amount of work to handle them is $O(R)$, where R is the number of occurrences of the pattern in the text. The cost cannot be $o(R)$ if we report all the occurrences.

Hence, if $f(n)$ concatenations are performed along all the process, our total search cost is $O(f(n) + R)$. The value of $f(n)$ depends on the compression algorithm. We have also to add a preprocessing cost to build the $S[\ ]$ table, which is $O(\sigma + m)$.

In all cases, the space complexity of our algorithms is $O(n + R)$, since we need to store the descriptions of the blocks already seen and the matches found. Notice that this n refers in fact to the size of the compression window, and the R to the matches present in that window only.

Finally, we consider the practical problem of uncompressing a neighborhood of the occurrences. In practice it is undesirable that we just give the text positions matching the pattern. It is much better to uncompress and show a neighborhood of the match. This neighborhood can be defined as the line holding the occurrence, the record (delimited by some given pattern), a fixed number of characters, etc.

Assume that we know a pattern position and want to show a neighborhood. We just decompress the surrounding blocks forward and backward, until from the plain text obtained we determine that the neighborhood has been decompressed. To decompress a block we have two cases: (a) the block is a letter, in which case we deliver the letter, (b) the block is a concatenation of other blocks, in which case we decompress each of those blocks in turn. This process takes $O(N)$ time at most (where N is the size of the decompressed neighborhood), since at each step we either obtain one character of N or split the final text to be obtained, and it is not possible to split it more than $O(N)$ times. This shows that it is practical to show a part of a Ziv-Lempel compressed file without necessarily uncompressing the whole file.

4 LZ78 Compression

4.1 Compression Algorithm

The Ziv-Lempel compression algorithm of 1978 (usually named LZ78 [32]) is based on a dictionary of blocks, in which we add every new block computed. At the beginning of the compression, the dictionary contains a single block b_0 of length 0. The current step of the compression is as follows: if we assume that a prefix $t_1 \ldots t_i$ of T has been already compressed in a sequence of blocks $Z = b_1 \ldots b_c$, all them in the dictionary, then we look for the longest prefix of the rest of the text $t_{i+1} \ldots t_u$ which is a block of the dictionary. Once we found this block, say b_k of length l_k, we construct a new block $b_{c+1} = (k, t_{i+l_k+1})$, we write the pair at the end of the compressed file Z, i.e $Z = b_1 \ldots b_c b_{c+1}$, and we add the block to the dictionary. It is easy to see that this dictionary is prefix-closed (i.e. any prefix of an element is also an element of the dictionary) and a natural way to represent it is a trie.

We give as an example the compression of the word *ananas* in Figure 4. The first block is $(0, a)$, and next $(0, n)$. When we read the next a, a is already the block 1 in the dictionary, but an is not in the dictionary. So we create a third block $(1, n)$. We then read the next a, a is already the block 1 in the dictionary, but as do not appear. So we create a new block $(1, s)$.

The compression algorithm is $O(u)$ in the worst case and efficient in practice if the dictionary is stored as a trie, which allows rapid searching of the new text

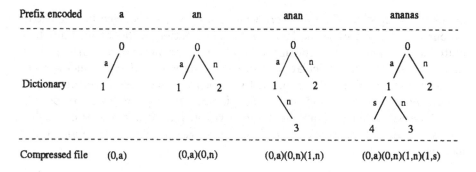

Fig. 4. Compression of the word *ananas* with the algorithm LZ78.

prefix (for each character of T we move once in the trie). The decompression needs to build the same dictionary (the pair that defines the block c is read at the c-th step of the algorithm), although this time it is not convenient to have a trie, and an array implementation is preferable. Compared to LZ77, the compression is rather fast but decompression is slow. LZ78 is used by Unix's *Compress* program.

Many variations on LZ78 exist, which deal basically with the best way to code the pairs in the compressed file, or with the best way to update the window. A particularly interesting variant is from Welch, called LZW [28]. In this case, the extra letter (second element of the pair) is not coded, but it is taken as the first letter of the next block (the dictionary is started with one block per letter). A variant over this is presented by Miller and Wegman [21] (which we call LZMW), where the new block is not the previous one plus the first letter of the new one, but simply the concatenation of the previous and the new one.

4.2 Pattern Matching in LZ78 Compressed Files

Our general algorithm for searching in a sequence of blocks $Z = b_1 \ldots b_n$ can be directly applied if we consider the new letter added after each block created by the LZ78 compression algorithm as a separate block. That is, each new pair (k, a) read at step c is taken as a reference to a previous block (b_k) followed by a literal block (a). Hence, we compute the description of the concatenation of b_k and a and add it as the new block b_c to our dictionary. At the same time, we update the state of the search using the description of b_c just computed. Of course, in practice we manage this one-letter block in a special way, to speed-up the block concatenation. We keep all the descriptions of the blocks b_k in an array which is directly accessed.

The algorithm we obtain is quite the same as in [2]. The main differences are that we obtain this algorithm as a particular case of a general string search algorithm for text that comes in blocks, that their algorithm is originally designed for LZW compression, and that we search all the occurrences of the pattern, not only the first one. Moreover, we present a practical implementation based on bit-parallelism, while [2] is a theoretical work that has not been implemented. To our knowledge ours is the first real implementation of this algorithm[2]. It is quite easy to adapt our algorithm to work on other variants of LZ78, such as LZW or LZMW. In particular we can easily adapt to different window management policies. The simplest one is that when the compressor memory is full, the dictionary is deleted and compression is restarted. Others try to remove the least interesting blocks from the dictionary, e.g. [12]. Our searcher can follow the same steps of the compressor along the search, using the same amount of memory.

4.3 Analysis

The theoretical complexity of the pattern matching algorithm is $O(n+R)$ (recall that, as we use bit-parallelism, we have $O(mn/w + R)$ time for long patterns). If $n = o(u)$, this is faster than searching in the uncompressed text. In practical terms, the algorithm is rather efficient since no extra work apart from one block concatenation and one update of the search is performed per element of the compressed file.

Our experimental results, however (Section 7), show that the algorithm takes in practice twice the time of a Shift-Or run on the uncompressed text. This is because Shift-Or is very simple, and although we process many characters of the uncompressed text in one shot, in practice the cost of each step is big enough to amortize any possible gain due to compression. A specific problem is the locality of reference: the compressed matching algorithm reads random positions in the array of block definitions, while the uncompressed algorithm works basically in-place. The caching mechanism of the computer largely favors this last approach.

However, there is a positive result. Searching the compressed file with this algorithm is twice as fast as decompressing it and then searching the uncompressed file. For this comparison we are assuming that the file is compressed with LZ77 (which is much faster than LZ78 to decompress) and consider the time of *gunzip*, which is an optimized decompression software. Hence, if the text collection is kept compressed (which is definitely of interest) then it is much faster to search directly the compressed files.

We have tried to further improve our algorithm. For instance, we have created a variant called Mark-LZ78. In this compression algorithm, we mark with a bit flag for each block if the block is a leaf of the dictionary trie or not, to avoid storing the block description if this block is not used anymore. However, as we show in the experiments, the performance does not improve.

5 LZ77 Compression

5.1 Compression Algorithm

The Ziv-Lempel compression algorithm of 1977 (usually named LZ77 [31]) is, in some sense, simpler than LZ78, since the basic idea is just to recognize two

---

[2] See, however, [18], in this very same conference.

repeated segments of the text and to mark the second as a reference (position in the text and length of the repeated part) to the first one. More formally, assume that a prefix $t_1 \ldots t_i$ of T has been already compressed in a sequence of blocks $Z = b_1 \ldots b_c$. We look for the longest prefix v of $t_{i+1} \ldots t_u$ which appears already in $t_1 \ldots t_i t_{i+1} \ldots t_{i+|v|-1}$. Once we have it, say that we find it starting at position $j \leq i$, we add a new block $(j, |v|)$ to the compressed file Z. A special case occurs if v is empty, in which case t_{i+1} is a new letter and we code it with a special block $(0, t_{i+1})$. With the same example *ananas*, we obtained: $(0, a)$ *nanas*; $(0, a)(0, n)$ *anas*; $(0, a)(0, n)(1, 3)$ *s*; $(0, a)(0, n)(1, 3)(0, s)$.

Notice that the above definition allows that the referenced block overlaps the one which is being compressed. Another variant avoids this for simplicity, i.e. v must be found in $t_1 \ldots t_i$. In this case the compression of *ananas* becomes: $(0, a)$ *nanas*; $(0, a)(0, n)$ *anas*; $(0, a)(0, n)(1, 2)$ *as*; $(0, a)(0, n)(1, 2)(1, 1)$ *s*; $(0, a)(0, n)(1, 2)(1, 1)(0, s)$.

Yet another variant codes the repeated block and then the letter which follows it in the still uncompressed text. There are many other variants as well, mainly related to how to represent the pairs in the compressed file and how to compress fast. In general, the position j is coded as the difference $i + 1 - j$, since the last occurrence of the block is used and v is normally restricted to not appear too far away from t_i.

LZ77 compresses more than LZ78, both in theory and in practice. From a theoretical point of view, the variant which allows overlaps can obtain a compressed file of $O(1)$ blocks in the best case, while the one not allowing overlaps obtains at most $O(\log u)$. LZ78, on the other case, cannot obtain less than $O(\sqrt{u})$. This is easily seen by considering the best-case file $T = a^u$. In practice it is also true that LZ77 compresses more than LZ78. LZ77 is implemented in the Gnu *gzip* program.

Compression is rather slow with LZ77. It is expensive in time and space to find the longest prefix of the uncompressed part of the file that appears already in the compressed part. In theory, the compression is $O(u)$ in time and space by the use of a suffix tree or a DAWG automaton [31, 30]. In practice, the search in done in a buffer window and an large hash table is normally used, as in *gzip*. An experimental comparison of different techniques to find the prefix can be found in [6]. The decompression algorithm, on the other hand, is very fast (faster than for LZ78) because to decompress a block is it just necessary to copy a part of the text and no dictionary has to be kept.

5.2   Pattern Matching in LZ77 Compressed Files

Our algorithm for LZ77 is an adaptation of the general algorithm on blocks, with a main difference. On LZ77 compressed files, when we want to process a new block, the situation shown in Figure 5 generally occurs: the new block references a sequence of r contiguous previously processed blocks, but it overlaps with the first and last one (u and v in the Figure). That is, the new block does not exactly correspond to previously processed blocks. Therefore, we do not have all the information on the blocks u and v that we need to concatenate the blocks.

We solve this by computing recursively the descriptions of the two blocks u and v with the same method. That is, we simulate that we are back in the text, where those blocks appeared, and compute their description (this may trigger more recursive invocations with the same purpose). When we finally obtain the descriptions of u and v, we concatenate all the referenced blocks to obtain the description of the new block. Another possibility is that the new block is completely inside another block already processed, in which case we have to recursively consider the blocks that define the referenced block.

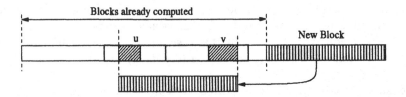

Fig. 5. Recursive computation of the description of a block in LZ77 compressed files.

We explain now a technique to concatenate the r blocks in low average time. Instead of computing Pref(B) and Suff(B) of the first block, then concatenating with the second, then to the third, until the r blocks are concatenated, we compute Suff(B) from the first block to the r-th and Pref(B) from the r-th block to the first one. We analyze this shortly.

5.3   Analysis and Improvements

We analyze now the many aspects of our algorithm and propose some improvements.

Block concatenation. If we use the proposed block concatenation technique, we have that in the worst case only the first m blocks can affect Suff(B) and only the last m blocks can affect Pref(B), so the worst case time for concatenating the blocks becomes $O(\min(u, mn))$.

We show now that on average only $O(\log m)$ blocks are processed until Suff(B) becomes stable. Each new block character we process will either extend the current suffixes of the set Suff(B) or make them disappear from the set. Each suffix is removed from the set with probability $1 - 1/\sigma$ (i.e. if the new character block cannot extend it). Before we read the block characters all the m pattern positions are in Suff(B), and therefore on average no pattern positions remain in the set after $O(\log m)$ block characters are read (after the i-th character is read, the pattern positions $m - i$ to $m - 1$ cannot be removed from the set, but their situation cannot change anyway).

Even if we consider all blocks of length 1 (the worst), we work on average $O(n \log m)$ because of concatenations. The same reasoning holds for Pref(B).

The only part of the block concatenation which cannot skip blocks is the computation of Matches(B). However, this adds up $O(R)$ time along all the search. Therefore, the total time for block concatenation is $O(\min(u, n \log m) + R)$ on average.

Finding the blocks. We consider now how to find the indices of the block that define a text position j. We keep an array with the blocks already seen. Binary searching the text position among these blocks adds $O(n \log n)$ to the cost. Instead, we keep a table of $O(n)$ entries where the element i points to the block where the text position $\lfloor iu/n \rfloor$ is defined. By accessing this table we directly arrive at the correct block with an average inaccuracy of $O(u/n)$, and a final binary search finds the correct position, for a total cost of $O(n \log(u/n))$ (in practice a linear search is faster for the final part). This gives good results in practice. Another alternative is that the compressor does not store the text position and length of the repeated part, but instead it gives the block numbers involved and the offsets inside u and v. Since a text position needs $O(\log u)$ bits and a block number plus an offset inside the block needs on average $\lceil \log_2 n \rceil + \lceil \log_2(u/n) \rceil = O(\log u)$ bits, the *order* of compression ratio should not worsen. We show in the experiments that this version of the algorithm (called Block-LZ77) is faster than the plain version, since no searching of the text position is necessary. However, compression ratios worsen significantly in practice due to round-offs.

Computing partial blocks. However, the really costly part of the algorithm is not here, but in the recursive computation of the partial blocks u and v. If we consider that each time we perform a recursive call we "split" the original block B at a new position, then it is clear that at most $|B|$ recursive calls can be done until we have split it in single characters and therefore we have found the definition of each one. This shows that the total cost of the recursive calls is $O(u)$ in the worst case. Our experiments suggest that this is also the average case, but we were not able to prove it.

Consider now the cost of the recursive invocations in the case where the new block B is strictly inside its referencing block. For instance, a letter which repeats inside a large block could trigger a long chain of recursive invocations until its real definition is found. In the worst case, we could have a block of size s which references one of size $s - 1$, and this one references another of size $s - 2$, and so on. We would work $O(s)$, but the size of the text at that point is $O(s^2)$. Hence, at text position i we cannot work more than $\sqrt{i}$, which gives a total worst-case cost of $O(n\sqrt{u})$, which is too high. This problem does not disappear if the compressor always stores the first occurrence of the repeated block instead of the last one, because we may not point to the first occurrence when we consider partial blocks.

Hence the total amount of work is $\omega(u)$ in the worst case whenever $n = \omega(\sqrt{u})$, and we conjecture that this is also the average case. See the left plot of Figure 6, where we have experimented with the *English* text described in Section 7. Least squares fitting shows that a good model for the number of recursive invocations per text character is $0.177 + 0.1 \ln u$ (with less than 0.5% error in the approximation). The experiment suggests that the algorithm is $O(u \log u)$ on average. This is, unfortunately, worse than uncompressing and searching. We present now some techniques to improve this situation.

Improvements. A first improvement we tried consisted in storing more information than simply one description per block. For instance, when we compute the description for the partial blocks u and v (which are not part of the original sequence of blocks), we could store instead of discarding them. If later another block needs the description of u and v, we have already computed them. Figure 6 (right plot) shows that the total amount of recursive calls is reduced using this technique, and we conjecture that in this case we work $O(u)$ (least squares fitting yields a complexity of $O(u^{0.99927})$). These blocks, however, cannot be easily stored in the array of blocks since they do not belong to the sequence. A hashing implementation gave bad results in practice, that is, the cost to add the new blocks outweighted the gains of having them already computed. This could change for longer texts, if the orders of the two algorithms are different.

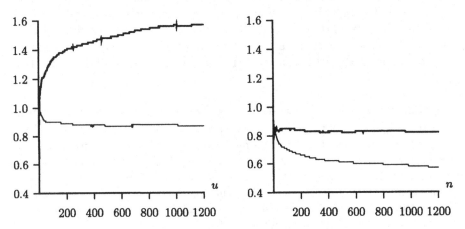

Fig. 6. Number of recursive invocations (thick line) and block concatenations (thin line) per text character, for natural language text. The left plot shows the basic algorithm and the right plot shows the improvement of adding the computed blocks.

Another improvement, which gave good practical results, was to try to compute *less* (instead of more) information. Our aim was to avoid the recursive computation of u and v. Hence, instead of computing their descriptions recursively, we pessimistically assume that they match all the pattern positions. If they are short enough we will not have a match even assuming this, and we could process them without actually obtaining their descriptions. Only when we find a (possible) match we backtrack to the point where it could have been started and compute correctly the involved blocks. For each block, we store whether it has been correctly or pessimistically computed. As we show in the experiments, this improves search time for patterns of length 15 or more in practice. However, the method is limited since we cannot skip more than m characters of T without having at least one character correctly computed, hence in the very best case we pay $O(u/m)$ with this speedup. We call this algorithm Skip-LZ77 (and combined with Block-LZ77 it yields Skip-Block-LZ77).

Final remarks. Even with all these improvements, the experiments show that this algorithm is much slower than decompressing (with *gunzip*) and searching (with Shift-Or). Although ours is the first algorithm to directly search in LZ77-compressed text, we believe that it is not possible in practice to beat a decompress-then-search approach. The root of this limitation lies in the need to recursively compute u and v. Another consequence of the existence of partial blocks is that, even if the compressor uses a window of fixed size to select the strings to repeat, we need to keep in memory all the previous blocks, since even if they are not directly referenced anymore, we may need to resort to them in case of partial blocks. We propose in the next section a slightly different compression scheme which gets rid of all the aspects of LZ77 compression that degrade the searching performance.

We finish this section with a couple of comments. First, as it is clear from the algorithm, we do not handle the case of overlapping compression, i.e. when the referenced block can overlap with the new block B. Although we could handle it, the result is the same in cost as if the compressor avoided such overlapping (i.e. performing many steps, where a step ends when an overlap occurs). Second, other variants of LZ77 are easily accommodated. Finally, we notice that a neighborhood of size N around the occurrences can be obtained using the general mechanism at $O(N\sqrt{u})$ cost (or, according to the empirical results, $O(N\log u)$ cost). This is because of the cost to find the definitions of the incomplete blocks.

6   A New Hybrid Compression Algorithm

It became clear in the previous section that the worst part of the cost of the LZ77 search algorithm was due to the cost of recursively computing partial blocks, and of finding the block corresponding to a text position. We design a new compression algorithm between LZ78 and LZ77, to have multiple-block compression (not just one block like in LZ78), but also to avoid the recursive situation which appears in searching LZ77-compressed files (Figure 5).

We propose the following algorithm. Assume that a prefix $t_1 \ldots t_i$ of T has been already compressed in a sequence of block $Z = b_1 \ldots b_c$. We look now for the longest prefix v of $t_{i+1} \ldots t_u$ which is represented by a sequence $b_r \ldots b_{r+h}$ already present in the compressed file. If there are many alternative choices for the same v, we take the one with the minimum of blocks (to reduce the cost of concatenations). And if still several possibilities occur, we take the first occurrence (the minimum in the number of the first block). We code this new block by (r, h). As in LZ77, if v is empty (i.e the letter t_{i+1} is new), we code a special block $(0, t_{i+1})$. With the same example *ananas*, we obtain: $(0, a)$ *nanas*; $(0, a)(0, n)$ *anas*; $(0, a)(0, n)(1, 1)$ *as*; $(0, a)(0, n)(1, 1)(1, 0)$ *s*; $(0, a)(0, n)(1, 1)(1, 0)(0, s)$.

The main advantage of this compression scheme is that it avoids the recursive case in the LZ77 pattern matching (Figure 5), because we know already that the new block corresponds directly to a concatenation of already processed blocks. Moreover, we do not need to search the text position in the blocks, since we can directly access the relevant blocks.

The compression can still be performed in $O(u)$ time by using a sparse suffix tree [16] where only the block beginnings are inserted and when we fall out of the trie we take the last node visited which corresponds to a block ending. Decompression is slower than for LZ77, since we need to keep track of the blocks already seen to be able to retrieve the appropriate text. Finally, the compression ratio is in principle worse than for LZ77 since we are limited in the text segments that we can use. On the other hand, the numbers to code are smaller since we code block positions in $O(\log n)$ bits instead of text positions in $O(\log u)$ bits. Moreover, if we use a simple trick, the compression is in general better than for LZ78 since we are not limited to using just one block. The trick is to represent the pairs $(r, 0)$ as $(2r)$, and the pairs $(r, h+1)$ as $(2r+1, h)$. This pays off because the second element of the pair is frequently zero.

The searching algorithm is like that of LZ77 except because we do not need to search for the blocks and we do not have to recursively find the partial blocks u and v (they simply do not exist now). From the analysis of the LZ77 pattern matching algorithm we have that we work $O(\min(u, n \log m) + R)$ on average and $O(\min(u, mn) + R)$ in the worst case (thanks to the improved algorithm to concatenate blocks). In practice, this algorithm performance is very close to LZ78 pattern matching. We also tried a marked version (called Mark-Hybrid) where for each block a bit is stored which tells whether or not the block will be used again, but as for LZ78, the search time does not improve in practice.

Unlike LZ77, we can use less memory if the compressor restricts the references to a window of the text. Since there are no recursive references, those blocks which are far away in the past need not be stored since they will not be referenced anymore. Hence, as in LZ78, we need the same memory as the compressor. A window of size N can be displayed in $O(N)$ time.

7 Experimental Results

We show in this section our empirical results on the behavior or our search and compression schemes. We first study the compression techniques and later the search performance.

We use mainly two files for the experiments. One is an *English* literary text (from B. Franklin) of 1.29 Mb, filtered to lower-case and with separators normalized. The other is the *DNA* chain of "h.influenzae", of 1.36 Mb. For comparative purposes, we also show the results on some files of the the Calgary Corpus[3]: two books (book*), six troff-formatted scientific articles (paper*) and three source program codes (prog*).

---

[3] ftp://ftp.cpsc.ucalgary.ca/pub/projects/text.compression.corpus/

7.1   Compression Performance

It is interesting to study the compression performance of the algorithms for two reasons: first, we propose a hybrid compression scheme which we have to evaluate in terms of compression ratios. Second, our search algorithms use a technique to code the pairs which speeds up search time but which is suboptimal: the numbers are stored in as many bytes as needed (using the highest bit to denote if there are more bytes or not).

We first compare the number of bits needed to code a file with our hybrid compression scheme against the same number for LZ77 and LZ78. We call this approach "bit-coding". This is aimed to give and idea of the expected compression performance when the file is compressed with a real technique (such as Elias [10] or Huffman codes). Many other improvements are possible. A deeper study of the best techniques for our hybrid compressor is deferred for future study.

Table 1 shows the results. The "Ideal" column counts exactly the bits used by each number stored in the compressed file, while both "Elias" columns count the number of bits needed to represent the numbers using these codes[4] [10]. The letters, on the other hand, are Huffman coded. For *English* and *DNA* we show in a second line the percentages for different variants of the compressors: Block-LZ77, Mark-LZ78 and Mark-Hybrid, respectively. With our Hybrid compression method, we obtain estimated compression ratios comparable to LZ77. The Hybrid and LZ77 compression is better than LZ78 except for *DNA*, where only two bits are necessary to code a letter. Block-LZ77, on the other hand, compresses quite badly.

We now perform a practical comparison using our byte-coding techniques against good LZ77 and LZ78 compressors, namely *gzip* and *Compress* respectively. This is to show how much compression are we loosing in order to ease the searching process.

Table 2 shows the compression ratios achieved. The percentages in the second row of *English* and *DNA* have the same meaning as before. Interestingly, *Compress* is better than *gzip* on *DNA*, which rarely happens on natural language texts. Our compression ratios show a penalty with respect to those of *gzip*. Our byte compression method is very simple, and these results show in which proportion our compression ratios could be improved by engineering techniques, keeping in mind that complicating the encoding of the numbers risks slowing down the pattern matching process.

7.2   Search Algorithms

We compare now the search time for our algorithms against the decompressing and searching approach. The experiments were run on a Sun UltraSparc-1 of 167 MHz, with 64 Mb of RAM, running Solaris 2.5.1. We consider user time, which is within 2% of accuracy with 95% confidence. Time is expressed in seconds everywhere in this section.

---

[4] Recall that Elias-γ precedes the number x by its length in unary, while Elias-δ uses Elias-γ to code that length that precedes the number.

File	Size	Ideal			Elias-γ			Elias-δ		
	(Kb)	LZ77	LZ78	Hybrid	LZ77	LZ78	Hybrid	LZ77	LZ78	Hybrid
English	1,324	29.67%	36.15%	29.28%	59.34%	64.01%	58.57%	48.96%	52.04%	46.17%
		52.45%	*38.01%*	*31.24%*	*104.9%*	*82.31%*	*62.48%*	*74.25%*	*54.71%*	*48.75%*
DNA	1,390	28.03%	25.30%	29.08%	56.06%	47.33%	58.18%	45.77%	37.71%	46.40%
		47.21%	*26.77%*	*31.15%*	*94.43%*	*67.62%*	*62.30%*	*73.14%*	*39.91%*	*49.03%*
book1	751	34.10%	40.70%	35.62%	68.20%	70.83%	71.25%	41.26%	44.96%	41.50%
book2	597	29.33%	40.21%	30.44%	58.66%	69.46%	60.89%	35.51%	44.41%	35.72%
paper1	52	32.33%	46.20%	34.29%	64.53%	77.01%	68.59%	41.05%	51.92%	41.91%
paper2	80	32.68%	43.00%	34.80%	65.27%	72.84%	69.60%	41.08%	48.28%	42.01%
paper3	45	35.10%	45.50%	38.12%	70.07%	76.23%	76.24%	44.84%	51.36%	46.55%
paper4	13	37.60%	47.95%	41.07%	74.74%	78.30%	82.15%	49.92%	54.81%	51.55%
paper5	12	39.85%	50.79%	41.74%	79.13%	82.42%	83.49%	52.63%	57.92%	52.39%
paper6	37	33.60%	47.72%	35.69%	67.03%	79.08%	71.38%	42.91%	53.72%	43.81%
progc	39	32.21%	47.99%	34.16%	64.24%	79.14%	68.32%	41.24%	53.96%	41.95%
progl	70	22.45%	39.10%	23.30%	44.82%	65.83%	44.92%	28.04%	43.85%	27.65%
progp	48	21.34%	40.36%	22.46%	42.54%	66.95%	46.60%	27.16%	45.33%	28.46%

Table 1. Estimated compression ratios with three different methods. For each number in the compressed file, if we note n the bits needed to code it, then *Ideal* counts only n, *Elias-γ* counts $2n$ and *Elias-δ* counts $n + 2\lceil \log_2 n \rceil$. The second line (in italics) of *English* and *DNA* correspond to Block-LZ77, Mark-LZ78 and Mark-Hybrid, respectively.

File	*gzip*	*Compress*	Byte-LZ77	Byte-LZ78	Byte-Hybrid
English	35.58%	38.90%	44.49%	54.41%	43.29%
			79.32%	*56.20%*	*45.24%*
DNA	30.44%	27.96%	41.07%	43.17%	42.23%
			75.24%	*44.90%*	*44.22%*
book1	40.76%	43.19%	53.21%	59.92%	53.30%
book2	33.83%	41.05%	45.60%	58.55%	46.53%
paper1	34.94%	47.17%	54.70%	66.17%	52.67%
paper2	36.19%	43.99%	54.65%	62.02%	52.10%
paper3	38.89%	47.63%	60.19%	67.92%	58.75%
paper4	41.66%	52.36%	69.20%	75.71%	68.24%
paper5	41.78%	55.04%	72.27%	79.47%	68.16%
paper6	34.72%	49.06%	56.84%	69.33 %	54.76%
progc	33.51%	48.32%	54.97%	67.99%	51.95%
progl	22.71%	37.89%	37.82%	55.30%	35.47%
progp	22.77%	38.90%	35.97%	57.20%	34.20%

Table 2. Compression ratios for classical compressors and our byte versions. The second (italics) lines of *English* and *DNA* correspond to Block-LZ77, Mark-LZ78 and Mark-Hybrid, respectively.

In general, searching a compressed text has the additional advantage over the uncompressed text that it performs less I/O. However, this is relevant if we compare compressed versus uncompressed searching. This is not what we compare here: we consider that the text is always compressed. Hence, we measure the cost of searching it without decompressing versus the cost of decompressing it and then searching. Clearly the last task can be done using an intermediate buffer in main memory, and therefore the I/O is the same in both cases.

Figure 7 compares the marked and unmarked versions of LZ78 and the Hybrid compressor. As it can be seen, there is no advantage in practice by the use of marking. Therefore, we do not further consider the marked versions. Another conclusion we take from the figure is that the searcher for Hybrid compression is slightly faster than for LZ78 on *English* but slower for *DNA*. This may be related to the good performance of the LZ78 compressor on *DNA*.

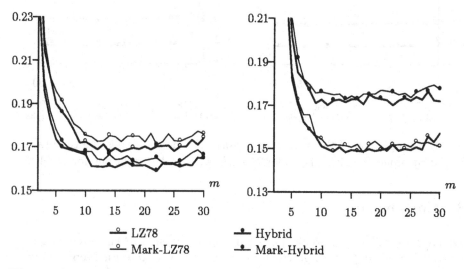

Fig. 7. Comparison between the marked and unmarked versions of LZ78 and Hybrid compressors. The left plot is for *English* text and the right one for *DNA*.

Figure 8 compares all the search algorithms together, as well as decompression (with *gunzip*) plus search time (with Shift-Or and BNDM [25], a bit-parallel searcher which is the fastest in practice together with [27]). It can be seen that Block-LZ77 improves significantly over LZ77, and that the Skip-LZ77 versions improve as the pattern length grows. However, all the LZ77 search algorithms are not competitive against decompressing and searching, especially on *DNA*. On the other hand, both the Hybrid and LZ78 search algorithms are twice as fast as decompressing and searching.

Table 3 compares the time to search a random 10-letter pattern on *English*, *DNA* and the selected files of the Calgary Corpus. We consider the time to decompress with *gunzip* and to search with Shift-Or (as seen, for $m = 10$ the time is very close to BNDM). We show the results for LZ78 and Hybrid only, as LZ77 has been shown to be much inferior.

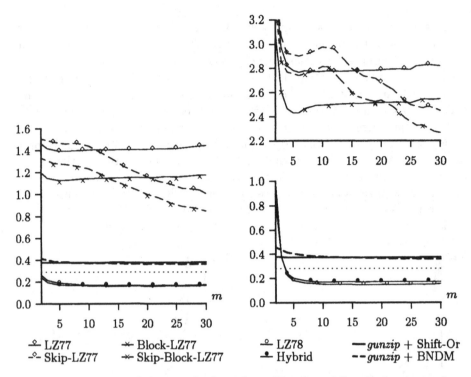

–⚬– LZ77	–✱– Block-LZ77	–⚬– LZ78	— *gunzip* + Shift-Or
–⚬– Skip-LZ77	–✱– Skip-Block-LZ77	–●– Hybrid	-- *gunzip* + BNDM

Fig. 8. Comparison of the search algorithms. The dotted line is the time taken by *gunzip* alone. The left plot is for *English* text and the right one for *DNA*.

8   Conclusions

We have focused in the problem of string matching on Ziv-Lempel compressed text. This is an important practical problem, as it is of interest keep the texts compressed and at the same time being able to efficiently search on them.

We presented a general paradigm to search in a text that is expressed as a sequence of blocks, which abstracts the main features of Ziv-Lempel compression. Then, we applied the technique to the different variants, i.e. LZ77 and LZ78. For LZ78, we are able to search in half the time of uncompressing and searching, while for LZ77 our algorithm, although much slower, is the first one proposed to search on LZ77 compressed text. This motivated us to present a new hybrid compression technique which allows to search as fast as in LZ78 but which keeps many of the features of LZ77 compression, being in practice similar in compression ratios.

Therefore, we are able to search in a compressed text faster than uncompressing and then searching. In general, on the other hand, searching on compressed text at the same speed of on uncompressed text seems difficult to achieve in practice because of a basic problem of locality of reference.

Future work involves studying better the performance of our hybrid compression, both in theory and in practice (especially on finding better methods to encode the numbers while keeping the good search times). We also plan to work

File	*gunzip*	Shift-Or	LZ78	Hybrid
English	28.80	8.90	17.24 (45.7%)	16.65 (44.2%)
DNA	28.10	9.21	15.10 (40.5%)	17.27 (46.3%)
book1	18.40	4.92	10.91 (46.8%)	11.42 (49.0%)
book2	12.40	4.14	8.01 (48.4%)	7.78 (47.0%)
paper1	1.80	1.67	1.88 (54.2%)	1.92 (55.3%)
paper2	2.40	1.76	2.07 (49.8%)	2.18 (52.4%)
paper3	1.80	1.60	1.73 (50.9%)	1.88 (55.3%)
paper4	1.20	1.48	1.50 (56.0%)	1.59 (59.3%)
paper5	0.80	1.42	1.52 (68.5%)	1.54 (69.4%)
paper6	1.90	1.53	1.69 (49.3%)	1.78 (51.9%)
progc	1.50	1.55	1.73 (56.7%)	1.75 (57.4%)
progl	1.90	1.72	1.88 (51.9%)	1.84 (50.8%)
progp	1.20	1.62	1.74 (61.7%)	1.70 (60.3%)

Table 3. Search times for different files, in 1/100-th of seconds. The percentages indicate the time of the compressed searching as a fraction of uncompressing plus Shift-Or searching.

more in understanding the behavior of the LZ77 search algorithm. Finally, we plan to allow for more flexible search, including features such as allowing classes of characters and Hamming errors (some work has been already done in [26]).

This is a field where important theoretical and practical development is necessary, and we have presented new results in both aspects. We hope that more improvements are to come.

References

[1] A. Amir and G. Benson. Efficient two-dimensional compressed matching. In *Proc. Second IEEE Data Compression Conference*, pages 279–288, March 1992.

[2] A. Amir, G. Benson, and M. Farach. Let sleeping files lie: Pattern matching in Z-compressed files. *Journal of Computer and System Sciences*, 52(2):299–307, 1996.

[3] R. Baeza-Yates. Text retrieval: Theory and practice. In *12th IFIP World Computer Congress*, volume I, pages 465–476. Elsevier Science, September 1992.

[4] R. Baeza-Yates and G. Gonnet. A new approach to text searching. *Communications of the ACM*, 35(10):74–82, October 1992.

[5] T. Bell, J. Cleary, and I. Witten. *Text Compression*. Prentice Hall, New Jersey, 1990.

[6] T. Bell and D. Kulp. Longest-match string searching for Ziv-Lempel compression. *Software– Practice and Experience*, 23(7):757–771, July 1993.

[7] J. Bentley, D. Sleator, R. Tarjan, and V. Wei. A locally adaptive data compression scheme. *Communications of the ACM*, 29:320–330, 1986.

[8] R. S. Boyer and J. S. Moore. A fast string searching algorithm. *Communications of the ACM*, 20(10):762–772, 1977.

[9] A. Czumaj, Maxime Crochemore, L. Gasieniec, S. Jarominek, Thierry Lecroq, W. Plandowski, and W. Rytter. Speeding up two string-matching algorithms. *Algorithmica*, 12:247–267, 1994.

[10] P. Elias. Universal codeword sets and representations of the integers. *IEEE Transactions on Information Theory*, 21:194–203, 1975.

[11] M. Farach and M. Thorup. String matching in Lempel-Ziv compressed strings. In *27th ACM Annual Symposium on the Theory of Computing*, pages 703–712, 1995.

[12] E. Fiala and D. Greene. Data compression with finite windows. *Communications of the ACM*, 32(4):490–505, 4 1989.

[13] L. Gasieniec, M.Karpinksi, W.Plandowski, and W. Rytter. Efficient algorithms for Lempel-Ziv encodings. In *Proc. SWAT'96*, 1996.

[14] R. N. Horspool. Practical fast searching in strings. *Software Practice and Experience*, 10:501–506, 1980.

[15] D. Huffman. A method for the construction of minimum-redundancy codes. *Proc. of the I.R.E.*, 40(9):1090–1101, 1952.

[16] J. Kärkkäinen and E. Ukkonen. Sparse suffix trees. In *COCOON'96*, pages 219–230, 1996. LNCS v. 1090.

[17] M. Karpinski, A. Shinohara, and W. Rytter. Pattern matching problem for strings with short descriptions. *Nordic Journal of Computing*, 4(2):172–186, 1997.

[18] T. Kida, M. Takeda, A. Shinohara, and S. Arikawa. Shift-and approach to pattern matching in lzw compressed text. In *Proc. CPM'99*, 1999. To appear.

[19] D. E. Knuth, J. H. Morris, Jr, and V. R. Pratt. Fast pattern matching in strings. *SIAM Journal on Computing*, 6(1):323–350, 1977.

[20] U. Manber. A text compression scheme that allows fast searching directly in the compressed file. *ACM Transactions on Information Systems*, 15(2):124–136, 1997.

[21] V. Miller and M. Wegman. Variations on a theme by Ziv and Lempel. In *Combinatorial Algorithms on Words*, volume 12 of *NATO ASI Series F*, pages 131–140. Springer-Verlag, 1985.

[22] A. Moffat. Word-based text compression. *Software Practice and Experience*, 19(2):185–198, 1989.

[23] E. Moura, G. Navarro, N. Ziviani, and R. Baeza-Yates. Direct pattern matching on compressed text. In *Proc. SPIRE'98*, pages 90–95. IEEE CS Press, 1998.

[24] E. Moura, G. Navarro, N. Ziviani, and R. Baeza-Yates. Fast searching on compressed text allowing errors. In *Proc. SIGIR'98*, pages 298–306. York Press, 1998.

[25] G. Navarro and M. Raffinot. A bit-parallel approach to suffix automata: Fast extended string matching. In *Proc. CPM'98*, LNCS v. 1448, pages 14–33, 1998.

[26] G. Navarro and M. Raffinot. A general practical approach to pattern matching over Ziv-Lempel compressed text. Technical Report TR/DCC-98-12, Dept. of Computer Science, Univ. of Chile, 1998.

[27] D. Sunday. A very fast substring search algorithm. *Communications of the ACM*, 33(8):132–142, August 1990.

[28] T. A. Welch. A technique for high performance data compression. *IEEE Computer Magazine*, 17(6):8–19, June 1984.

[29] I. Witten, R. Neal, and J. Cleary. Arithmetic coding for data compression. *Communications of the ACM*, 30(6):520–541, 1987.

[30] M. Zipstein. Data compression with factor automata. *Theor. Comput. Sci.*, 92(1):213–221, 1992.

[31] J. Ziv and A. Lempel. A universal algorithm for sequential data compression. *IEEE Trans. Inf. Theory*, 23:337–343, 1977.

[32] J. Ziv and A. Lempel. Compression of individual sequences via variable length coding. *IEEE Trans. Inf. Theory*, 24:530–536, 1978.

Pattern Matching in Text Compressed by Using Antidictionaries

Yusuke Shibata, Masayuki Takeda, Ayumi Shinohara, and Setsuo Arikawa

Department of Informatics, Kyushu University 33
Fukuoka 812-8581, Japan
{yusuke, takeda, ayumi, arikawa}@i.kyushu-u.ac.jp
http://www.i.kyushu-u.ac.jp

Abstract. In this paper we focus on the problem of compressed pattern matching for the text compression using antidictionaries, which is a new compression scheme proposed recently by Crochemore et al. (1998). We show an algorithm which preprocesses a pattern of length m and an antidictionary M in $O(m^2 + \|M\|)$ time, and then scans a compressed text of length n in $O(n + r)$ time to find all pattern occurrences, where $\|M\|$ is the total length of strings in M and r is the number of the pattern occurrences.

1 Introduction

Compressed pattern matching is one of the most interesting topics in the combinatorial pattern matching, and many studies have been undertaken on this problem for several compression methods from both theoretical and practical viewpoints. See Table 1. One important goal of compressed pattern matching is to achieve a linear time complexity that is proportional not to the original text length but to the compressed text length.

Recently, Crochemore *et al.* proposed a new compression scheme: *text compression using antidictionary* [8]. Contrary to the compression methods that make use of dictionaries, which are particular sets of strings occurring in texts, the new scheme exploits an *antidictionary* that is a finite set of strings that do not occur as factors in text, i.e. that are *forbidden*. Let $a_1 \ldots a_n \in \{0,1\}^+$ be the text to be compressed. Suppose we have read a prefix $a_1 \ldots a_j$ at a certain moment. If the string $a_i \ldots a_j b$ ($i \leq j$, $b \in \{0,1\}$) is a forbidden word, namely, is in the antidictionary, then the next symbol a_{j+1} cannot be b. In other words, the next symbol a_{j+1} is predictable. Based on this idea, the compression method removes such predictable symbols from the text. The compression and the decompression are performed by using the automaton accepting the set of strings in which no forbidden words occur as factors.

In this paper we focus on the problem of compressed pattern matching for the text compression using antidictionaries. We present an algorithm that solves the problem in $O(m^2 + \|M\| + n + r)$ time using $O(m^2 + \|M\|)$ space, where m and n are the pattern length and the compressed text length, respectively, $\|M\|$

M. Crochemore, M. Paterson (Eds.): CPM'99, LNCS 1645, pp. 37–49, 1999.
© Springer-Verlag Berlin Heidelberg 1999

Table 1. Compressed pattern matching.

compression method	compressed pattern matching algorithms
run-length	Eilam-Tzoreff and Vishkin [11]
run-length (two dim.)	Amir, Landau, and Vishkin [6]; Amir and Benson [2,3]; Amir, Benson, and Farach [5]
LZ77	Farach and Thorup [12]; Gąsieniec, Karpinski, Plandowski, and Rytter [14]
LZW	Amir, Benson, and Farach [4]; Kida, Takeda, Shinohara, Miyazaki, and Arikawa [17]; Kida, Takeda, Shinohara, and Arikawa [16]
straight-line program	Karpinski, Rytter, and Shinohara [15]; Miyazaki, Shinohara, and Takeda [20]
Huffman	Fukamachi, Shinohara, and Takeda [13]; Miyazaki, Fukamachi, Takeda, and Shinohara [19]
finite state encoding	Takeda [22]
word based encoding	Moura, Navarro, Ziviani, and Baeza-Yates [9,10]
pattern substitution	Manber [18]; Shibata, Kida, Fukamachi, Takeda, A. Shinohara, T. Shinohara, Arikawa [21]

denotes the total length of strings in antidictionary M, and r is the number of pattern occurrences. Since M is a part of the compressed representation of text, the text scanning time is $O(\|M\| + n + r)$, which is linear in the compressed text length $\|M\| + n$, when ignoring r. Moreover, in the case where a set of text files share a common antidictionary [8], we can regard the $O(\|M\|)$ time processing of M as a preprocessing. Then the $O(n + r)$ time text scanning will be fast in practice. The proposed algorithm thus has desirable properties.

2   Preliminaries

Strings x, y, and z are said to be a *prefix*, *factor*, and *suffix* of the string $u = xyz$, respectively. The sets of prefixes, factors, and suffixes of a string u are denoted by *Prefix(u)*, *Factor(u)*, and *Suffix(u)*, respectively. A prefix, factor, and suffix of a string u is said to be *proper* if it is not u. The length of a string u is denoted by $|u|$. The empty string is denoted by ε, that is, $|\varepsilon| = 0$. The ith symbol of a string u is denoted by $u[i]$ for $1 \leq i \leq |u|$, and the factor of a string u that begins at position i and ends at position j is denoted by $u[i : j]$ for $1 \leq i \leq j \leq |u|$. The reversed string of a string u is denoted by u^R. The total length of strings of a set S is denoted by $\|S\|$. For strings x and y, denote by $Occ(x, y)$ the set of occurrences of x in y. That is,

$$Occ(x, y) = \{ |x| \leq i \leq |y| \mid x = y[i - |x| + 1 : i] \}.$$

The next lemma follows from the periodicity lemma.

Lemma 1. *If $Occ(x, y)$ has more than two elements and the difference of the maximum and the minimum elements is at most $|x|$, then it forms an arithmetic progression, in which the step is the smallest period of x.*

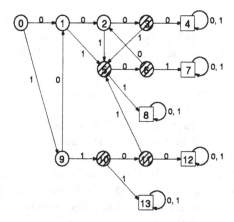

Fig. 1. Automaton $\mathcal{A}(M)$ for $M = \{0000, 111, 011, 0101, 1100\}$. Circles and squares denote the final and the nonfinal states, respectively. Shaded circles denote the predict states.

3   Text Compression Using Antidictionary

In this section we describe the text compression scheme recently proposed by Crochemore *et al.* [8].

3.1   Method

Let $B = \{0, 1\}$. Suppose that $T \in B^+$ be the text to be compressed. A *forbidden word* for T is a string $u \in B^+$ that is not a factor of T. A forbidden word is said to be *minimal* if it has no proper factor that is forbidden. An *antidictionary* for T is a set of minimal forbidden words for T.

Let M be an antidictionary for T. Then the text T is in the set $B^* \backslash B^* M B^*$. The automaton accepting the set $B^* \backslash B^* M B^*$ can be built from M in $O(\|M\|)$ time in a similar way to the construction of the Aho-Corasick pattern matching machine [1]. We denote the automaton by

$$\mathcal{A}(M) = (Q, B, \delta, \varepsilon, M),$$

where $Q = Prefix(M)$ is the set of states; B is the alphabet; δ is the state transition function from $Q \times B$ to Q defined as

$$\delta(u, a) = \begin{cases} u, & \text{if } u \in M; \\ \text{longest string in } Q \cap \textit{Suffix}(ua), & \text{otherwise;} \end{cases}$$

ε is the initial state; M is the set of final states. Figure 1 shows the automaton $\mathcal{A}(M)$ for $M = \{0000, 111, 011, 0101, 1100\}$, which is an antidictionary for text $T = 11010001$.

The encoder and the decoder in this compression scheme are obtained directly from the automaton $\mathcal{A}(M)$. The encoder $\mathcal{E}(M)$ is a generalized sequential machine based on $\mathcal{A}(M)$ with output function $\lambda : Q \times B$ defined by

$$\lambda(u, a) = \begin{cases} a, & \text{if } Deg(u) = 2; \\ \varepsilon, & \text{otherwise,} \end{cases}$$

where $Deg(u) = |\{a \in B | \delta(u, a) \notin M\}|$. The decoder $\mathcal{D}(M)$ is a generalized sequential machine obtained by swapping the input label and the output label on each arc of the encoder $\mathcal{E}(M)$. Figure 2 illustrates the move of the encoder $\mathcal{E}(M)$ based on $\mathcal{A}(M)$ of Fig. 1 which takes as input $T = 11010001$ and emits 110. It should be noted that, any prefix of 1101000100 with length greater than 6 is compressed into the same string 110. For a decompression we therefore need the length of T together with the encoded string itself. Formally, the compressed representation of T is a triple $\langle M, b_1 \ldots b_n, N \rangle$, where M is an antidictionary, $b_1 \ldots b_n$ is output from the encoder, and N is the length of T.

Let us denote by $MF(T)$ the set of all minimal forbidden words for T. In the case of binary alphabet we have $|MF(T)| \le 2 \cdot |T| - 2$ as shown in [7]. To shorten the representation size of the above triple, we need a way to build a 'good' antidictionary as a subset of $MF(T)$. Crochemore et al. presented in [8] a simple method in which antidictionary is the set of forbidden words of length at most k, where k is a parameter. It is reported in [8] that the compression ratio in practice is comparable to pkzip.

$$
\begin{array}{llcccccccc}
\text{input:} & 1 & 1 & 0 & 1 & 0 & 0 & 0 & 1 \\
\text{state:} & 0 \rightarrow 9 \rightarrow & 10 \rightarrow & 11 \rightarrow & 5 \rightarrow & 6 \rightarrow & 2 \rightarrow & 3 \rightarrow & 5 \\
\text{output:} & 1 & 1 & \varepsilon & \varepsilon & \varepsilon & \varepsilon & 0 & \varepsilon
\end{array}
$$

Fig. 2. Move of encoder $\mathcal{E}(M)$ for $T = 11010001$.

3.2   Decoder without ε-Moves

Note that the decoder $\mathcal{D}(M)$ mentioned above has ε-moves. For a simple presentation of our algorithm, we shall define a generalized sequential machine $\mathcal{G}(M)$ obtained by eliminating the ε-moves from the decoder $\mathcal{D}(M)$.

Let us partition the set Q into four disjoint subsets M, Q_0, Q_1, and Q_2 by

$$Q_i = \{u \in Q \backslash M \mid Deg(u) = i\} \qquad (i = 0, 1, 2).$$

A state p in Q_1 is called a *predict state* because of the uniqueness of outgoing arc when ignoring the arcs into states in M. Namely, there exists exactly one symbol a such that $\delta(p, a) \notin M$. We denote such symbol a by *NextSymbol(p)*, and denote by *NextState(p)* the state $\delta(p, a)$.

Consider, for $p \in Q_1$, the sequence $p_1, p_2, \ldots$ of states in Q_1 defined by $p_1 = p$ and $p_{i+1} = NextState(p_i)$ $(i = 1, 2, \ldots)$. There are two cases: One is the case that there exists an integer $m > 0$ such that, for $i = 1, 2, \ldots, m - 1$, $p_i \in Q_1$, and $p_m \in Q_0 \cup Q_2$. The other is the case of no such integer m, namely, the sequence continues infinitely. Let us call the sequence the *predict path* of p, and denote by *Terminal*(p) the last state p_m. In the infinite case, let *Terminal*$(p) = \bot$, where $\bot$ is a special state not in Q. (Therefore, *Terminal*$(p) \in Q_0 \cup Q_2 \cup \{\bot\}$.) The finite/semi-infinite string spelled out by the predict path of $p \in Q_1$ is denoted by *Sequence*(p). It is easy to see that:

Lemma 2. *For any $p \in Q_1$, there exist $u, v \in B^*$ with $|uv| < |Q_1|$ such that*

$$Sequence(p) = u\, v\, v\, \cdots .$$

Now we are ready to define a generalized sequential machine $\mathcal{G}(M)$, where the set of states is $Q_0 \cup Q_2 \cup \{\bot\}$; the state transition function is $\delta_{\mathcal{G}} : Q_2 \times B \to Q_0 \cup Q_2 \cup \{\bot\}$ defined by

$$\delta_{\mathcal{G}}(u, a) = \begin{cases} Terminal(\delta(u, a)), & \delta(u, a) \in Q_1; \\ \delta(u, a), & \text{otherwise;} \end{cases}$$

the output function is $\lambda_{\mathcal{G}} : Q_2 \times B \to B^+ \cup B^\infty$ defined by

$$\lambda_{\mathcal{G}}(u, a) = \begin{cases} a \cdot Sequence(\delta(u, a)), & \delta(u, a) \in Q_1; \\ a, & \text{otherwise,} \end{cases}$$

where B^∞ denotes the set of semi-infinite strings over B. Figure 3 shows the decoder $\mathcal{G}(M)$ obtained in this way from the automaton $\mathcal{A}(M)$ of Fig. 1.

Decompression algorithm using $\mathcal{G}(M)$ is shown in Fig. 4. It should be emphasized that, if the decoder $\mathcal{G}(M)$ enters a state q and then reads a symbol a such that $\lambda_{\mathcal{G}}(q, a)$ is semi-infinite, the symbol is the last symbol of the output from the encoder $\mathcal{E}(M)$. In this case the decoder $\mathcal{G}(M)$ halts after emitting an appropriate length prefix of $\lambda_{\mathcal{G}}(q, a)$ according to the value of N.

4   Main Result

Generally, most of text compression methods can be recognized as mechanisms to factorize a text into several blocks as $\mathcal{T} = u_1 u_2 \ldots u_n$ and to store a sequence of 'representations' of blocks u_i. In the LZW compression, for example,

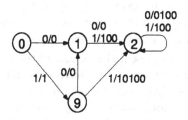

Fig. 3. Decoder $\mathcal{G}(M)$ for $M = \{0000, 111, 011, 0101, 1100\}$.

```
Input.      A compressed representation ⟨M, b₁...bₙ, N⟩ of a text T = T[1 : N].
Output.     Text T.
begin
    ℓ := 0;
    q := ε;
    for i := 1 to n − 1 do begin
        u := λ_G(q, bᵢ);
        q := δ_G(q, bᵢ);
        ℓ := ℓ + |u|;
        print u
    end;
    u := λ_G(q, bₙ);
    print the prefix of u with length N − ℓ
end.
```

Fig. 4. Decompression by $\mathcal{G}(M)$.

the representation of a block u_i is just an integer which indicates the node of dictionary trie representing the string u_i. In the case of the compression using antidictionaries, the way of representation of block is slightly complicated.

Consider how to simulate the move of the KMP automaton for a pattern $\mathcal{P}$ running on the uncompressed text T. Let $\delta_{KMP} : \{0, 1, \ldots, m\} \times B \to \{0, 1, \ldots, m\}$ be the state transition function of the KMP automaton for $\mathcal{P} = \mathcal{P}[1 : m]$. We extend δ_{KMP} to the domain $\{0, 1, \ldots, m\} \times B^*$ in the standard manner. We also define the function λ_{KMP} on $\{0, 1, \ldots, m\} \times B^*$ by

$$\lambda_{KMP}(j, u) = \{1 \leq i \leq |u| \mid \mathcal{P} \text{ is a suffix of string } \mathcal{P}[1 : j] \cdot u[1 : i] \}.$$

We want to devise a pattern matching algorithm which takes as input a sequence of representations of blocks $u_1, u_2, \ldots, u_n$ of T and reports all occurrences of $\mathcal{P}$ in T in $O(n + r)$ time, where $r = |Occ(\mathcal{P}, T)|$. Then we need a mechanism for obtaining in $O(1)$ time the value $\delta_{KMP}(j, u)$ and a linear size representation of the set $\lambda_{KMP}(j, u)$. In the case of the LZW compression such mechanism can be realized in $O(m^2 + n)$ time using $O(m^2 + n)$ space as stated in [4] and [17]. Similar idea can also be applied to the case of text compression by antidictionaries, except that block u_i, which will be an input to the second arguments of δ_{KMP} and λ_{KMP}, is represented in a different manner.

In our case a block u_i is represented as a pair of the current state q of $\mathcal{G}(M)$ and the first symbol b_i of u_i. Therefore we have to keep the state transitions of $\mathcal{G}(M)$. An overview of our algorithm is shown in Fig. 5. The algorithm makes $\mathcal{G}(M)$ run on $b_1 \ldots b_n$ to know inputs $u_1, u_2, \ldots, u_n$ to the KMP automaton being simulated. Figure 6 illustrates the move of the algorithm searching the compressed text 110 for the pattern $\mathcal{P} = 0001$.

We have the following theorems which will be proved in the next section.

Input.   A compressed representation $\langle M, b_1 b_2 ... b_n, N \rangle$ of a text $\mathcal{T} = \mathcal{T}[1 : N]$,
and a pattern $\mathcal{P} = \mathcal{P}[1 : m]$.
Output.   All positions at which $\mathcal{P}$ occurs in $\mathcal{T}$.
begin
    /* *Preprocessing* */
    Construct the KMP automata and the suffix tries for $\mathcal{P}$ and $\mathcal{P}^R$;
    Construct the automaton $\mathcal{A}(M)$ from M;
    Construct the predict path graph from $\mathcal{A}(M)$;
    Perform the processing required for $\delta_\mathcal{G}$, δ_{KMP}, and λ_{KMP} (See Section 5.);

    /* *Text scanning* */
    $\ell := 0$;
    $q := \varepsilon$;
    $state := 0$;
    for $i := 1$ **to** $n - 1$ **do begin**
        $u := \lambda_\mathcal{G}(q, b_i)$;
        $q := \delta_\mathcal{G}(q, b_i)$;
        for each $p \in \lambda_{\mathrm{KMP}}(state, u)$ **do**
            Report a pattern occurrence that ends at position $\ell + p$;
        $state := \delta_{\mathrm{KMP}}(state, u)$;
        $\ell := \ell + |u|$
    end;
    $u := \lambda_\mathcal{G}(q, b_n)$;
    for each $p \in \lambda_{\mathrm{KMP}}(state, u)$ such that $\ell + p \le N$ **do**
        Report a pattern occurrence that ends at position $\ell + p$
end.

Fig. 5. Pattern matching algorithm.

Theorem 1. *The function which takes as input $(q, a) \in Q_2 \times B$ and returns in $O(1)$ time the value $\delta_\mathcal{G}(q, a)$, can be realized in $O(\|M\|)$ time using $O(\|M\|)$ space.*

Theorem 2. *The function which takes as input a triple $(j, q, a) \in \{0, \ldots, m\} \times Q_2 \times B$ and returns in $O(1)$ time the value*

$$\delta_{\mathrm{KMP}}(j, u) \qquad (u = \lambda_\mathcal{G}(q, a)),$$

can be realized in $O(\|M\| + m^2)$ time using $O(\|M\| + m^2)$ space.

Theorem 3. *The function which takes as input a triple $(j, q, a) \in \{0, \ldots, m\} \times Q_2 \times B$ and returns in $O(1)$ time a linear size representation of the set*

$$\lambda_{\mathrm{KMP}}(j, u) \qquad (u = \lambda_\mathcal{G}(q, a)),$$

can be realized in $O(\|M\| + m^2)$ time using $O(\|M\| + m^2)$ space.

Then we have the following result.

input :		1	1	0	
state of $\mathcal{G}(M)$:	0 $\longrightarrow$ 9	$\longrightarrow$	2	$\longrightarrow$ 2	
u :		1	10100	0100	
state of KMP automaton :	0 $\longrightarrow$ 0	$\longrightarrow$	2	$\longrightarrow$ 2	
output :		$\emptyset$	$\emptyset$	{8}	

Fig. 6. Move of pattern matching algorithm when $\mathcal{T} = 110100010$ and $\mathcal{P} = 0001$.

Theorem 4. *The problem of compressed pattern matching for the text compression using antidictionaries can be solved in $O(\|M\| + n + m^2 + r)$ time using $O(\|M\| + m^2)$ space.*

5   Algorithm in Detail

This section gives a detailed presentation of the algorithm to prove Theorems 1, 2, and 3.

5.1   Proof of Theorem 1

For a realization of $\delta_{\mathcal{G}}$, we have to find, for each $q \in Q_0 \cup Q_2 \cup \{\perp\}$, the pairs $(p, b) \in Q_2 \times B$ such that $\delta(p, b) = p' \in Q_1$ and $Terminal(p') = q$. First of all, we mention the graph consisting of the predict paths, which plays an important role in this proof.

Consider the subgraph of $\mathcal{A}(M)$ in which the arcs are limited to the outgoing arcs from predict nodes. We add auxiliary nodes $v = \langle p, b \rangle$ and new arcs labelled b from v to $q \in Q_1$ such that $p \in Q_2$, $b \in B$, and $\delta(p, b) = q$ to the subgraph. We call the resulting graph *predict path graph*. Figure 7 shows the predict path graph obtained from $\mathcal{A}(M)$ of Fig. 1.

The predict path graph illustrates, for $(p, b) \in Q_2 \times B$, the string $\lambda_{\mathcal{G}}(p, b)$ as a path which starts at the auxiliary node $\langle p, b \rangle$, passes through nodes in Q_1, and either finally encounters a node in $Q_0 \cup Q_2$, or flows into a loop consisting only of nodes in Q_1. A connected component of the predict path graph falls into two classes: (a) a tree which has as root a node in $Q_0 \cup Q_2$ and has as leaves

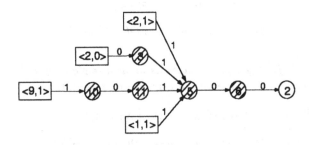

Fig. 7. Predict path graph. Rectangles denote the auxiliary nodes.

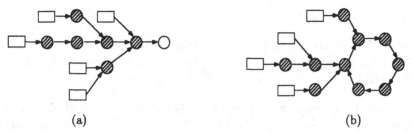

Fig. 8. Connected components of predict path graph.

auxiliary nodes, and (b) a loop with trees, each of which has as root a node on the loop and has leaves auxiliary nodes. See Fig. 8.

Now we are ready to prove Theorem 1. Construction of $\delta_\mathcal{G}$ is as follows: First, we set $\delta_\mathcal{G}(p,b) = \delta(p,b)$ for every $(p,b) \in Q_2 \times B$ with $\delta(p,b) \in Q_0 \cup Q_2$. Next, for every node $q \in Q_0 \cup Q_2$ of the predict path graph, we traverse the tree that has q as root. Note that the leaves of the tree are auxiliary nodes $\langle p, b \rangle$ such that $Terminal(\delta(p,b)) = q$, and we can set $\delta_\mathcal{G}(p,b) = q$. Finally, for every node q on loops of the predict path graph, we traverse the tree that has q as root. The leaves of the tree are auxiliary nodes $\langle p, b \rangle$ such that $Terminal(\delta(p,b)) = \bot$, and hence we set $\delta_\mathcal{G}(p,b) = \bot$. The total time complexity is linear in the number of nodes of the predict path graph, i.e. $O(\|M\|)$. The proof is now complete.

5.2   Proof of Theorem 2

In the following discussions, we are frequently faced with the need to get some value as a function of u, the strings that are spelled out by the paths from auxiliary nodes. Even when the value for each path can be computed in time proportional to the path length, the total time complexity is not $O(\|M\|)$ since more than one path can share common arcs.

Suppose that the value for each path can be computed by making an automaton run on the path in the reverse direction. Then, we can compute the values for such paths by traversing every tree in the depth-first-order using a stack. Since this method enables us to 'share' the computation for a common suffix of two strings, the total time complexity is linear in the number of arcs, i.e. $O(\|M\|)$. This technique plays a key role in the following proofs.

For an integer j with $0 \le j \le m$ and for a factor u of $\mathcal{P}$, let us denote by $N_1(j, u)$ the largest integer k with $0 \le k \le j$ such that $\mathcal{P}[j - k + 1 : j] \cdot u$ is a prefix of $\mathcal{P}$. Let $N_1(j, u) = nil$, if no such integer exists. Then, we have:

$$\delta_{\mathrm{KMP}}(j, u) = \begin{cases} N_1(j, u) + |u|, & \text{if } u \text{ is a factor of } \mathcal{P} \text{ and } N_1(j, u) \ne nil; \\ \delta_{\mathrm{KMP}}(0, u), & \text{otherwise.} \end{cases}$$

We assume that the second argument u of N_1 is given as a node of the suffix trie for $\mathcal{P}$. Amir et al. [4] showed the following fact.

Lemma 3 (Amir et al. 1996). *The function which takes as input $(j, u) \in \{0, \ldots, m\} \times Factor(\mathcal{P})$ and returns the value $N_1(j, u)$ in $O(1)$ time, can be realized in $O(m^2)$ time using $O(m^2)$ space.*

We have also the next lemma.

Lemma 4. *The function which takes as input $(q, a) \in Q_2 \times B$ and returns $u = \lambda_G(q, a)$ as a node of the suffix trie for $\mathcal{P}$ when $u \in Factor(\mathcal{P})$, can be realized in $O(\|M\| + m^2)$ time using $O(\|M\| + m^2)$ space.*

Proof. We use the technique mentioned above. We can ignore the infinite strings. That is, we can ignore the trees in which a root is on a loop. Consider the problem of determining whether u^R is a factor of $\mathcal{P}^R$. It can be solved in $O(\min\{|u|, m\})$ time using the suffix trie for $\mathcal{P}^R$. If u^R is a factor of $\mathcal{P}^R$, the node u of the suffix trie for $\mathcal{P}$ can be determined directly from the node u^R of the suffix trie for $\mathcal{P}^R$ assuming a trivial one-to-one mapping between the two suffix tries, which can be computed in $O(m^2)$ time.

Lemma 5. *The function which takes as input $(q, a) \in Q_2 \times B$ such that $u = \lambda_G(q, a)$ is finite and returns in $O(1)$ time the value $\delta_{\mathrm{KMP}}(0, u)$, can be realized in $O(\|M\| + m)$ time using $O(\|M\| + m)$ space.*

Proof. We use the technique mentioned above again. We have to consider the problem of finding the length of longest suffix of u that is also a prefix of $\mathcal{P}$. This is equivalent to finding the length of longest prefix of u^R that is also a suffix of $\mathcal{P}^R$. It is solved in $O(\min\{|u|, m\})$ time using the suffix tree for $\mathcal{P}^R$. We can ignore the trees in which a root is on a loop.

Theorem 2 follows from the lemmas above.

5.3   Proof of Theorem 3

According to whether a pattern occurrence covers the boundary between the strings $\mathcal{P}[1:j]$ and u, we can partition the set $\lambda_{\mathrm{KMP}}(j, u)$ into two disjoint subsets as follows.

$$\lambda_{\mathrm{KMP}}(j, u) = \lambda_{\mathrm{KMP}}(j, \tilde{u}) \cup X(u),$$

where

$$X(u) = \{|\mathcal{P}| \leq i \leq |u| \mid \mathcal{P} \text{ is a suffix of } u[1:i]\},$$

and $\tilde{u}$ is the longest prefix of u that is also a proper suffix of $\mathcal{P}$. Let

$$Y(j, \ell) = Occ(\mathcal{P}, \mathcal{P}[1:j] \cdot \mathcal{P}[m - \ell + 1:m]) \ominus j,$$

where $\ominus$ denotes the element-wise subtraction. It is easy to see $\lambda_{\mathrm{KMP}}(j, \tilde{u}) = Y(j, |\tilde{u}|)$. It follows from Lemma 1 that the set $Y(j, \ell)$ has the following property:

Lemma 6. *If $Y(j, \ell)$ has more than two elements, it forms an arithmetic progression, where the step is the smallest period of $\mathcal{P}$.*

Lemma 7. *The function which takes as input* $(j, \ell) \in \{0, \ldots, m\} \times \{0, \ldots, m\}$ *and returns in* $O(1)$ *time an* $O(1)$ *space representation of the set* $Y(j, \ell)$, *can be realized in* $O(m^2)$ *time using* $O(m^2)$ *space.*

Proof. It follows from Lemma 6 that $Y(j, \ell)$ can be stored in $O(1)$ space as a pair of the minimum and the maximum values in it. The table storing the minimum values of $Y(j, \ell)$ for all (j, ℓ) can be computed in $O(m^2)$ time as stated in [4]. (Table N_2 defined in [4] satisfies $\min(Y(j, \ell)) = m - N_2(j, \ell)$.) By reversing the pattern $\mathcal{P}$, the table the maximum values is also computed in $O(m^2)$ time. The smallest period of $\mathcal{P}$ is computed in $O(m)$ time.

Lemma 8. *The function which takes as input* $(q, a) \in Q_2 \times B$ *and returns in* $O(1)$ *time the value* $|\tilde{u}|$ *with* $u = \lambda_g(q, a)$, *can be realized in* $O(\|M\| + m)$ *time using* $O(\|M\| + m)$ *space.*

Proof. We shall consider the problem of finding the length of longest suffix of u^R that is also a proper prefix of $\mathcal{P}^R$. This can be solved by using the KMP automaton for $\mathcal{P}^R$. But we have to consider the case where u is semi-infinite. In the finite string case, we make the automaton start at the root of tree with initial state. But in the infinite string case, we must change the value of the initial state. Let v be the string spelled out by the loop starting at the root of the tree being considered. We must pay attention to the case where a pattern suffix is also a prefix of the string v^ℓ with $\ell > 0$. To determine the correct value of the initial state at the root node, we make the automaton go around the loop exactly ℓ times and stop it at the root node that is the starting point, where ℓ is the smallest integer with $\ell \cdot |v| > |\mathcal{P}|$. The state of the automaton at that moment is the desired value.

Lemma 9. *The function which takes as input* $(q, a) \in Q_2 \times B$ *and returns in* $O(1)$ *time a linear size representation of the set* $X(u)$ *with* $u = \lambda_g(q, a)$, *can be realized in* $O(\|M\| + m)$ *time using* $O(\|M\| + m)$ *space.*

Proof. By using the KMP automaton for the reversed pattern, we mark the predict nodes at which the pattern begins. Suppose that every predict node has a pointer to the nearest proper ancestor that is marked. Such pointers are realized using $O(\|M\|)$ time and space. This enables us to get the elements of $X(u)$ in $O(|X(u)|)$ time.

Theorem 3 follows from the lemmas above.

6   Concluding Remarks

In this paper we focused on the problem of compressed pattern matching for the text compression using antidictionaries proposed recently Crochemore *et al.* [8]. We presented an algorithm which has a linear time complexity proportional to the compressed text length, when we exclude the pattern preprocessing. We are now implementing the algorithm to evaluate its performance from practical viewpoints. In [16] we showed that the Shift-And approach is effective in the compressed pattern matching for the LZW compression. We think that the Shift-And approach will be substituted for the KMP automaton approach presented in this paper and show a good performace in practice when the pattern length m is not so large, say $m \leq 32$.

For a long pattern we can also consider the following method. Let k be the length of the longest forbidden word in the antidictionary. By using the syncronizing property [8], we obtain:

Lemma 10. *If $|\mathcal{P}| \geq k - 1$, then $\delta(u, \mathcal{P}) = \delta(\varepsilon, \mathcal{P})$ for any state u in Q such that $\delta(u, \mathcal{P}) \notin M$.*

Let $p = \delta(\varepsilon, \mathcal{P})$. Since $p \in M$ implies that $\mathcal{P}$ cannot occur in $\mathcal{T}$, we can assume $p \notin M$. If p is in Q_1, then let $q = Terminal(p)$. Otherwise, let $q = p$. We can monitor whether the state of $\mathcal{A}(M)$ is in state p by using the function $\delta_{\mathcal{G}}$ to check $\mathcal{G}(M)$ is in state q. If so, we shall confirm it. Our preliminary experiments suggest that this search method is efficient in practice.

References

1. A. V. Aho and M. Corasick. Efficient string matching: An aid to bibliographic search. *Comm. ACM*, 18(6):333–340, 1975.
2. A. Amir and G. Benson. Efficient two-dimensional compressed matching. In *Proc. Data Compression Conference'92*, page 279, 1992.
3. A. Amir and G. Benson. Two-dimensional periodicity and its application. In *Proc. 3rd Ann. ACM-SIAM Symp. on Discrete Algorithms*, pages 440–452, 1992.
4. A. Amir, G. Benson, and M. Farach. Let sleeping files lie: Pattern matching in Z-compressed files. *Journal of Computer and System Sciences*, 52:299–307, 1996.
5. A. Amir, G. Benson, and M. Farach. Optimal two-dimensional compressed matching. *Journal of Algorithms*, 24(2):354–379, 1997.
6. A. Amir, G. M. Landau, and U. Vishkin. Efficient pattern matching with scaling. *Journal of Algorithms*, 13(1):2–32, 1992.
7. M. Crochemore, F. Mignosi, and A. Restivo. Minimal forbidden words and factor automata. In L. Brim, J. Gruska, and J. Zlatuska, editors, *Proc. 23rd Internationial Symp. on Mathematical Foundations of Computer Science*, volume 1450 of *Lecture Notes in Computer Science*, pages 665–673. Springer-Verlag, 1998.
8. M. Crochemore, F. Mignosi, A. Restivo, and S. Salemi. Text compression using antidictionaries. Technical Report IGM-98-10, Institut Gaspard-Monge, 1998.
9. E. S. de Moura, G. Navarro, N. Ziviani, and R. Baeza-Yates. Direct pattern matching on compressed text. In *Proc. 5th International Symp. on String Processing and Information Retrieval*, pages 90–95. IEEE Computer Society, 1998.
10. E. S. de Moura, G. Navarro, N. Ziviani, and R. Baeza-Yates. Fast sequencial searching on compressed texts allowing errors. In *Proc. 21st Ann. International ACM SIGIR Conference on Research and Development in Information Retrieval*, pages 298–306. York Press, 1998.
11. T. Eilam-Tzoreff and U. Vishkin. Matching patterns in strings subject to multi-linear transformations. *Theoretical Computer Science*, 60(3):231–254, 1988.

12. M. Farach and M. Thorup. String-matching in Lempel-Ziv compressed strings. In *Proc. 27th Ann. ACM Symp. on Theory of Computing*, pages 703–713, 1995.

13. S. Fukamachi, T. Shinohara, and M. Takeda. String pattern matching for compressed data using variable length codes. Submitted, 1998.

14. L. Gąsieniec, M. Karpinski, W. Plandowski, and W. Rytter. Efficient algorithms for Lempel-Ziv encoding. In *Proc. 4th Scandinavian Workshop on Algorithm Theory*, volume 1097 of *Lecture Notes in Computer Science*, pages 392–403. Springer-Verlag, 1996.

15. M. Karpinski, W. Rytter, and A. Shinohara. An efficient pattern-matching algorithm for strings with short descriptions. *Nordic Journal of Computing*, 4:172–186, 1997.

16. T. Kida, M. Takeda, A. Shinohara, and S. Arikawa. Shift-And approach to pattern matching in LZW compressed text. In *Proc. 10th Ann. Symp. on Combinatorial Pattern Matching*, Lecture Notes in Computer Science. Springer-Verlag, 1999. to appear.

17. T. Kida, M. Takeda, A. Shinohara, M. Miyazaki, and S. Arikawa. Multiple pattern matching in LZW compressed text. In *Proc. Data Compression Conference '98*, pages 103–112. IEEE Computer Society, 1998.

18. U. Manber. A text compression scheme that allows fast searching directly in the compressed file. In *Proc. 5th Ann. Symp. on Combinatorial Pattern Matching*, volume 807 of *Lecture Notes in Computer Science*, pages 113–124. Springer-Verlag, 1994.

19. M. Miyazaki, S. Fukamachi, M. Takeda, and T. Shinohara. Speeding up the pattern matching machine for compressed texts. *Transactions of Information Processing Society of Japan*, 39(9):2638–2648, 1998. (in Japanese).

20. M. Miyazaki, A. Shinohara, and M. Takeda. An improved pattern matching algorithm for strings in terms of straight-line programs. In *Proc. 8th Ann. Symp. on Combinatorial Pattern Matching*, volume 1264 of *Lecture Notes in Computer Science*, pages 1–11. Springer-Verlag, 1997.

21. Y. Shibata, T. Kida, S. Fukamachi, M. Takeda, A. Shinohara, T. Shinohara, and S. Arikawa. Byte pair encoding: a text compression scheme that accelerates pattern matching. Technical Report DOI-TR-161, Department of Informatics, Kyushu University, April 1999.

22. M. Takeda. Pattern matching machine for text compressed using finite state model. Technical Report DOI-TR-142, Department of Informatics, Kyushu University, October 1997.

On the Structure of Syntenic Distance

David Liben-Nowell

Department of Computer Science
Cornell University
Ithaca, NY 14853 USA
dln@cs.cornell.edu

Abstract. This paper examines some of the rich structure of the syntenic distance model of evolutionary distance, introduced by Ferretti, Nadeau, and Sankoff. The syntenic distance between two genomes is the minimum number of fissions, fusions, and translocations required to transform one into the other, ignoring gene order within chromosomes. We prove that the previously unanalyzed algorithm given by Ferretti et al is a 2-approximation and no better, and that, further, it always outperforms the algorithm presented by DasGupta, Jiang, Kannan, Li, and Sweedyk. We also prove the same results for an improved version of the Ferretti et al algorithm. We then prove a number of properties which give insight into the structure of optimal move sequences. We give instances in which any move sequence working solely within connected components is nearly twice optimal, and a general lower bound based on the spread of genes from each chromosome. We then prove a monotonicity property for the syntenic distance, and bound the difficulty of the hardest instance of any given size. We briefly discuss the results of implementing these algorithms and testing them on real synteny data.

1 Introduction

Numerous models for measuring the evolutionary distance between two species have been proposed in the past. These models are often based upon high-level (non-point) mutations which rearrange the order of genes within a chromosome. The distance between two genomes (or chromosomes) is defined as the minimum number of moves of a certain type required to transform the first into the second. A move for the *reversal distance* [1] is the replacement of a segment of a chromosome by the same segment in reversed order. For the *transposition distance* [2], a legal move consists of removing a segment of a chromosome and reinserting it at some other location in the chromosome.

In [6], Ferretti, Nadeau, and Sankoff propose a somewhat different sort of measure of genetic distance, known as *syntenic distance*. This model abstracts away from the order of the genes within chromosomes, and handles each chromosome as an unordered set of genes. The legal moves are *fusions*, in which two chromosomes join into one, *fissions*, in which one chromosome splits into two, and reciprocal *translocations*, in which two chromosomes exchange sets of genes. In practice, the order of genes within chromosomes is often unknown,

M. Crochemore, M. Paterson (Eds.): CPM'99, LNCS 1645, pp. 50–65, 1999.

and this model allows the computation of the distance between species regardless. Additional justification follows from the observation that interchromosomal evolutionary events occur with relative rarity with respect to intrachromosomal events. (For some discussion of this and related models, see [5, 8].)

Ferretti et al propose using synteny as a measure of the distance between genomes, and present a heuristic to approximate this distance. Although they give some experimental data on its performance, no formal analysis of this approximation algorithm is given. Identifying a performance guarantee for this algorithm has remained an open question since.

In [3], DasGupta, Jiang, Kannan, Li, and Sweedyk show a number of results on the syntenic distance problem. They prove that computing the syntenic distance between genomes is NP-hard, and provide a simple polynomial-time 2-approximation. They also prove a number of other useful structural results.

Our results. As with many NP-complete problems, reasoning about the syntenic distance is difficult. We are able, however, to show some results on the structure of the problem, and analyze previously unanalyzed heuristics, including the original algorithm of Ferretti et al. These results give interesting insight into the rich structure of optimal move sequences. The structural properties aid in reasoning about the syntenic distance, and may lead to improved approximation algorithms.

Using results from [3], we prove a general lower bound for the syntenic distance between two genomes. Roughly, if for many chromosomes c in one genome, genes from c appear in many of the chromosomes of the other genome, then the instance is hard to solve. This lower bound may be helpful in developing improved approximation algorithms, since it implies that for the class of instances in which this scattering occurs, previously proposed algorithms are less than a factor of 2 away from optimal.

We prove a *monotonicity theory* for syntenic distance, showing a natural ordering on the difficulty of problem instances. We define the *syntenic diameter of order n $D_y(n)$* (in the spirit of the reversal and transposition diameters [7]) as the maximum number of moves required to solve an instance of size n. Monotonicity identifies a worst instance of size n, and implies that $D_y(n)$ is exactly the number of moves required to solve this instance. We prove that this worst instance requires between $2n - 3 - \log_{4/3}(2n - 3)$ and $2n - 3$ moves, using our lower bound. We leave open the question of providing tighter bounds on this distance, though we conjecture that the minimum number of moves required to solve this instance is exactly $2n - 3$.

Instance-by-instance comparison of two heuristics is a valuable notion that is rarely explored. This type of analysis leads to very strong results in comparing the performance of two approximation algorithms, even those with the same approximation ratio. Using this technique, we analyze the previously unanalyzed approximation algorithm given by Ferretti et al, settling the open question of finding a performance guarantee for this algorithm. We prove that this algorithm is never worse than the approximation algorithm presented by DasGupta et al in [3], immediately giving a performance guarantee of 2. We also show that there are instances in which the algorithm performs $2 - \epsilon$ away from optimal.

We also consider the algorithm resulting from making the fixes necessary to handle the instances in which the original algorithm performs $2 - \epsilon$ away from optimal. We prove the same results about this modified algorithm: it is also a 2-approximation that always outperforms the DasGupta et al algorithm, and there are instances in which it performs a factor of $2 - \epsilon$ away from optimal.

Call the *connected components* of an instance the connected components of the intersection graph of the chromosomes. We prove the surprising result that there are instances in which the optimal move sequence *must* connect two unconnected components, and any move sequence that fails to do so is in fact $2 - \epsilon$ away from optimal. This implies that any approximation algorithm that works only with components (as all currently proposed algorithms do) is doomed to be no better than a 2-approximation. This raises the new problem of *connected synteny*, in which the move sequence is constrained to work only within connected components. The above results indicate that the algorithms presented in [3] and [6] (and the modified version of the latter) are only 2-approximations for connected synteny, as well.

We also discuss a preliminary implementation of the syntenic distance model and all of the algorithms discussed above. We discuss the results of running all three algorithms on eight sets of real synteny data from the Institut National de la Recherche Agronomique (INRA) Comparative Homology Database [4].

2 Notational Preliminaries and Previous Heuristics

The syntenic distance model is as follows: a *chromosome* is a subset of a set of n *genes*, and a *genome* is a collection of k chromosomes. A genome can be transformed by any of the following moves (for S, T, U, and V non-empty sets of genes): (1) a *fusion* $(S, T) \longrightarrow U$, where $U = S \cup T$; (2) a *fission* $S \longrightarrow (T, U)$, where $T \cup U = S$; or (3) a *translocation* $(S, T) \longrightarrow (U, V)$, where $U \cup V = S \cup T$. The *syntenic distance* between genomes $\mathcal{G}_1$ and $\mathcal{G}_2$ is then given by the minimum number of moves required to transform $\mathcal{G}_1$ into $\mathcal{G}_2$.

The *compact representation* of an instance of synteny is described in [6] and formalized in [3]. This representation makes the goal of each instance uniform and thus eases reasoning about move sequences. For an instance in which we are attempting to transform genome $\mathcal{G}_1$ into genome $\mathcal{G}_2$, we relabel each gene a contained in a chromosome of $\mathcal{G}_1$ by the numbers of the chromosomes of $\mathcal{G}_2$ in which a appears. Formally, we replace each of the k sets S in $\mathcal{G}_1$ with $\bigcup_{\ell \in S} \{i \mid \ell \in G_i\}$ (where $\mathcal{G}_2 = G_1, G_2, \ldots, G_n$) and attempt to transform these sets into the collection $\{1\}, \{2\}, \ldots, \{n\}$. As an example of the compact representation (given in [6]), consider the instance

$$
\begin{array}{ll}
\mathcal{G}_1 = \{x, y\}, & \text{(Chromosome 1)} \\
\phantom{\mathcal{G}_1 = } \{p, q, r\}, & \text{(Chromosome 2)} \\
\phantom{\mathcal{G}_1 = } \{a, b, c\} & \text{(Chromosome 3)}
\end{array}
\qquad
\begin{array}{ll}
\mathcal{G}_2 = \{p, q, x\}, & \text{(Chromosome 1)} \\
\phantom{\mathcal{G}_2 = } \{a, b, r, y, z\} & \text{(Chromosome 2).}
\end{array}
$$

The compact representation of $\mathcal{G}_1$ with respect to $\mathcal{G}_2$ is $\{1, 2\}, \{1, 2\}, \{2\}$ and the compact representation of $\mathcal{G}_2$ with respect to $\mathcal{G}_1$ is $\{1, 2\}, \{1, 2, 3\}$. For an instance of synteny in this compact notation, we will write $\mathcal{S}(n, k)$ to refer to the instance where there are n elements and k sets in the compact representation. Let $D(\mathcal{S}(n, k))$ be the minimum number of moves required to solve a synteny instance $\mathcal{S}(n, k)$.

We will say that two sets S_1 and S_2 are *connected* if $S_1 \cap S_2 \neq \emptyset$, and that both are in the same *component*. For a gene ℓ, let $\mathsf{count}(\ell)$ be the number of chromosomes in which ℓ appears.

In [6], Ferretti et al present the approximation algorithm reproduced in Fig. 1, which we denote by $\mathcal{F}$. (Two genes are *syntenic* iff they appear in the same chromosome.) Although they provide some empirical evidence on the algorithm's performance, they do not give any formal analysis.

Select an uneliminated gene ℓ to work on, under the following priorities:

Priority (A). Any ℓ for which $\mathsf{count}(\ell) = 1$.

Priority (B). Any ℓ for which $\mathsf{count}(\ell) = 2$.

Priority (C). If all $\mathsf{count}(\ell) > 2$, pick ℓ which minimizes $\mathsf{count}(\ell)$ and, if there are several such, which minimizes $\mathsf{count}(\ell')$ for some ℓ' in the chromosome remaining from the last operation involving ℓ. If there are several such, choose ℓ so that after it is operated on, $\sum_\ell \mathsf{count}(\ell)$ is minimized.

For the ℓ selected above, do one of the following operations:

Operation (1). If $\mathsf{count}(\ell) = 1$ and some of the genes syntenic with ℓ appear in no other chromosomes, effect a fission to create a separate chromosome $\{\ell\}$.

Operation (2). If $\mathsf{count}(\ell) = 1$ and all genes ℓ' syntenic with ℓ appear in $\mathsf{count}(\ell') \geq c_{\min} > 1$ chromosomes, effect a translocation to obtain a separate chromosome $\{\ell\}$. The second chromosome involved in the translocation is one that contains some gene ℓ' syntenic with ℓ, with $\mathsf{count}(\ell') = c_{\min}$, and, if there are several, with a maximal number of genes syntenic with ℓ.

Operation (3). If $\mathsf{count}(\ell) > 1$, effect $\mathsf{count}(\ell) - 2$ fusions followed by one translocation[a], again to produce a separate $\{\ell\}$.

---

[a] This translocation could actually be a fusion if no other genes are present in the component.

Fig. 1. The approximation algorithm $\mathcal{F}$ [6].

Let $\mathcal{H}$ denote the approximation algorithm defined in [3]: for each connected component containing n_i elements and k_i sets, perform $k_i - 1$ fusions to produce one set with all n_i elements, then $n_i - 1$ fissions to produce the n_i singletons. Thus in an instance with p components, $\mathcal{H}$ requires $n + k - 2p$ moves. DasGupta et al show that this algorithm is a 2-approximation, a tight bound (the algorithm performs a factor of 2 away from optimal on the instance $\{1\}, \{1, 2\}, \ldots, \{1, 2, \ldots, n\}$). To derive the performance guarantee for $\mathcal{H}$, Das-

Gupta et al prove the following *component bound*: if an instance of synteny $S(n,k)$ has p components, then $D(S(n,k)) \geq n - p$.

3   An Analysis of $\mathcal{F}$

In this section, we prove a number of results about $\mathcal{F}$. We first show that $\mathcal{F}$ is *never* worse than $\mathcal{H}$, and is therefore a 2-approximation. We then show that the factor of 2 is tight by giving a class of instances in which $\mathcal{F}$ performs a factor of 2 away from optimal. (In Sect. 4, we give a modification of $\mathcal{F}$ that handles this class of instances optimally, and analyze the modified algorithm.)

Theorem 1. *For any instance $S(n,k)$ of synteny, $|\mathcal{F}(S(n,k))| \leq |\mathcal{H}(S(n,k))|$.*

Proof. Suppose that $\mathcal{F}$ generates a move sequence σ on $S(n,k)$. Suppose that in σ there are m_1 fissions (from Operation (1)), m_2 translocations (from Operations (2) and (3)), and m_3 fusions (from Operation (3)).

Every translocation generated by Operation (2) is of the form $(S \cup \{\ell\}, T) \longrightarrow (S \cup T, \{\ell\})$ where $\ell \notin S \cup T$ and, for some gene ℓ', $\ell' \in S \cap T$. Every translocation generated by Operation (3) is of the form $(S \cup \{\ell\}, T \cup \{\ell\}) \longrightarrow (S \cup T, \{\ell\})$ where $\ell \notin S \cup T$. Note that in either case, at the time that $\{\ell\}$ is produced, it appears nowhere else in the genome (i.e., $\mathsf{count}(\ell) = 1$).

We create a new move sequence σ' which differs from σ in that each translocation $(S_1 \cup S_2, T_1 \cup T_2) \longrightarrow (S_1 \cup T_1, S_2 \cup T_2)$ is replaced by the two-move sequence $(S_1 \cup S_2, T_1 \cup T_2) \longrightarrow S_1 \cup S_2 \cup T_1 \cup T_2 \longrightarrow (S_1 \cup T_1, S_2 \cup T_2)$.

By the form of the translocations and this translation, we have the following facts:

- Each of the newly-created fusions is within a connected component (the input sets are connected by ℓ' for Operation (2) and ℓ for Operation (3)).
- Each of the newly-created fissions produces a singleton $\{\ell\}$ for a gene ℓ that appears nowhere else in the genome.

Now we examine the fusions and fissions in σ. Each original fusion (from Operation (3)) is also within a component (the two input sets are connected by ℓ), and each fission (in Operation (1)) produces a singleton of a gene that appears nowhere else in the genome. Thus, every fusion in σ' fuses two sets in the same component, and every fission in σ' produces a singleton set with an element that appears nowhere else in the genome.

Clearly we can rearrange σ' to completely solve each component before beginning the next, since there are no intercomponent dependencies. Further, inside each component we can put all the fusions before all the fissions, since the fissions merely remove the last instance of an element from a larger set. In other words, after rearrangement, σ' does *exactly* what $\mathcal{H}$ does: within each component, it fuses all the sets into one massive set, and then fissions off individual elements one at a time. Note that $|\sigma'| = m_1 + 2 \cdot m_2 + m_3 = m_2 + |\sigma|$, and thus $|\sigma| = |\sigma'| - m_2 = |\mathcal{H}(S(n,k))| - m_2$. In other words, $\mathcal{F}$ performs m_2 moves better than $\mathcal{H}$ on each input. $\square$

Corollary 2. $\mathcal{F}$ *is a 2-approximation.*

Proof. Immediate from Theorem 1 and the fact that $\mathcal{H}$ is a 2-approximation. □

We now show the corresponding lower bound for $\mathcal{F}$:

Lemma 3. *For any $\epsilon > 0$, there exists an instance $\mathcal{S}(n, k)$ with $|\mathcal{F}(\mathcal{S}(n, k))| \geq (2 - \epsilon) \cdot D(\mathcal{S}(n, k))$.*

Proof. Select any n such that $1/(n-1) \leq \epsilon$. We give a synteny instance $\mathcal{S}(n, n)$ such that $D(\mathcal{S}(n, n)) = n - 1$ and $|\mathcal{F}(\mathcal{S}(n, n))| = 2n - 3$. Then the ratio between the result of $\mathcal{F}$ and the optimal is $(2(n-1) - 1)/(n-1)$, i.e., only $1/(n-1)$ better than two times optimal.

The instance $\mathcal{S}(n, n)$ consists of $\{1, 2, \ldots, n\}$ and $n - 1$ copies of $\{n\}$. Here is an $n - 1$ move sequence solving the instance:

[$n - 1$ moves]   For $i = 1$ to $n - 1$, translocate the ith singleton $\{n\}$ with the remaining elements of the large set to produce the singleton $\{i\}$.

Each move removes one of the $n - 1$ genes appearing only in the large set while absorbing another of the singleton $\{n\}$ sets, so that after $n - 1$ of these moves all the ns have been joined.

Now, we examine what $\mathcal{F}$ does on this input. Genes $1, 2, \ldots, n-1$ are exactly symmetric in this instance, so we assume without loss of generality that $\mathcal{F}$ selects them in ascending order.

[$n - 2$ moves]   For $i = 1$ to $n - 2$, count(i) = 1. The gene $n - 1$ is syntenic with i and appears in no other chromosome, so by Operation (1) we fission off the singleton $\{i\}$. This leaves $\{n - 1, n\}$ and $n - 1$ copies of $\{n\}$.

[1 move]   count($n - 1$) = 1, so select it. By Operation (2), translocate $\{n - 1, n\}$ and $\{n\}$ to produce $\{n - 1\}$ and $\{n\}$. This leaves $n - 1$ copies of $\{n\}$.

[$n - 2$ moves]   Fuse the $n - 1$ copies of $\{n\}$ by Operation (3).

Thus $\mathcal{F}$ requires $n - 2$ fissions, 1 translocation, and $n - 2$ fusions, or $2n - 3$ moves total.   □

4   A Possible Improvement to $\mathcal{F}$

Note that the non-optimality of $\mathcal{F}$ on the instance in Lemma 3 is only as the result of applications of Operation (1). When the applications of this operation have been completed, the resulting genome is $\{n - 1, n\}$ and $n - 1$ copies of $\{n\}$. $\mathcal{F}$ takes $n - 1$ more moves after this point, which, by the component bound, is optimal. The non-optimality of Operation (1) is sufficient to cause $\mathcal{F}$ to be a factor of 2 away from optimal. The difficulty with $\mathcal{F}$ results from overzealous applications of Operation (1) when Operation (2) could do some good. (Notice from Theorem 1 that the more translocations $\mathcal{F}$ does, the better its performance.) Call $\mathcal{F}'$ the algorithm resulting from making the following fixes to $\mathcal{F}$ to deal with this problem:

- Apply Operation (1) only if *all* of the genes syntenic with ℓ appear in no other chromosomes.
- Apply Operation (2) if *any* gene syntenic with ℓ appears in another chromosome. The second chromosome involved in the translocation is selected as in $\mathcal{F}$, but ignoring those genes ℓ' syntenic with ℓ such that count$(\ell') = 1$.

Note that $\mathcal{F}'$ performs optimally on the bad instance for $\mathcal{F}$ in Lemma 3: the genes are still selected in the same order, but each of the first $n-1$ fissions becomes a translocation, and the instance is solved after these translocations.

Theorem 4. *For any instance $S(n,k)$ of synteny, $|\mathcal{F}'(S(n,k))| \leq |\mathcal{H}(S(n,k))|$.*

Proof. Analogous to the proof of Theorem 1.   □

Corollary 5. *$\mathcal{F}'$ is a 2-approximation.*   □

The following lemma shows, however, that $\mathcal{F}'$ is also no better than a 2-approximation.

Lemma 6. *For any $\epsilon > 0$, there exists an instance $S(n,k)$ with $|\mathcal{F}'(S(n,k))| \geq (2-\epsilon) \cdot D(S(n,k))$.*

Proof. We give an instance $S(\alpha i+1, \alpha i+i+1)$ with $D(S(\alpha i+1, \alpha i+i+1)) = \alpha i+i$ and $\mathcal{F}'(S(\alpha i+1, \alpha i+i+1)) = 2\alpha i - 1$ for $3 \leq \alpha < i$. Selecting $\alpha \geq 3$ and $i > \alpha$ so that $\epsilon \geq (2i+1)/(\alpha i + i)$ then gives us an instance in which $\mathcal{F}'$ performs $(2-\epsilon)$ away from optimal.

Consider the instance $S(\alpha i + 1, \alpha i + i + 1)$ consisting of the following sets, for $3 \leq \alpha < i$:

- $S = \{1, 2, \ldots, \alpha i + 1\}$
- $X_{j,q} = \{j\}$, for $1 \leq j \leq i$ and $1 \leq q \leq \alpha$
- i copies of $Z = \{i+1, i+2, \ldots, \alpha i + 1\}$.

Here is a move sequence solving $S(\alpha i + 1, \alpha i + i + 1)$ in $\alpha i + i$ moves:

[$i-1$ moves]	Fuse the i copies of Z, leaving S, the $X_{j,q}$s, and Z.
[1 move]	Translocate S and Z to produce $\{1, 2, \ldots, \alpha i\}$ and $\{\alpha i + 1\}$.
[$(\alpha - 1)i$ moves]	Translocate $(\alpha - 1)$ of the singletons for each of the genes $1, 2, \ldots, i$ with the set $\{1, 2, \ldots, \alpha i\}$ to produce the singletons $\{i+1\}, \{i+2\}, \ldots, \{\alpha i\}$. This leaves $\{1, 2, \ldots, i\}$ and $\{1\}, \{2\}, \ldots, \{i\}$.
[i moves]	For $j = 1$ to $i-1$, translocate $\{j\}$ with the large set to remove j from it. This leaves two copies of $\{i\}$. Fuse these to solve the instance.

We now examine what $\mathcal{F}'$ does on this instance. Notice that

$$\text{count}(\ell) = \begin{cases} \alpha + 1 & \text{for } \ell \in \{1, 2, \ldots, i\} \\ i + 1 & \text{for } \ell \in \{i+1, i+2, \ldots, \alpha i + 1\}. \end{cases}$$

Since $1, 2, \ldots, i$ are completely symmetric in this instance, without loss of generality we assume that the algorithm picks them in ascending order. Similarly, $i + 1, i + 2, \ldots, \alpha i + 1$ are symmetric, and we assume without loss of generality that they are selected in ascending order, as well.

[α moves] $\alpha < i$, so we first select $\ell = 1$. Applying Operation (3), we fuse all singletons $\{1\}$ and then translocate with S to produce $\{1\}$ and $\{2, 3, \ldots, \alpha i + 1\}$.

[α moves]   Select $\ell = 2$. Apply Operation (3) as above to produce $\{2\}$ and $\{3, 4 \ldots, \alpha i + 1\}$.

$\quad\vdots$ $\quad\vdots$

[α moves]   Select $\ell = i$. Apply Operation (3) as above to produce $\{i\}$ and $\{i + 1, i + 2, \ldots, \alpha i + 1\}$. The only remaining sets are $i + 1$ copies of the set Z.

[i moves]   Select $\ell = i+1$. Apply Operation (3) to fuse i copies of Z, and then translocate the last two copies to produce $\{i + 1\}$ and $\{i + 2, i + 3, \ldots, \alpha i + 1\}$.

[$(\alpha - 1)i - 1$ moves]   Fission the remaining set $\{i + 2, i + 3, \ldots, \alpha i + 1\}$ into singletons, by Operation (1).

$\mathcal{F}'$ thus has to complete $\alpha i + i + (\alpha - 1)i - 1 = 2\alpha i - 1$ moves to solve this instance.   □

Note that $\mathcal{F}$ does poorly on these instances, as well — bad choices of the genes by Priority (C) are sufficient to cause the non-optimality, and $\mathcal{F}$ selects genes in the same way as $\mathcal{F}'$.

5   Moves between Connected Components

It seems intuitive that when attacking an instance of synteny consisting of two distinct connected components, the optimal move sequence would never fuse these components together. Both $\mathcal{H}$ and $\mathcal{F}$ (and $\mathcal{F}'$) work within connected components, in fact. However, the following theorem shows that this approach is doomed to be no better than a 2-approximation.

Theorem 7. *For any algorithm $\mathcal{A}$ that works only within connected components, and for any $\epsilon > 0$, there exists an instance $\mathcal{S}(n, k)$ where $|\mathcal{A}(\mathcal{S}(n, k))| \geq (2 - \epsilon) \cdot D(\mathcal{S}(n, k))$.*

Proof. We construct an instance of synteny $\mathcal{S}(n, n)$ solvable in $n - 1$ moves, but for which $\mathcal{A}$ will require $2n - 4$ moves. Selecting n so that $\epsilon \geq 2/(n - 1)$ yields an instance where $\mathcal{A}$ is $2 - \epsilon$ away from optimal.

Consider the instance $\mathcal{S}(n, n)$ consisting of $\{1, 2, \ldots, n - 1\}$ and $n - 1$ copies of $\{n\}$. First we observe that there is a move sequence solving this instance in $n - 1$ moves:

[1 move]   Translocate $\{1, 2, \ldots, n - 1\}$ and $\{n\}$ to produce $\{1\}$ and $\{2, 3, \ldots, n - 1, n\}$.

[$n-2$ moves]   For $i = 2$ through $n - 1$, perform a translocation of
the set $\{i, i+1, \ldots, n\}$ and $\{n\}$ to produce $\{i\}$ and
$\{i+1, i+2, \ldots, n\}$.

For any algorithm $\mathcal{A}$ working only within components, however, the moves
that $\mathcal{A}$ can make are severely limited. Since $\{1, 2, \ldots, n-1\}$ is a component all
by itself, there is no choice but to complete $n - 2$ fissions. The $n - 1$ copies of $\{n\}$
also form an entire component by themselves. Thus the only possible moves are
to complete $n - 2$ fusions to create a unique singleton. Therefore, $\mathcal{A}$ completes
$2n - 4$ moves on this instance.   □

It is now natural to define the *connected synteny problem*, to find the minimum number of moves required to transform one genome into another with
all moves constrained to work only within a single component. We will use
$\hat{D}(\mathcal{S}(n, k))$ to denote the minimum number of moves within components required
to solve a synteny instance $\mathcal{S}(n, k)$.

Obviously, $\hat{D}(\mathcal{S}(n, k)) \geq D(\mathcal{S}(n, k))$. Because $\mathcal{H}$, $\mathcal{F}$, and $\mathcal{F}'$ all generate
move sequences that work within components, these algorithms are also 2-
approximations for the connected synteny problem. In each of the examples
in which these algorithms are $2 - \epsilon$ away from optimal, the optimal move sequence works only within components. (In fact, there is only one component in
each example.) Thus $\mathcal{H}$, $\mathcal{F}$, and $\mathcal{F}'$ are all 2-approximations for this problem,
and no better. Whether it is easier to approximate connected synteny, however,
remains an open question.

6   Non-redundancy and Monotonicity

In this section, we show that, for any instance, there is an optimal move sequence
containing no moves that produce two sets with non-empty intersection. We also
prove a monotonicity property for syntenic distance.

We first need to introduce an extension to our notation to handle the case of
empty sets as input. If $S_1, \ldots, S_k$ is a collection of sets and, for some i, $S_i = \emptyset$, we
understand the synteny instance $\mathcal{S}(n, k) = S_1, \ldots, S_k$ to represent the synteny
instance $T(n, k - 1) = S_1, \ldots, S_{i-1}, S_{i+1}, \ldots, S_k$.

Lemma 8. *If there is a move sequence* $\sigma = (\sigma_1, \sigma_2, \ldots, \sigma_m)$ *solving* $S_1, \ldots, S_i \cup$
$\{a\}, \ldots, S_k$ *where* $a \notin S_i$ *(with* S_i *possibly empty), then there exists a move
sequence* σ' *solving* $S_1, \ldots, S_i, \ldots, S_k$ *in at most* m *steps.*

Proof (by induction on m). For the base case ($m = 1$), σ_1 must solve the instance. We have two cases (a cannot appear in more than one additional set,
since otherwise no single move could solve the instance):

- The element a also appears in some set $S_{j \neq i}$.
  σ_1 must take $S_i \cup \{a\}$ and S_j as input, and produce the singleton $\{a\}$ as
  output. Otherwise, two copies of the gene a remain, or the copy of a is
  bundled up with some other element(s). This first restriction implies that σ_1
  cannot be a fission.

If σ_1 is the fusion $(S_j, S_i \cup \{a\}) \longrightarrow \{a\}$, it must be the case that $S_j = \{a\}$ and $S_i = \emptyset$. Thus $S_1, \ldots, S_k$ is already in the target form, and in the new instance we are done without making any move.

If σ_1 is a translocation, a must occur in only one of the output sets, for otherwise it appears twice and the instance is not solved. Thus $\sigma_1 = (S_i \cup \{a\}, S_j) \longrightarrow (S_i \cup [S_j - \{a\}], \{a\})$. We can replace this by $\sigma_1' = (S_i, S_j) \longrightarrow (S_i \cup [S_j - \{a\}], \{a\})$ to solve the instance $S_1, \ldots, S_k$.

- a does not appear elsewhere in the genome.

  Then the last move need not involve the singleton $\{a\}$. If it does not, then it must be the case that $S_i = \emptyset$. (Otherwise after the last move of the sequence a is in a non-singleton and the instance has not been solved.) In this case, simply doing the last move will solve $S_1, \ldots, S_k$.

  If the last move does involve $S_i \cup \{a\}$, it is not a fusion since any fusing would couple a with some other element. (a would have to be coupled with some element $b \neq a$, since a does not appear elsewhere in the genome.)

  If σ_1 is a fission, then it must produce a singleton set $\{a\}$ and some other set not containing a in order to solve the instance. Since $a \notin S_i$, this means that $\sigma_1 = S_i \cup \{a\} \longrightarrow (\{a\}, S_i)$. If we replace $S_i \cup \{a\}$ by S_i in the instance, the instance is already in the target form and we can skip this move.

  If σ_1 is a translocation, it must be $(S_j, S_i \cup \{a\}) \longrightarrow (U, \{a\})$ for some set U, or else (as with the fusion case) the instance would not be solved. If $a \in U$ then the instance is not solved, since a appears twice. Therefore it must be the case that $U = S_i \cup S_j$. To solve the new instance, we can simply do the fusion $(S_i, S_j) \longrightarrow S_i \cup S_j$ and we are done.

For the inductive case ($m \geq 2$), first we handle the cases when σ_1 is any move that does not involve the set $S_i \cup \{a\}$. For ℓ and j distinct from i:

- $\sigma_1 = (S_\ell, S_j) \longrightarrow S_\ell \cup S_j$. Then $\sigma_{2\ldots m}$ solves $S_r (1 \leq r \leq k, r \neq \ell, r \neq j, r \neq i), S_i \cup \{a\}, S_\ell \cup S_j$. By the inductive hypothesis, we have a move sequence σ' solving $S_r (1 \leq r \leq k, r \neq \ell, r \neq j, r \neq i), S_i, S_\ell \cup S_j$ in at most $m - 1$ moves.

- $\sigma_1 = S_\ell \longrightarrow (U, V)$. Then $\sigma_{2\ldots m}$ solves $S_r (1 \leq r \leq k, r \neq \ell, r \neq i), S_i \cup \{a\}, U, V$. By the inductive hypothesis, we have a move sequence σ' solving $S_r (1 \leq r \leq k, r \neq \ell, r \neq i), S_i, U, V$ in at most $m - 1$ moves.

- $\sigma_1 = (S_\ell, S_j) \longrightarrow (U, V)$. Then $\sigma_{2\ldots m}$ solves $S_r (1 \leq r \leq k, r \neq \ell, r \neq j, r \neq i), S_i \cup \{a\}, U, V$. By the inductive hypothesis, we have a move sequence σ' solving $S_r (1 \leq r \leq k, r \neq \ell, r \neq j, r \neq i), S_i, U, V$ in at most $m - 1$ moves.

In each case, doing σ_1 and σ' solves $S_1, \ldots, S_k$ in at most m moves. We now consider the cases in which σ_1 takes $S_i \cup \{a\}$ as input.

- Suppose σ_1 is a fission, and that $S_i = S_{i_1} \cup S_{i_2}$.

  If $\sigma_1 = S_i \cup \{a\} \longrightarrow (S_{i_1} \cup \{a\}, S_{i_2})$, then $\sigma_{2\ldots m}$ solves the instance $S_r (1 \leq r \leq k, r \neq i), S_{i_1} \cup \{a\}, S_{i_2}$. By the inductive hypothesis, there is a σ' solving $S_r (1 \leq r \leq k, r \neq i), S_{i_1}, S_{i_2}$ in at most $m - 1$ steps. Then doing $S_i \longrightarrow (S_{i_1}, S_{i_2})$ followed by σ' solves $S_1, \ldots, S_k$ in at most m steps.

If $\sigma_1 = S_i \cup \{a\} \longrightarrow (S_{i_1} \cup \{a\}, S_{i_2} \cup \{a\})$, then $\sigma_{2\ldots m}$ solves the instance $S_r(1 \leq r \leq k, r \neq i), S_{i_1} \cup \{a\}, S_{i_2} \cup \{a\}$ in $m - 1$ moves. By the inductive hypothesis applied to $\sigma_{2\ldots m}$ and $S_{i_1} \cup \{a\}$, there is a σ' solving $S_r(1 \leq r \leq k, r \neq i), S_{i_1}, S_{i_2} \cup \{a\}$ in at most $m - 1$ steps. Applying the inductive hypothesis again, this time to σ' and $S_{i_2} \cup \{a\}$, we have that there is a σ'' solving $S_r(1 \leq r \leq k, r \neq i), S_{i_1}, S_{i_2}$ in at most $m - 1$ steps. Then doing $S_i \longrightarrow (S_{i_1}, S_{i_2})$ followed by σ'' solves $S_1, \ldots, S_k$ in at most m steps.

- Suppose that σ_1 is the fusion $(S_i \cup \{a\}, S_\ell) \longrightarrow S_i \cup \{a\} \cup S_\ell$. Then $\sigma_{2\ldots m}$ solves the instance $S_r(1 \leq r \leq m, r \neq \ell, r \neq i), S_i \cup \{a\} \cup S_\ell$ in $m - 1$ steps. If $a \in S_\ell$, then $S_i \cup \{a\} \cup S_\ell = S_i \cup S_\ell$. Thus $\sigma_{2\ldots m}$ solves $S_r(1 \leq r \leq m, r \neq \ell, r \neq i), S_i \cup S_\ell$, and doing $(S_i, S_\ell) \longrightarrow S_i \cup S_\ell$ and $\sigma_{2\ldots m}$ solves $S_1, \ldots, S_k$ in m steps.

  If $a \notin S_\ell$, then by the inductive hypothesis, there is a σ' solving $S_r(1 \leq r \leq m, r \neq \ell, r \neq i), S_i \cup S_\ell$ in $m - 1$ steps. To solve $S_1, \ldots, S_k$, we do the fusion $(S_i, S_\ell) \longrightarrow S_i \cup S_\ell$ and run σ', which requires at most m steps.

- Suppose σ_1 is a translocation using the set $S_i \cup \{a\}$ and S_ℓ, where $S_i = S_{i_1} \cup S_{i_2}$ and $S_\ell = S_{\ell_1} \cup S_{\ell_2}$. Then σ_1 must look like one of the following:

  (1) $(S_\ell, S_i \cup \{a\}) \longrightarrow (S_{\ell_1} \cup S_{i_1}, S_{\ell_2} \cup S_{i_2} \cup \{a\})$

  (2) $(S_\ell, S_i \cup \{a\}) \longrightarrow (S_{\ell_1} \cup S_{i_1} \cup \{a\}, S_{\ell_2} \cup S_{i_2} \cup \{a\})$.

  In either case we replace this move by the translocation $\sigma'_1 = (S_\ell, S_i) \longrightarrow (S_{\ell_1} \cup S_{i_1}, S_{\ell_2} \cup S_{i_2})$.

  In case (1), if $a \in S_{\ell_2}$, then $\sigma_{2\ldots m}$ solves $S_r(1 \leq r \leq k, r \neq \ell, r \neq i), S_{\ell_1} \cup S_{i_1}, S_{\ell_2} \cup S_{i_2}$ in $m - 1$ steps, since $S_{\ell_2} \cup S_{i_2} \cup \{a\} = S_{\ell_2} \cup S_{i_2}$. Then we can do σ'_1 and $\sigma_{2\ldots m}$ to solve $S_1, \ldots, S_k$ in m steps. If $a \notin S_{\ell_2}$, then $\sigma_{2\ldots m}$ solves $S_r(1 \leq r \leq k, r \neq \ell, r \neq i), S_{\ell_1} \cup S_{i_1}, S_{\ell_2} \cup S_{i_2} \cup \{a\}$ in $m - 1$ steps. By the inductive hypothesis, there is a move sequence σ' solving $S_r(1 \leq r \leq k, r \neq \ell, r \neq i), S_{\ell_1} \cup S_{i_1}, S_{\ell_2} \cup S_{i_2}$ in at most $m - 1$ steps. Gluing this together with σ'_1 yields a sequence solving $S_1, \ldots, S_k$ in at most m moves.

  In case (2), if $a \in S_{\ell_1}$ then $S_{\ell_1} \cup S_{i_1} \cup \{a\} = S_{\ell_1} \cup S_{i_1}$ and this move is actually $(S_\ell, S_i \cup \{a\}) \longrightarrow (S_{\ell_1} \cup S_{i_1}, S_{\ell_2} \cup S_{i_2} \cup \{a\})$, which is exactly case (1). Otherwise, $a \notin S_{\ell_1}$. If $a \in S_{\ell_2}$, for exactly the same reason as above (with the roles of S_{ℓ_1} and S_{ℓ_2} reversed), we are again in case (1). Thus the only interesting case is when $a \notin S_{\ell_1}$ and $a \notin S_{\ell_2}$. In this case, $\sigma_{2\ldots m}$ solves $S_r(1 \leq r \leq k, r \neq \ell, r \neq i), S_{\ell_1} \cup S_{i_1} \cup \{a\}, S_{\ell_2} \cup S_{i_2} \cup \{a\}$ in $m - 1$ moves. By the inductive hypothesis applied to $\sigma_{2\ldots m}$ and $S_{\ell_1} \cup S_{i_1} \cup \{a\}$, we have a move sequence σ' solving $S_r(1 \leq r \leq k, r \neq \ell, r \neq i), S_{\ell_1} \cup S_{i_1}, S_{\ell_2} \cup S_{i_2} \cup \{a\}$ in at most $m - 1$ moves. Applying the inductive hypothesis again, to σ' and $S_{\ell_2} \cup S_{i_2} \cup \{a\}$, we have a move sequence σ'' solving $S_r(1 \leq r \leq k, r \neq \ell, r \neq i), S_{\ell_1} \cup S_{i_1}, S_{\ell_2} \cup S_{i_2}$ in at most $m - 1$ moves. This is exactly the result of doing the translocation σ'_1, so doing σ'_1 and σ'' solves $S_1, \ldots, S_k$ in at most m moves.   □

Define a *redundant move* as any move creating two sets S and T such that $S \cap T \neq \emptyset$. (Note that only fissions and translocations can be redundant, because fusions do not create two sets.)

We need the following result on reordering from [3] to prove a theorem on redundancy: for $S(n, k)$ an instance of synteny and $\sigma = (\sigma_1, \ldots, \sigma_m)$ any move solving the instance with m_1 fusions, m_2 translocations, and m_3 fissions, there exists a move sequence σ' solving the instance in $m' \leq m$ moves in which every fusion precedes every translocation precedes every fission, using $m_1' \leq m_1$ fusions, $m_2' \leq m_2$ translocations, and $m_3' \leq m_3$ fissions. (DasGupta et al actually state this lemma for the case where σ is optimal, but the proof extends to a general σ straightforwardly.) We refer to a move sequence in which the fusions precede the translocations precede the fissions as in *canonical order*.

Theorem 9. *For any synteny instance $S(n, k)$, there exists an optimal move sequence making no redundant moves.*

Proof. Let $\sigma = (\sigma_1, \ldots, \sigma_m)$ be a canonically-ordered optimal move sequence solving $S(n, k)$. There are no redundant fusions at all (by the definition of a redundant move). Any redundant fission must yield two copies of at least one gene a, say $S_1 \cup S_2 \cup \{a\} \longrightarrow (S_1 \cup \{a\}, S_2 \cup \{a\})$. But then there are two copies of the gene a, and since all succeeding moves are also fissions, the number of as can only increase, and therefore the instance will not be solved.

Then the only possible redundant moves are translocations of the form $(T_1 \cup T_2 \cup V, U_1 \cup U_2 \cup W) \longrightarrow (T_1 \cup U_1 \cup V \cup W, T_2 \cup U_2 \cup V \cup W)$ for some non-empty overlap $V \cup W$, with $V \cap (T_1 \cup T_2) = \emptyset$ and $W \cap (U_1 \cup U_2) = \emptyset$. Then by repeatedly applying the transformation described in Lemma 8 to σ for every element of $V \cup W$, we can solve the instance resulting from replacing this redundant move by the translocation $(T_1 \cup T_2 \cup V, U_1 \cup U_2 \cup W) \longrightarrow (T_1 \cup U_1 \cup V \cup W, T_2 \cup U_2)$ in at most as many moves. Repeating this sequentially for every redundant move in σ yields a move sequence of length at most m with no redundant moves. $\square$

Note that the canonicalizing process does not create redundancies: with a non-redundant move sequence as input, it produces a non-redundant canonical move sequence as output. Thus we can convert any move sequence σ into a non-redundant canonical move sequence by consecutively applying canonicalization, redundancy elimination, and canonicalization again.

Theorem 10 (Monotonicity). *Let $S_1, \ldots, S_k$ and $T_1, \ldots, T_k$ be two collections of sets where, for all $1 \leq i \leq k$, $T_i \subseteq S_i$. Let $n = |\bigcup_i S_i|$ and $n' = |\bigcup_i T_i|$. Let $S(n, k) = S_1, \ldots, S_k$ and let $T(n', k) = T_1, \ldots, T_k$. Then $D(S(n, k)) \geq D(T(n', k))$.*

Proof (by induction on $\delta = \sum_i |S_i - T_i|$).

Base case ($\delta = 0$). Then each $S_i = T_i$, and $T(n', k) = S(n, k)$, so their distances are trivially equal.

Inductive case ($\delta \geq 1$). Let $\sigma = (\sigma_1, \sigma_2, \ldots, \sigma_m)$ be an optimal move sequence solving $S(n, k)$. Let j be the minimum index such that $S_j \supset T_j$ and let a be any element in $S_j - T_j$. By applying the transformation described in Lemma 8, we can convert σ into a σ' solving $S_1, S_2, \ldots, S_j - \{a\}, \ldots, S_k$ in at most m steps. This instance is one element "closer" to $T(n', k)$, so, by the inductive hypothesis, we can solve $T(n', k)$ in at most m steps. $\square$

7   A Lower Bound on Synteny

In this section, we give a lower bound on syntenic distance when many elements
appear in many sets. The intuition for the bound is the following: suppose that,
in the compact representation, many genes appear in many chromosomes. This
occurs exactly when, in the non-compact representation, for many chromosomes
c in genome $\mathcal{G}_1$, genes from c appear in many of the chromosomes in genome $\mathcal{G}_2$.
This can only occur if many evolutionary events "scattered" c from $\mathcal{G}_1$ to $\mathcal{G}_2$. If
this occurs for many chromosomes c, then many events must have occurred for
many chromosomes, and thus the distance between the genomes must be large.

To prove this lower bound, we need the following restricted form of the synteny problem, defined in [3]. Define the *linear synteny* problem as the synteny
problem in which all move sequences are constrained as follows:

- The first $k - 1$ moves must be fusions or severely restricted translocations.
  One of the input sets is initially designated as the *merging set*. Each of the
  first $k - 1$ moves takes the current merging set Δ as input, along with one
  unused input set S, and produces a new merging set Δ'. If some element
  a appears nowhere in the genome except in Δ and S, then the move is the
  translocation $(\Delta, S) \longrightarrow (\Delta', \{a\})$, where $\Delta' = (\Delta \cup S) - \{a\}$. If there is
  no such element a, then the move simply fuses the two sets: $(\Delta, S) \longrightarrow \Delta'$,
  where $\Delta' = \Delta \cup S$.
- If Δ is the merging set after the $k - 1$ fusions and translocations, then each
  of the next $|\Delta| - 1$ moves simply fissions off a singleton $\{a\}$ and produces
  the new merging set $\Delta' = \Delta - \{a\}$.

Let $\tilde{D}(\mathcal{S}(n, k))$ be the length of the optimal linear move sequence. Note that
if a linear move sequence performs m_1 fusions in the first $k - 1$ moves, then the
move sequence contains $k - m_1 - 1$ translocations. After the $k - 1$ fusions and
translocations are complete, there are $n - k + m_1 + 1$ elements left in the merging
set, since exactly one element is eliminated by each translocation. Therefore,
$n - k + m_1$ fissions must be performed to eliminate the remaining elements.
Thus the length of the linear move sequence is $n + m_1 - 1$ moves. (Every move
either is a fusion or removes one element, and all but the last element must be
removed.)

Theorem 11. *For any instance of synteny* $\mathcal{S}(n, k)$,

$$\tilde{D}(\mathcal{S}(n, k)) \geq n - 1 + \max_{1 \leq c \leq k-1} \left[c - \left| \{\ell \mid \mathsf{count}(\ell) \leq c + 1\} \right| \right].$$

Proof. Consider an arbitrary c between 1 and $k-1$, and consider any linear move
sequence solving $\mathcal{S}(n, k)$. In the first c moves, only genes ℓ such that $\mathsf{count}(\ell) \leq
c + 1$ can be eliminated. (Any ℓ with $\mathsf{count}(\ell) > c + 1$ remains present in at least
one unused input set, since the first c moves can only merge $c + 1$ sets.)

Thus, in the first c moves we have at most $\left| \{\ell \mid \mathsf{count}(\ell) \leq c + 1\} \right|$ transloca-
tions, and therefore at least $c - \left| \{\ell \mid \mathsf{count}(\ell) \leq c + 1\} \right|$ fusions. Thus the instance
requires at least $n - 1 + c - \left| \{\ell \mid \mathsf{count}(\ell) \leq c + 1\} \right|$ moves to solve. $\square$

DasGupta et al prove that for any instance $S(n, k)$ of synteny, $\tilde{D}(S(n, k)) \leq D(S(n, k)) + \log_b D(S(n, k))$, for some constant b. (In the full version of their paper, DasGupta et al show that we can take $b = 4/3$.) This gives us the following bound on the general synteny problem:

Corollary 12. *For any instance of synteny $S(n, k)$,*

$$D(S(n, k)) + \log_{4/3}(D(S(n, k)))$$

$$\geq n - 1 + \max_{1 \leq c \leq k-1} \left[c - \left| \{ \ell \mid \text{count}(\ell) \leq c + 1 \} \right| \right].$$

$\square$

This bound may help in the development of improved approximation algorithms for the (linear) synteny problem. In particular, for a significant class of instances, $\mathcal{H}$ is better than a factor of 2 away from the linear optimal solution:

Corollary 13. *For any instance $S(n, k)$ of synteny in which $n \geq k$, if there exists some c such that $c - \left| \{ \ell \mid \text{count}(\ell) \leq c + 1 \} \right| \geq \beta n + 1$ for $\beta \in (0, 1]$, then*

$$\frac{|\mathcal{H}(S(n, k))|}{\tilde{D}(S(n, k))} < \frac{2}{\beta + 1}.$$

Proof. Suppose that $S(n, k)$ has p components. Then

$$\frac{|\mathcal{H}(S(n, k))|}{\tilde{D}(S(n, k))} \leq \frac{n + k - 2p}{n - 1 + \beta n + 1} \leq \frac{2n - 2p}{(1 + \beta)n} < \frac{2n}{(1 + \beta)n} = \frac{2}{1 + \beta}.$$

$\square$

8   Syntenic Diameter

In this section, we consider the notion of the hardest instance of a given size, and give bounds on how hard it is. We define the *syntenic diameter of order n* as

$$D_y(n) \overset{\text{def}}{=} \max_{S(n,n)} D(S(n, n)),$$

the number of moves required to solve the worst instance of up to n elements and n sets. We also define the *complete n-instance $\mathcal{K}_n(n, n)$* of synteny, which consists of n copies of the set $\{1, \ldots, n\}$.

Lemma 14. $D_y(n) = D(\mathcal{K}_n(n, n))$.

Proof. Immediate from monotonicity.

$\square$

Lemma 15. $\tilde{D}(\mathcal{K}_n(n, n)) = 2n - 3$.

Proof. All genes appear n times, so

$$n - 2 - \left| \{\ell \mid \text{count}(\ell) \leq n - 1\} \right| = n - 2.$$

By Theorem 11, then, $\tilde{D}(\mathcal{K}_n(n,n)) \geq 2n - 3$. We easily have that $\tilde{D}(\mathcal{K}_n(n,n)) \leq 2n - 3$: complete $n - 2$ fusions to leave two copies of $\{1, \ldots, n\}$, complete 1 translocation to eliminate n and leave $\{1, \ldots, n - 1\}$, and then $n - 2$ fissions to solve the instance. This is a linear move sequence of length $2n - 3$ solving $\mathcal{K}_n(n,n)$.   □

Theorem 16. $2n - 3 \geq D(\mathcal{K}_n(n,n)) \geq 2n - 3 - \log_{4/3}(2n - 3).$

Proof. Clearly for any synteny instance $D(\mathcal{S}(n,k)) \leq \tilde{D}(\mathcal{S}(n,k))$. Then, from the bound on linear synteny proved by DasGupta et al, we have that

$$\tilde{D}(\mathcal{K}_n(n,n)) \geq D(\mathcal{K}_n(n,n)) \geq \tilde{D}(\mathcal{K}_n(n,n)) - \log_{4/3} D(\mathcal{K}_n(n,n))$$
$$\geq \tilde{D}(\mathcal{K}_n(n,n)) - \log_{4/3} \tilde{D}(\mathcal{K}_n(n,n)).$$

By Lemma 15, we have

$$2n - 3 \geq D(\mathcal{K}_n(n,n)) \geq 2n - 3 - \log_{4/3}(2n - 3).$$

□

Note that this is almost tight, with only a $\log_{4/3}(2n - 3)$ window for the syntenic diameter. Even more strongly, however, we conjecture that there is no way to solve $\mathcal{K}_n(n,n)$ in any faster way than the linear sequence described in the proof of Lemma 15:

Conjecture 17. $D(\mathcal{K}_n(n,n)) = 2n - 3.$

9   A Preliminary Implementation

We have implemented all of the heuristics discussed above ($\mathcal{H}$, $\mathcal{F}$, and $\mathcal{F}'$) in Standard ML. The full implementation is approximately 750 lines of code. We have run these algorithms on eight sets of real synteny data, found on the INRA Comparative Homology Database [4]. We make the following observations based upon the results of these tests:

- In all cases, $\mathcal{F}'$ performed at least as well as $\mathcal{F}$; on one of the eight data sets, $\mathcal{F}'$ outperformed $\mathcal{F}$ by one move.
- In most cases (5 of 8), the component bound was actually attained by both $\mathcal{F}$ and $\mathcal{F}'$. In the 6th case, $\mathcal{F}'$ achieved the component bound and $\mathcal{F}$ was within one move of it. In each of these six cases, the instance can be solved using only translocations.

The latter point may raise some question about the validity of the model (that it is too easy to solve too many instances, and thus that the model fails to be informative about relative distances among groups of species), or may indicate that there is simply insufficient synteny data presently available.

10   Conclusions and Future Work

We have proven a number of interesting structural results for syntenic distance, including monotonicity and the fact that improving the approximation ratio for this problem will require an algorithm that works among components. These results may help in solving the obvious remaining open question:

– Is there an approximation algorithm for syntenic distance that achieves an approximation ratio strictly better than 2?

The lower bound from Theorem 11 may be useful in improving the approximation ratio. Other interesting open questions include:

– Can connected synteny be approximated any better than general synteny?
– Can we improve the bound on $D(\mathcal{K}_n(n, n))$ and prove Conjecture 17?

Acknowledgements. We thank Jon Kleinberg for extensive and fruitful discussions on numerous aspects of this paper and the syntenic distance problem. We also thank Chris Jeuell, Manish Sambhu, and Wes Weimer for reading and commenting on previous versions of this paper.

References

[1] Vineet Bafna and Pavel A. Pevzner. Genome rearrangements and sorting by reversals. In *34th IEEE Symposium on Foundations of Computer Science*, pages 148–157, 1993.

[2] Vineet Bafna and Pavel A. Pevzner. Sorting by transpositions. In *Proceedings of the 6th Annual ACM-SIAM Symposium on Discrete Algorithms*, pages 614–623, 1995.

[3] Bhaskar DasGupta, Tao Jiang, Sampath Kannan, Ming Li, and Elizabeth Sweedyk. On the complexity and approximation of syntenic distance. In *1st Annual International Conference on Computational Molecular Biology*, pages 99–108, 1997.

[4] Institut National de la Recherche Agronomique. Comparative homology database. http://locus.jouy.inra.fr/cgi-bin/lgbc/mapping/common/taxonomy.pl.

[5] Jason Ehrlich, David Sankoff, and Joseph H. Nadeau. Synteny conservation and chromosome rearrangements during mammalian evolution. *Genetics*, 147(1):289–296, September 1997.

[6] Vincent Ferretti, Joseph H. Nadeau, and David Sankoff. Original synteny. In *7th Annual Symposium on Combinatorial Pattern Matching*, pages 159–167, 1996.

[7] Pavel Pevzner and Michael Waterman. Open combinatorial problems in computational molecular biology. In *Proceedings of the Third Israel Symposium on Theory of Computing and Systems*, pages 158–173, January 1995.

[8] David Sankoff and Joseph H. Nadeau. Conserved synteny as a measure of genomic distance. *Discrete Applied Mathematics*, 71:247–257, 1996.

Physical Mapping with Repeated Probes: The Hypergraph Superstring Problem

Serafim Batzoglou[1] and Sorin Istrail[2]

[1] MIT Laboratory for Computer Science, Cambridge, MA 02139, USA
serafim@theory.lcs.mit.edu
[2] Sandia National Laboratories, Massively Parallel Computer Research Lab
MS1110, Albuquerque, NM 87185-1110
scistra@cs.sandia.gov

"A problem for the next century."
Paul Erdös

Abstract. We focus on the combinatorial analysis of physical mapping with repeated probes. We present computational complexity results, and we describe and analyze an algorithmic strategy. We are following the research avenue proposed by Karp [9] on modeling the problem as a combinatorial problem – the Hypergraph Superstring Problem – intimately related to the Lander-Waterman stochastic model [16]. We show that a sparse version of the problem is MAXSNP-complete, a result that carries over to the general case. We show that the *minimum Sperner decomposition* of a set collection, a problem that is related to the Hypergraph Superstring problem, is NP-complete. Finally we show that the Generalized Hypergraph Superstring Problem is also MAXSNP-hard. We present an efficient algorithm for retrieving the PQ-tree of optimal zero repetition solutions, that provides a constant approximation to the optimal solution on sparse data. We provide experimental results on simulated data.

1 Introduction and Previous Work

Physical mapping using hybridization data involves the construction of genomic maps based on the information contained in the clone-probe hybridization matrix. The mapping technique has to cope with combinatorial difficulties that are specific to the hybridization data. There are errors like chimerism, false negatives or false positives, that come from the limitations in experimental accuracy. Errors introduce specific combinatorial problems whose solutions could provide good mapping hypotheses. Usually these optimization problems are NP-hard and various heuristics – based on generalizations of the *Consecutive Ones Property* (C1P) [14] – have been designed to cope with them e.g., [7], [4]. Another important combinatorial dimension of the mapping problem arises from the the fact that most probes have multiple occurrences on the genomic region to be mapped. The literature dealing with algorithms for mapping in the presence of

M. Crochemore, M. Paterson (Eds.): CPM'99, LNCS 1645, pp. 66–77, 1999.

repeated probes is quite limited. In this paper we consider the combinatorial difficulties of physical mapping with repeated probes, we identify some computational bottlenecks, and we propose algorithms that exhibit various degrees of measurable success.

The fundamental modeling paper of the area is the paper by Lander and Waterman [5] in which the widely accepted Lander-Waterman model is introduced and analyzed; see also [13], [3] and [11] for further mathematical and statistical analyses. According to the Lander-Waterman model, clones are distributed uniformly along the genomic region, and probes are distributed according to a Poisson distribution.

The only published algorithmic work focussing on mapping with repeated probes seems to be [6], although further recent work devoted to the problem is in progress [10], [18]. In [6] algorithmic strategies are proposed, based on the Lander-Waterman model by attempting to approximate the likelihood function, leading to NP-complete optimization problems that are reasonably tractable in practice. The algorithmic strategy proposed there uses local search 3-opt Lin-Kernigan type heuristics. No approximation algorithms with a provable guarantee were obtained. Based on this work, Karp [9] proposed the problem of designing approximation algorithms with guaranteed error bounds for the shortest superstring of a set collection – in our present terminology, the Hypergraph Superstring Problem. This optimization problem is a combinatorial problem intimately related to the Lander-Waterman model, capturing the search for minimal explanations of the hybridization data. This combinatorial problem was introduced before (see [19], [21], [8]) and it is notoriously difficult [8], [12]. We are interested here in the sparse version of the problem, consistent with biologically relevant data of the Lander-Waterman model.

Kou proves in a paper devoted to information retrieval and file organization [20] that a variant of the C1P – modeling multiple storage of records – is NP-complete. In our terminology the result is that the Hypergraph Superstring Problem for strict Sperner hypergraphs is NP-complete. In [8], non-tight upper and lower bounds were obtained for the hypergraph superstring length for the special case of the hypergraph being the power set of a finite set. [17] gives a comprehensive overview of the problem.

A clone-probe hybridization matrix is a 0/1 matrix with rows representing clones, columns representing probes, and a 1 in position (i, j) if and only if probe j is incident to clone i. Any permutation of the columns of such a matrix results in the same clone/probe incidence relationship. A collection of clones has the Consecutive Ones Property (C1P)[14] if there is a permutation of the columns of the hybridization matrix that allows each row (clone) to be of the form $0 \cdots 01 \cdots 10 \cdots 0$ - in a consecutive ones form. The obvious biological relevance of the C1P is that each clone spans a connected region of the genome. A clone-probe hybridization matrix containing "perfect" data, i.e., containing no errors and only unique probes, is a matrix that obeys the C1P. An important property for a heuristic mapping algorithm is to retrieve the C1P in the absence of errors [4]. This is one of the properties that our mapping algorithms achieve.

A feature of the Lander-Waterman model is the *Sperner property* of a set collection: no set is included in the other. Indeed, as the number of probes increases, the set of clones of the Lander-Waterman model has the Sperner property with high probability. The *PQ*-tree algorithm [14] that retrieves the C1P uses a framework that hierarchically decomposes the initial collection of sets into subcollections that avoid sets included in unions of other sets.

The C1P property of a hybridization matrix ensures that there are no repeated probes. The *Sperner decomposition* of a set collection satisfying the C1P, and the optimal merging of sets in such a collection to obtain a *PQ*-tree are relatively easy computing tasks. Both tasks become computationally intractable for very sparse instances of data with repeated probes To get insight into the new combinatorial difficulties, consider the intersection graph *IG* of a set collection. The vertices are the sets of the collection, and an edge exists between two vertices when the corresponding sets intersect. In the C1P case, the strict Sperner collections are sets of disjoint paths (SDP) in *IG*, while in the Hypergraph Superstring Problem they are general graphs. These facts point out to the importance of strict Sperner collections, as building blocks in the hierarchical decomposition of the Hypergraph Superstring Problem. As we will see in this paper, both the Sperner decomposition as well as the optimal merging of the sets in a strict Sperner collection are MAXSNP- /NP-complete tasks.

In all the above discussion the implicit assumption has been that a probe never appears more than once in a particular clone. This is a simplifying assumption that is justifiable probabilistically by the Lander-Waterman model, as the Poisson parameter λ governing probe distribution decreases. However, this property is not necessarily guaranteed in practice. In fact the genome deviates from the Lander-Waterman model by means of certain sequence patterns that are repeated and could cause higher than expected probe repetition. An alternative model therefore, is to seek the minimal explanation of the hybridization data in the form of a *multiset superstring* that allows for possible repetition of probes in a single clone. We prove that this problem is also MAXSNP-complete.

We present and test the *GREEDY-MERGE* algorithm that is based on Sperner decomposition of hypergraphs, with the following provable performance: (1) it retrieves the PQ-tree of all optimal zero-repetition superstrings; (2) on strict Sperner hypergraphs it is provably a 1.5625-approximation algorithm;(3) it provides a 2-approximation for hypergraphs with a restricted Sperner decomposition. The algorithm has cubic worst-case time complexity, and is much faster on sparse, biologically relevant data. We test the algorithm on data generated according to the Lander-Waterman model and found that it approximates the length of the initial (correct) superstring within a factor of 1.1 in most problems involving 100-200 clones, 200-400 probes, and 1.5 to 4.9 average probe repetition.

2 Background

2.1 Physical Mapping

DNA molecules are very long sequences over an alphabet of four letters, or nucleotides: $\{A, G, C, T\}$. The study of a genomic region involves breaking it into smaller pieces that can be sequenced by present technologies. *Physical Mapping* involves reassembling the true arrangement of the pieces on the initial genomic region, and then sequencing the smallest subset of pieces that cover the region. The *cloning* procedure incorporates the pieces of DNA into biological hosts. Each such copy is a *clone*. Through self-replication, a large number of copies of each clone are obtained. The result is a clone library containing many copies of pieces of the initial genomic region. The reconstruction process is based on data indicating "overlap" between clones. One method of detecting overlaps is through the hybridization of short sequences, called *probes*. Hybridization occurs when a probe sequence is complementary to a subsequence of a clone. If the probe has a unique occurrence on the initial genomic region and if two clones are hybridized by the same probe then they overlap. This assumes ideal experimental conditions, i.e., no errors. So, unique probes detect overlap. However, in general probes are complementary to multiple places on the genomic region so detecting overlap is ambiguous. The information contained in the hybridization data can be summarized as follows. Let the clones be $\{C_1, \ldots, C_n\}$ and the probes be $\{P_1, \ldots, P_m\}$. Let the matrix H be defined by $H[i,j] = 1$ if probe P_j hybridizes to clone C_i, and $H[i,j] = 0$ otherwise. The problem studied in this paper is that of using hybridization data given in the matrix H to reassemble the clones such as to reconstruct the initial genomic region. Let us note that the process of breaking the DNA into pieces and selecting probes, even in a perfect cloning and hybridization experimental scenario, might result in loss of information. Therefore, we may not be able to obtain the exact reconstruction. To well-define the problem, we aim at obtaining the maximal mapping information consistent with H.

2.2 The Lander-Waterman Model

We will first define the Lander-Waterman model and then formulate a combinatorial problem in terms of hypergraphs, an appropriate framework for probe/clone hybridization data. Then superstrings are introduced in order to search for the minimal number total repetition of the probes needed to explain the hybridization data.

The Lander-Waterman Model

1. A *clone* is an interval of length 1 contained in the interval $[0, N]$. The left endpoints of the clones are independent random variables, uniformly distributed over $[0, N - 1]$.

2. Probes $1, \ldots, m$ are distributed along the interval $[0, N]$ according to independent Poisson processes of rate λ. That is, a probe occurs at a short interval of length dx with probability λdx, and disjoint intervals are independent.

2.3   The Hypergraph Superstring Problem

Hypergraphs. A *hypergraph* is a pair $H = (X, \mathcal{S})$, where X is a finite set, and $\mathcal{S} = \{S_1, \ldots, S_m\}$ is a family of subsets of X. The sets S_i are called *hyperedges*. The following definitions apply to hypergraphs as well to families of sets. A hypergraph is B-*bounded* if all of its hyperedges have at most B elements. A hypergraph is a *chain* if $\mathcal{S} = \{S_1, \ldots, S_m\}$ and $S_1 \subseteq S_2 \subseteq \cdots \subseteq S_m$. A hypergraph is called *antichain*, or Sperner, if no S_i is included in S_j, for every $i, \neq j, 1 \leq i, j \leq m$. A hypergraph is called *strict Sperner* if no hyperedge is included in the union of the other hyperedges, or equivalently every hyperedge has a *characteristic* element.

A *Sperner decomposition* of a hypergraph $H = (X, \mathcal{S})$ is a decomposition of $\mathcal{S}$ into subfamilies of sets called *levels* $\mathcal{S}_1, \ldots, \mathcal{S}_t$ such that: (1) the levels partition $\mathcal{S}$, i.e. $\mathcal{S} = \mathcal{S}_1 \cup \cdots \cup \mathcal{S}_m$ and $\mathcal{S}_i \cap \mathcal{S}_j = \emptyset, 1 \leq i \neq j \leq t$; (2) $\mathcal{S}_i$ is a strict Sperner family of sets for every i, $1 \leq i \leq t$ and (3) $\bigcup \mathcal{S}_t \subseteq \bigcup \mathcal{S}_2 \subseteq \cdots \subseteq \bigcup \mathcal{S}_t$.

Consider the clone-probe hybridization matrix of a Lander-Waterman process. Let P be the set of probes, and let $C = \{C_1, \ldots C_m\}$ be the clones viewed as sets of probes. Then $H_{LW} = (P, C)$ is the associated hypergraph. According to the Lander-Waterman model, the arrivals of the left endpoints of the clones are distributed according to a Poisson process of rate $\frac{m}{N-1}$. If $|P|$ is large enough, with high probability no clone is a subclone of any other clone. Then H_{LW} is a Sperner hypergraph. The average number of probes per clone is $\lambda |P|$.

Multiset Superstrings. A string $\sigma = \sigma_1 \cdots \sigma_r$, is a multiset superstring of any subset of $U(\sigma) = \{S : 1 \leq \beta \leq \eta \leq r : S = \{\sigma_\beta, \sigma_{\beta+1}, \ldots, \sigma_\eta\}\}$.

Set Superstrings. A string σ is a set superstring (or simply, superstring) of any subset of $V(\sigma) = \{S : \forall \beta \leq i < j \leq \eta \quad \sigma_i \neq \sigma_j, S = \{\sigma_\beta, \ldots, \sigma_\eta\}\}$

For $S \in U(\sigma)$ or $S \in V(\sigma)$ we define $\beta_\sigma(S), \eta_\sigma(S)$ so that $S = \{\sigma_{\beta_\sigma(S)}, \ldots, \sigma_{\eta(S)}\}$. We say that σ *expresses* S if $S \in U(\sigma)$ ($S \in V(\sigma)$, also denoted by $S \in \sigma$. A multiset (set) superstring σ is *non-repeting* if no letter in σ occurs more than once.

Now we are ready to define our main computational problems:

The Hypergraph Set Superstring Problem: *Given* a Hypergraph $H = (X, \mathcal{S})$ *find* a superstring $\sigma = \sigma_1 \ldots \sigma_n$ for H of minimal length n.

The Hypergraph Multiset Superstring Problem: *Given* a Hypergraph $H = (X, \mathcal{S})$ *find* a multiset superstring $\sigma = \sigma_1 \ldots \sigma_n$ for H of minimal length n.

Remark. Let us remark that the corresponding Graph Superstring Problem, where the hyperedges have exactly two elements can be solved in time linear in the number of edges in the graph. The minimum superstring coincides with the Eulerian path if the graph has such a path. In the general case, it coincides with the minimum size collection of Eulerian paths that cover all the edges.

Our problem, the Hypergraph Superstring problem, is therefore a hypergraph generalization of the Eulerian path problem in graphs.

The Sperner Decomposition of a Hypergraph Problem: *Given* a Hypergraph $H = (X, S)$ and an integer $k > 0$, *decide* whether there exists a Sperner decomposition into k levels.

3 Computational Complexity of the Hypergraph Superstring Problems

We show that the hypergraph set superstring, and the hypergraph multiset superstring problems are MAXSNP-hard. We prove these results with an *L*-reduction from *TSP(1,2)* on bounded degree undirected graphs. The same reduction proves both problems to be MAXSNP-hard. We are thus strengthening Kou's result by showing that the same problem is MAXSNP-hard, which implies that it is computationally intractable to approximate within better than a multiplicative constant of optimal. We also show that computing a Sperner Decomposition of a hypergraph is a hard computational task: it is NP-complete to decide whether a two-level decomposition exists and more generally, to find the Sperner Decomposition with a minimal number of levels.

Theorem 1. *The Hypergraph Set Superstring Problem and the Hypergraph Multiset Superstring Problem are MAXSNP-hard even for 5-bounded strict Sperner hypergraphs.*

Proof. We use an *L*-reduction (intuitively a linear reduction, refer to [1]) from *TSP(1,2)* on undirected graphs, on instances where the graph formed by length-one edges has bounded degree. *TSP(1,2)* is the traveling salesman problem with distances $1, 2$. That is, given a complete graph G with edges of distance 1 and 2, find the shortest Hamiltonian path on the graph.[1] This problem has been shown to be *MAXSNP*-complete even if restricted to instances where the graph formed by the length-one edges has bounded degree [2].

Let $H_G = (V, E)$ be a graph of bounded degree D specifying an instance of *TSP(1,2)*. That is, H_G contains the edges of cost 1 in the corresponding *TSP(1,2)* graph G. For every $v \in V = \{1, \dots, n\}$, with associated edges $(v, u_1), \dots, (v, u_d)$ where $d \leq D$, define hyperedge $S_v = \{v, \{v, u_1\}, \dots, \{v, u_d\}\}$. The hypergraph H is then (X, S) where $X = \bigcup_{v \in V} S_v$ and $S = \{S_v | v \in V\}$. Clearly the above reduction can be performed in logarithmic space. Notice that the resulting set collection is Sperner because every set S_v has a distinguishing element $v \in S_v$. Moreover, $\forall v : |S_v| \leq D + 1$.

We will show that there is a Hamiltonian path on the graph G of *TSP(1,2)* of cost $n - 1 + k$ if and only if there is a (multiset, or set) superstring σ for S of length $m + k + 1$ where $m = |E|$. Since H_G is a graph of degree bounded by D, $m \leq D \times n$ is linear in n. This will establish that the above reduction is an *L*-reduction.

---

[1] That is, the shortest path that visits each node exactly once.

Say there is a Hamiltonian path of cost $n - 1 + k$. Since all edges have costs 1 or 2, we know the path uses $n - 1 - k$ edges from H and k edges of cost 2. Construct σ of cost $m + k + 1$ as follows: σ arranges the sets S_v in the order the nodes v are arranged on the path. Whenever an edge (u, v) in H_G is used on the path, S_u and S_v overlap in one element in σ. Then,

$$|\sigma| = \sum_{v=1}^{s} |S_v| - (n - 1 - k) = m + k + 1$$

Conversely, say that σ is a superstring of length $m + k + 1 = \sum_{v=1}^{n} |S_v| - (n - k - 1)$. Construct a path by reading in σ each vertex in the order it appears. Since σ is shorter than $\sum_{v=1}^{n} |S_v|$ by $(n - 1 - k)$ there is a total overlap of $(n - 1 - k)$ between the sets on the superstring. Since no two sets contain more than one common element, there are $(n - 1 - k)$ sets that overlap. These sets have a common edge. This establishes a total of $(n - 1 - k)$ edges from H_G used in the path, and hence a path of cost $(n - 1 + k)$.

Theorem 2. *The Sperner Decomposition of a Hypergraph Problem is NP-complete. In particular, distinguishing between 2 and 3 levels for the minimum Sperner decomposition of a hypergraph is NP-complete, even for 3-bounded hypergraphs with size ≤ 1 hyperedge intersections.*

Proof. (Sketch). Given a hypergraph $H = (X, \mathcal{S})$ and a partition of $\mathcal{S}$ into $\mathcal{S}_1, \mathcal{S}_2$, we can check efficiently the properties for a Sperner decomposition. Therefore, the Sperner Decomposition in k levels problem is in *NP*. We will show *NP*-hardness by a reduction from *3SAT*.

Let $\phi = \psi_1 \bigvee \ldots \bigvee \psi_m$ be a 3-CNF formula, with variables $x_1, \ldots, x_n$. We construct a hypergraph $\mathcal{S}_\phi$. Figure 1 shows the main part of the construction.

Two or three boxes connected by a line network correspond to one hyperedge. Any "o" contained in a box is a unique element in X. An "o" or "s" contained only in one box is contained only in one set. Such a set has to be in layer 1, because the union of layer 1 contains the union of layer 2. A set containing elements all belonging to sets in layer 1, has to be in a layer $\neq 1$.

Associate layer 1 with *TRUE* and layer 2 with *FALSE*. Then the top part of Figure 1 containing the three sets labeled *TRUE*, *TRUE*, and *FALSE*, should be self-explanatory. It follows that any two sets labeled x and $\bar{x}$ in Figure 1 are in different layers, in any 2-layer Sperner decomposition.

Assign either all the x-sets, or all the $\bar{x}$-sets to layer 1 for each variable x, thereby constructing a truth assignment. Among the x-sets and the $\bar{x}$-sets, notice in Figure 1 that there are some containing an s-element. These sets are meant to correspond to literals in the clauses of ϕ.

For each variable x with k_x occurrences of literal x and k'_x occurrences of literal $\bar{x}$ construct k_x x-sets with an s-element, and k'_x $\bar{x}$-sets with an s-element.

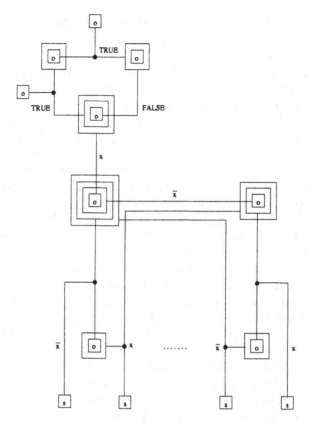

Fig. 1. Gadget for truth assignment

Finally, three s-elements collapse to one if and only if the corresponding literals are in the same clause ψ_i. Therefore there is one s-element for each clause.

Clearly a truth assignment satisfying every clause translates to a 2-level Sperner decomposition. Conversely, a 2-level Sperner decomposition correctly assigns truth value: $\forall x$ all the x-sets are in the same level, complement to the one with the $\bar{x}$-sets. Moreover, every s-element belongs to three sets one of which in level 1, thereby satisfying the corresponding clause.

4   Algorithms

We designed a collection of algorithms that incrementally deal with more complex hypergraph structures. They provide a collection of subroutines from which the *SPERNER-GREEDY-MERGE* algorithm is constructed. The algorithm *SPERNER-GREEDY-MERGE* retrieves the *Consecutive Ones Property* for a hybridization matrix, which hints on the strength of the algorithm to deal with all different kinds of imperfections in physical mapping data. Moreover, *SPERNER-*

GREEDY-MERGE has approximation guarantees on sparse, biologically relevant data. Complete details of the algorithms are included in the Appendix sent to the Program Committee.

The *Merge-Sequence-Pair* procedure is the basic building block of the algorithms. The algorithm merges pairs of already merged set collections. We say that a sequence of sets $\mathcal{A} = [A_1, \ldots, A_r]$ is a superstring collection for a set collection $\mathcal{S} = \{C_1, \ldots, C_s\}$ if for each i, $1 \leq i \leq s$ there are j_i, k_i, $1 \leq j_i \leq k_i \leq n$ such that $C_i = \bigcup_{j_i \leq l \leq k_i} A_l$, and A_l, A_m are disjoint for all $j_i \leq l < m \leq k_i$. If $\mathcal{A}$ and $\mathcal{B}$ are superstring collections for clone (set) collections $C_1, \ldots, C_{s_A}$ and $D_1, \ldots, D_{s_B}$, then *Merge-Sequence-Pair* finds the optimal way of merging the two set sequences $\mathcal{A}$ and $\mathcal{B}$ into $Merge(\mathcal{A}, \mathcal{B})$, a superstring collection for $\{C_1, \ldots, C_{s_A}, D_1, \ldots, D_{s_B}\}$. *Merge-Sequence-Pair* requires that $\{C_1, \ldots, C_{s_A}, D_1, \ldots, D_{s_B}\}$ is Sperner, and respects the order of the sets in set sequences $\mathcal{A}$ and $\mathcal{B}$. *Merge-Sequence-Pair* was designed to provide a way to merge efficiently, in an incremental greedy way, large collections of sets into one Q-node from which superstrings of the set collections can be obtained.

The *SPERNER-GREEDY-MERGE* algorihtm uses the *Merge-Sequence-Pair* algorithm in a greedy way to construct superstrings. That is, all possible *Merge-Sequence-Pair* operations are performed, each time performing the one that yields the greatest overlap between the two structures that are merged. Each of the initial structures (superstring collections) consists of one clone from the data set. The *SPERNER-GREEDY-MERGE* algorithm assumes that the clone collection is Sperner. At the first step of the algorithm all the clone intersection sizes are computed, and among the clone pairs that provide maximum intersections, one is chosen arbitrarily. This pair (call it C, D) is merged into a set sequence consisting of three sets, $C \backslash D, C \cup D, D \backslash C$. At each step, all new overlaps between the newly merged set sequence and the existing ones are computed. The pair to be merged is chosen arbitrarily among the ones with maximum overlap. The algorithm runs till there is no possible merge with non-zero overlap. In the case that there is a non-repeating superstring for the initial set of clones, *SPERNER-GREEDY-MERGE* retrieves the *PQ*-tree of all possible non-repeating superstrings.

The *GREEDY-MERGE* is dealing with Sperner levels, accommodating inclusions from higher levels of the Sperner decomposition. *GREEDY-MERGE* retrieves the C1P-property for arbitrary hypergraphs. It is a generalization of the *PQ*-tree C1P algorithm; it preserves the merges that are necessary for retrieving the consecutive ones property, performing them in a greedy fashion according to maximum overlaps. The *GREEDY-MERGE* algorithm uses the *SPERNER-GREEDY-MERGE* algorithm as a subroutine.

The algorithm *2-PHASE-GREEDY* is an approximation algorithm that works well on the strict Sperner hypergraphs. It achieves a 1.5625 worst-case ratio to the optimal solution. This algorithm is based on the *SPERNER-GREEDY-MERGE* algorithm, with some additional restrictions on the order in which the *Merge-Sequence-Pair* operations are performed.

5   Experimental Results

We implemented the *SPERNER-GREEDY-MERGE* algorithm and ran it on randomly generated data. The data were generated according to the Lander-Waterman model, where clones are intervals of length 1 distributed uniformly along the interval $[0, N]$.[2] The interval $[0, N]$ was divided in $1000N$ discrete positions and probes were distributed along $[0, N]$ according to a Poisson process, except that for each clone C, a probe p was allowed to occur only once. Any occurrences of p in C after the first, were discarded. This distribution is very similar to a pure Poisson distribution if, as in our case, the mean arriving time of a probe is much greater than the length of a clone, which is 1 in our case. The hypergraph that was given as input to *SPERNER-GREEDY-MERGE* consisted of all the maximal generated clones.

Table 1 displays some results of running the algorithm while varying N, the length of the interval where the clones are distributed; n, the number of clones used for generating the data, m, the number of probes used for generating the data; and λ for exponential distribution of the arriving time of probes. p is the average number of probes after generating the data, r_{avg} is the actual average number of repetitions of probes, approximately $= \lambda N$, and r_{max} is the average over all generated sequences, maximum number of repetitions of a single probe. L_0 is the average length of the generated sequences, and L_{GM} is the average length of the sequences or sequence fragments produced by *SPERNER-GREEDY-MERGE*. To facilitate presentation, the performance is presented in percentage of optimal that correspond to the ratio L_0/L_{GM}. That is, when we say that the performance is 95.9% as on the table below in the experiment running with $N = 20$ and 300 probes, we mean that *SPERNER-GREEDY-MERGE* produces on average a superstring collection of total length $1.0428 \times$ [length of the initial sequence].

N	n	m	p	r_{avg}	r_{max}	L_0	L_{GM}	Performance
5	200	200	159.2	1.6	3.9	259.1	292.7	88.7%
10	100	200	118.3	1.4	3.8	165	163.2	100%
10	100	200	145	1.5	3.7	216.5	238.8	90.7%
10	100	200	159	1.7	4.7	268	319.5	84.2%
20	100	200	186.7	2.4	6.8	451.3	453.8	99.5%
20	100	200	192.8	3	7.1	555.3	585.5	94.9%
20	100	200	196.4	3.4	7.8	660.3	699	94.5%
20	100	300	275.5	2.4	6.5	638	665.5	95.9%
30	100	300	293	3.3	8.5	951	913	100%
30	150	300	293	3.3	8.5	969	1041	93.1%
40	200	400	397.5	4.9	12.5	1886.5	1937.5	97.4%

Table 1. Results on data generated according to the Lander-Waterman model.

---

[2] The clone beginnings are distributed along $[0, N - 1]$ with uniform probability.

As can be seen, the major factor that seems to hurt the performance of the algorithm is the *coverage* of the gene, i.e. the average number of clones that cover each point in the interval $[0, N]$. This indicates that a hypergraph that is Sperner decomposable in a few layers is easier to handle than one that is decomposable in many layers. This experimental observation is consistent with our intuition that the Sperner Decomposition problem captures the essence of the difficulty of computing minimal superstrings. High probe repetition also hurts the performance of the algorithm, as expected. The performance of the algorithm increases with the number of probes. Therefore the algorithm is expected to produce good results given that a sufficient number of probes is used in the experiment. Finally the performance seems unaffected as the number of clones increases. Occasionally the algorithm produces a shorter superstring than the initial superstring. This would correspond to experimental conditions where either too few clones, or too few probes are used, resulting in under-specified instances of the problem.

6   Future Work

Further research will focus on returning to the Lander-Waterman model to relate the worst-case algorithmic approximability performance, to the probabilistic analysis of the algorithmic performance in the stochastic model. The mapping difficulties introduced by repeated probes as reported by the genomic centers for Human Chromosomes, e.g., the Human Y Chromosome, [15] seem well captured by the combinatorial structure of our algorithms. We are planning a detailed experimental analysis of the performance of our algorithms on real data.

On the theoretical side, it is an open question to prove a stronger inapproximability result for *MIN-HYPERGRAPH-SUPERSTRING*, or to demonstrate a constant approximation algorithm for the general problem.

Acknowledgements

We would like to thank Mike Waterman for discussions on the problem.

The first author would like to thank his advisor, Bonnie Berger, for discussions on the problem and for initiating the communication between the two authors. The first author is supported by a Merck Fellowship.

The second author wants to thank Lee Istrail for the information that Paul Erdös was going to give a lecture to him and his fellow finalists of the Mathematics Olympiad, at the invitation of the University of New Mexico, the host of the olympiad finals. This led to a one day long, unforgettable visit of Erdös at Sandia Labs.

This work was supported by the Applied Mathematical Sciences program, U.S. Department of Energy, Office of Energy Research, and was performed at Sandia National Laboratories, operated for the U.S. Department of Energy under contract No. DE-AC04-94AL85000.

References

[1] Papadimitriou C.H. Approximability. In *Computational Complexity*. Addison-Wesley Publishing Company, 1994.

[2] Papadimitriou C.H. and Yannakakis M. The traveling salesman problem with distances one and two. *Math. of Operations Research*, pages 1–12, 1993.

[3] Green E. D. and Green P. Sequence-tagged site (sts) content mapping of human chromosomes: Theoretical considerations and early experiences. *PCR Methods and Applications*, 1:77–90, 1991.

[4] Greenberg D.S. and Istrail S. Physical mapping by sts hybridization: Algorithmic strategiews and the challange of software evaluation. *Journal of Computational Biology*, 2, Number 2:219–274, 1995.

[5] Lander E.S. and Waterman M.S. Genomic mapping by fingerprinting random clones: A mathematical analysis. *Genomics*, 2:231–239, 1988.

[6] Alizadeh F., Karp M. R., Newberg L. A., and Weisser D. K. Physical mapping of chromosomes: A combinatorial problem in molecular biology. *Algorithmica*, 13:52–76, 1995.

[7] Alizadeh F., Karp R.M., Weisser D.K., and Zweig G. Physical mapping of chromosomes using unique probes. *Manuscript*, 1995.

[8] Lipski W. Jr. On strings containing all subsets as substrings. *Discrete Mathematics*, 21:253–259, 1978.

[9] Karp R. M. Mapping the genome: Some combinatorial problems arising in molecular biology. *Symposium on Discrete Algorithms*, SODA 93:278–285, 1993.

[10] Waterman M.S. *Personal communication about the work of Simon Tavare. Ocrober, 1997.*

[11] Nelson D. O. and Speed T. P. Statistical issues in constructing high resolution physical maps. *Statistical Science*, 9, No. 3:334–354, 1994.

[12] Erdos Paul. *Personal Communication*, 1993.

[13] Arratia R., Lander E. S., Tavare S., and Waterman M. S. Genomic mapping by anchoring random clones: A mathematical analysis. *Genomics*, 11:806–827, 1991.

[14] Booth K. S. and Lueker G. S. Testing for the consecutive ones property, interval graphs and planarity using pq-tree algorithms. *J. Comput. Sys. Sci.*, 13:335–379, 1976.

[15] Foote S., Vollrath D., Hilton A., and Page D. The human y chromosome: Overlapping dna clones spanning the euchromatic region. *Science*, pages 60–66, October 1992.

[16] Lander E. S. and Waterman M. S. Genomic mapping by fingerprinting random clones: A mathematical analysis. *Genomics*, 2, Number 2:219–274, 1988.

[17] Waterman M. S. In *Introduction to Computational Biology*. Chapman and Hall, 1995.

[18] Shamir. *Personal communication, October 1997.*

[19] Ghosh S.P. Consecutive storage of relevant records with redundancy. *Communications of the ACM*, 18:464–471, 1975.

[20] Kou A. T. Polynomial complete consecutive information retrieval problems. *SIAM J. Computing*, 6, No.1:67–75, 1977.

[21] Lipski W. Information storage and retrieval – mathematical foundations ii. *Theoretical Computer Science*, 3:183–212, 1976.

Hybridization and Genome Rearrangement

Nadia El-Mabrouk[1] and David Sankoff[2]

[1] Département d'informatique et de recherche opérationnelle, Université de Montréal,
CP 6128 succursale Centre-Ville, Montréal, Québec, H3C 3J7.
mabrouk@iro.umontreal.ca

[2] Centre de recherches mathématiques, Université de Montréal,
CP 6128 succursale Centre-Ville, Montréal, Québec, H3C 3J7.
sankoff@ere.umontreal.ca

Abstract. We infer post-hybridization rearrangements in a hybrid genome, given the gene orders on its chromosomes and some knowledge of the two parent genomes. We study this in two biologically and computationally different contexts, genome fusion and interspecific fertilization. Exact algorithms are furnished for some cases, and a heuristic based on the Hannenhalli-Pevzner theory for another.

1 Introduction

An important mechanism for the rapid emergence of a new, qualitatively different species is the hybridization of two existing species. These parent species will generally be fairly closely related, but may have very different phenotypic expressions. There are actually several types of biological processes that give rise to hybrids, and these are perhaps most widespread in the plant kingdom. In this paper, we explore two such processes – genome fusion and interspecific fertilization. In the first case we give an exact, linear time algorithm for reconstructing the ancestral hybrid from knowledge of the modern genome and data about which gene came from which parent species. We then introduce additional data, on parental species gene order, and try to reconstruct two stages of hybrid genome evolution, intra- and intergenomic (referring to the haploid components originating from the two parents). We adapt the techniques of Hannenhalli and Pevzner [2,3] in a heuristic for separating these stages and give upper and lower bounds for the optimal transition point between them.

In the case of interspecific fertility, we hypothesize that a key stage in the stabilization of the hybrid genome can be found by calculating the median of three diploid genomes, the two parents and the hybrid. We refer to a reduction of this problem [5] to the Traveling Salesman Problem.

Definitions

A **genome** G is a collection of N **chromosomes** $G_1, \cdots, G_N$. A chromosome is a string of **signed** ($+$ or $-$) elements from a set $\mathcal{E}$ of **genes**. Each gene in $\mathcal{E}$ appears exactly once in the set of N chromosomes.

M. Crochemore, M. Paterson (Eds.): CPM'99, LNCS 1645, pp. 78–87, 1999.
© Springer-Verlag Berlin Heidelberg 1999

For string $X = x_1, \cdots, x_m$, we write $-X$ for the inverted string $-x_m, \cdots, -x_1$. We define the following **rearrangement operations** as in Figure 1: **Inversion**, (or reversal) where any proper substring of a chromosome is inverted. (Inverting the entire chromosome only invokes an alternate notation for the identical chromosome, and does not constitute a rearrangement operation.) **Translocation**, where two chromosomes (one or both inverted), exchange prefixes of any length. A **fusion** is a translocation where one of the prefixes is the entire chromosome and the other prefix is null. A **fission** is a translocation where one of the starting chromosomes is the null string. Our analyses of translocations implicitly include fusions and fissions.

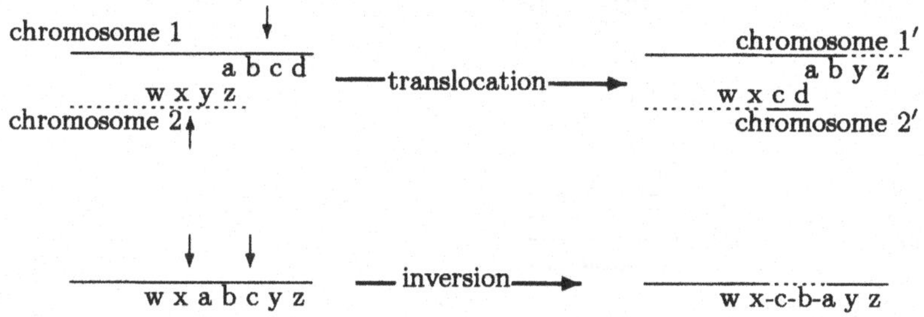

Fig. 1. Schematic view of genome rearrangement processes. Letters represent positions of genes. Vertical arrows at left indicate boundaries of affected substrings. Translocation exchanges prefixes of two chromosomes. Inversion reverses the order and sign of genes in a substring (dotted segment).

2 Resolution of Tetraploidy; Ancestral Synteny Unknown

One form of hybridization of two karyotypically distinct species sees the fusion of two genomes followed by a series of chromosomal rearrangement events until the hybrid genome is finally stabilized as a diploid (e.g. [1]). The two homologous versions of each gene, one from each parent species, may diverge functionally to create a gene family. From the moment of hybridization till the present, the two parent species may also undergo chromosomal rearrangement. Thus we have direct access to neither the ancestral hybrid genome nor the two contributing strains. In this section we provide a method for reconstructing the ancestral hybrid, given the order of the genes on its chromosomes as well as data (obtained, for example, from sequence analysis) on which of these genes originated from each of the parent species.

2.1 Formalization

Consider two genomes A and B having disjoints sets of genes, $\mathcal{E}(A)$ and $\mathcal{E}(B)$, respectively. Let G be a third genome with N chromosomes and gene set $\mathcal{E} =$

$\mathcal{E}(A) \cup \mathcal{E}(B)$. Given only $\mathcal{E}(A)$, $\mathcal{E}(B)$, and G, including how the genes are distributed and ordered on the N chromosomes of G, the problem is to find $d(G)$, the minimal number of inversions and translocations necessary to transform G into an **ancestral hybrid genome** H (with any number of chromosomes) satisfying the following condition: each chromosome of H contains genes from A only, or from B only. See Figure 2.

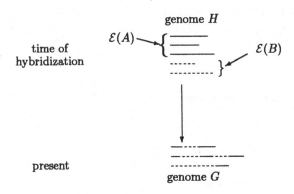

Fig. 2. Evolution of a hybrid genome resulting from genome fusion when gene origins, but not ancestral genome organization, is known. Genome H is to be reconstructed from knowledge of genome G, and ancestral gene sets $\mathcal{E}(A)$ and $\mathcal{E}(B)$ only.

2.2 Algorithm

The following procedure solves this problem exactly in time linear in the number of genes. The output attains the lower bound of the type found by Watterson *et al.*, except for certain special cases.

- In each chromosome G_i of G, amalgamate each substring of consecutive A-origin genes to form an **A-segment**. Similarly form the **B-segments**. A-segments and B-segments alternate along the length of the chromosome, separated by **breakpoints**.
- Transform each chromosome with an odd number $b_i > 1$ of breakpoints to a chromosome consisting of a single A-segment and a single B-segment by means of $\frac{b_i - 1}{2}$ inversions as follows.
  - While there remain at least 3 breakpoints, invert the fragment between the first and third breakpoints. Two A-segments are thus made adjacent and two B-segments are made adjacent.
  - Erase the breakpoints between the two adjacent A-segments and between the two adjacent B segments, thus reducing the number of breakpoints by two.

- Transform each chromosome with an even number $b_i > 2$ of breakpoints to a chromosome consisting of either two A-segments and a single B-segment, or two B's and one A, by means of $\frac{b_i-2}{2}$ inversions as follows.
  - While there remain at least 4 breakpoints, invert the fragment between the first and third breakpoints. Two A-segments are thus made adjacent and two B-segments are made adjacent.
  - Erase the breakpoints between the two adjacent A-segments and between the two adjacent B segments, thus reducing the number of breakpoints by two.
- Form as many pairs of ABA and BAB chromosomes as possible. Two translocations performed on each pair suffice to produce a homogeneous A chromosome and a homogeneous B chromosome, allowing the erasure of all four breakpoints.
- Suppose some 2-breakpoint chromosomes remain and they are all ABA. They may be amalgamated two by two, each time with a translocation that produces a homogeneous A chromosome and an ABA chromosome, and allows the erasure of two breakpoints, until only one ABA remains.
- Suppose instead of the previous step, the only 2-breakpoint chromosomes remaining are BAB. They may be amalgamated two by two, each time with a translocation that produces a homogeneous B chromosome and an BAB chromosome, and allows the erasure of two breakpoints, until only one BAB remains.
- If there are any 1-breakpoint chromosomes, form as many pairs of them as possible.
  - If there are no 2-breakpoint chromosomes, transform each of the pairs of one-breakpoint chromosomes into one homogeneous A chromosome and one homogeneous B by means of a single translocation, and erase the two breakpoints.
  - If there is a 2-breakpoint chromosome, transform all but one of the pairs of chromosomes into one homogeneous A chromosome and one homogeneous B by means of a single translocation, and erase the two breakpoints. Then two translocations suffice to transform the remaining pair and the 2-breakpoint chromosome into three homogeneous chromosomes, and to erase all four remaining breakpoints.
- If 1- or 2-breakpoint chromosomes remain there are several cases:
  - If all that remains is a single 1-breakpoint chromosome, one translocation (fission) is required to produce two homogeneous chromosomes and remove the breakpoint.
  - If all that remains is a single 1-breakpoint chromosome and a single 2-breakpoint one, two translocations (one a fission) are required to produce two homogeneous chromosomes and to remove all three breakpoints.
  - If all that remains is a single 2-breakpoint chromosome, two translocations (fissions) are required to produce two homogeneous chromosomes and to remove both breakpoints.

The output from this algorithm consists of homogeneous A chromosomes and homogeneous B chromosomes only, and the number of steps is $\lceil \sum_i b_i \rceil / 2 + \Psi$, where $\Psi = 0$ except if the last step of the algorithm must be activated, i.e., when there are no chromosomes G_i of forms $A \cdots B$ or $B \cdots A$, and an unequal number of chromosomes of forms $A \cdots A$ and $B \cdots B$. Here, $\Psi = 1$.

Note that there are generally many equally good solutions to this problem. In the next section, we reformulate the problem in order to pin down the structure of the ancestral genome somewhat. This will require additional data on the parent genomes and some assumptions about the amount of evolution in the hybrid compared to the purebred descendants of the parents.

3   Resolution of Tetraploidy; Ancestral Synteny and Gene Order Inferred

A second version of the hybridization problem uses the modern configurations A, B and G of the two parent genomes and the hybrid genome, respectively, to infer the three ancestral genomes A', B' and G' at the moment of hybridization, as on the left of Figure 3. Note that G' consists of the chromosomes in A' plus the chromosomes in B'.

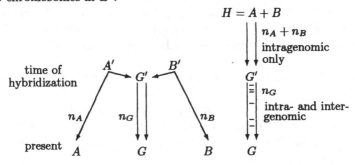

Fig. 3. Localization of ancestral hybrid immediately before intergenomic translocations

As a first step, we infer the total number n of evolutionary steps required to produce G from a construct H consisting of the chromosomes of A and the chromosomes of B, as on the right of Figure 3. We assume that G' is one of the intermediate steps in this evolution so that $n = n_A + n_B + n_G$, where n_X is the number of steps from genome X' to genome X, for $X \in \{A, B, G\}$.

Under the assumption that one of the first translocations to occur in the stabilization of the hybrid will be an **intergenomic** one, involving chromosomes from both A' and B', we could locate G' at the last step on the path from H to G before the first intergenomic translocation, as on the right of the figure. Unfortunately, the optimal path is not unique, and there will generally be one optimum whose *first* step is intergenomic, so that $n_A + n_B = 0$ and $n = n_G$. This may indeed be biologically meaningful in some contexts where hybrids evolve more rapidly than their parents. In other cases we may prefer to look for the path where n_G is as small as possible, to allow for a maximum of evolution in the parent species.

It is this latter problem we investigate in this section. First we sketch the method of Hannenhalli and Pevzner [2,3], hereafter "H-P", for finding the minimum number of translocations and inversions necessary to transform one genome into another, and show how a heuristic for finding a minimal n_G solution to the hybridization problem may be grafted onto their algorithm. We then show how to calculate, relatively quickly, an upper bound for this heuristic based on one step of the algorithm. Finally, we construct a lower bound based on a breakpoint argument.

3.1 The H-P Algorithm and a Heuristic for n_G

We will only sketch the H-P procedure, which is rather complex, and give additional details for those aspects which are modified in our heuristic. The first step in the comparison of two multi-chromosomal genomes through translocations and inversions is to reduce it to a problem of comparing two single chromosome genomes through inversion only. These latter genomes are constructed essentially by concatenating the individual chromosomes in the original genomes end-to-end in an arbitrary order. (Additional dummy genes, called **caps** must be appropriately inserted at the ends of the original chromosomes of both genomes). Translocation in an original genome becomes inversion in the new one. In the string representing a chromosome each gene $+x$ is replaced by the pair $x^t x^h$, and $-x$ by $x^h x^t$.

To find the minimum inversions $d(H, G)$ necessary to transform one single-chromosome genome H to another, G, H-P constructs a **cycle graph**, a bicolored graph $\mathcal{G}(V, E)$ with vertex set V containing x^t and x^h for all genes in $\mathcal{E}$, where black edges connect neighboring vertices in H, and gray edges connect neighboring vertices in G. Each vertex is thus adjacent to exactly one black edge and one gray edge. $\mathcal{G}$ therefore has a unique decomposition into disjoint alternating cycles. We set $b(\mathcal{G}) = |\mathcal{E}| - 1$, the number of black edges of $\mathcal{G}$, and $c(\mathcal{G})$ to be the number of cycles of $\mathcal{G}$. Note that $c(\mathcal{G})$ is maximal when $G = H$. The **size of cycle** C is the number of black (or gray) edges in C. The inversion distance between H and G is then:

$$d(H, G) = b(\mathcal{G}) - c(\mathcal{G}) + h(\mathcal{G}) + f(\mathcal{G}) \qquad (1)$$

where $h(\mathcal{G})$ is the number of **hurdles** in $\mathcal{G}$, and $f(\mathcal{G}) = 1$ if $\mathcal{G}$ is a **fortress** and $f = 0$ if not. (These concepts will be discussed below.)

A key concept in the algorithm is the **oriented component**. A gray edge in a cycle is oriented if the inversion disrupting the two adjacent black edges, i.e.,

a adjacent to b in H, b adjacent to c in G, c adjacent to d in H

becomes

a adjacent to c in H, c adjacent to b in G, b adjacent to d in H, replaces the cycle by two cycles. An oriented cycle is one containing at least one oriented gray edge. Two cycles whose containing gray edges that "cross", e.g., gene i adjacent to gene j in Cycle 1, gene k adjacent to gene t in Cycle 2 in G, but ordered i, k, j, t in H, are connected. A component of $\mathcal{G}(V, E)$ is a subset of the cycles, built recursively from one cycle, at each step adding all the remaining cycles connected to any of those already in the construction.

An oriented component has at least one oriented cycle. Hurdles are a particular class of unoriented components. The entire graph $\mathcal{G}(V, E)$ is a fortress if a certain configuration of hurdles obtains.

The H-P algorithm proceeds by decreasing $h - c$, the number of hurdles minus the number of cycles at each step. It handles each oriented component independently. If component C has γ_C black edges, and κ_C cycles, the algorithm proceeds to find a series of $\gamma_C - \kappa_C$ inversions that reduces the component to a set of γ_C 1-cycles.

Hurdles are treated somewhat differently. There is no inversion which will immediately increase the number of cycles in such a component. Instead, certain hurdles undergo an inversion which changes them into oriented components, decreasing the number of hurdles by one and leaving the number of cycles unchanged – hurdle "cutting". Other pairs of hurdles are merged by means of an inversion that *decreases* the cycle count by one, but also decreases the number of hurdles by two.

In each case, after the first inversion in a hurdle or pair of hurdles, the resulting configuration is an oriented component which may be reduced as above.

Unoriented components which are not hurdles will eventually become oriented through inversions operating on other components, and may then be reduced accordingly.

Thus the execution of the H-P algorithm involves repeatedly choosing an oriented cycle and performing an inversion around an oriented gray edge, thus increasing the number of cycles by one, except for the first inversion whenever hurdles must be cut or merged. The strategy for our heuristic focuses on the successive choices of cycles and edges within cycles. The idea is to stop the reduction of an oriented component when there is no choice of cycle and edge within the cycle which corresponds to an intragenomic translocation or inversion (i.e. involving genes from A only or genes from B only). Similarly, if either the conversion of a hurdle to an oriented component, or the pairing of two hurdles, corresponds to an intergenomic transfer, it is postponed.

This procedure is validated by the fact that each oriented component may be reduced independent of whatever inversions apply outside the component. Eventually, when no more intragenomic translocations are possible, we have reached a locally maximum value of $n_A + n_B$ (local minimum for n_G), the postponed reductions are re-started and the algorithm proceeds to an optimum solution.

3.2   An Upper Bound for the Heuristic

Suppose the decomposition of $\mathcal{G}(V, E)$ contains monogenomic oriented components $C_1, \cdots, C_r$ (each involving genes from a single genome only, A or B).

The decomposition may also contain other components. If component C_i has γ_{C_i} black edges and κ_{C_i} cycles, the r components will be reduced by $y = \sum_{i=1}^{r} \gamma_{C_i} - \kappa_{C_i}$ inversions. Then

$$d(H, G) - y \geq n_G,$$

where n_G is the value found by the heuristic.

This bound can be improved in three stages:

- By including, in the calculation of y, at least one monogenomic oriented cycle (if one exists) contained in each bi-genomic oriented component.
- By including, for each bi-genomic oriented component not satisfying the previous criterion, an intragenomic inversion (if one exists) around an oriented gray edge in a bi-genomic oriented cycle.
- By repeating the above steps on certain hurdles whose treatment does not depend on the previous analysis of other hurdles.

3.3   A Lower Bound for n_G

Label the genes in G according to whether they come from A or B as in Section 2, and form segments of contiguous A's and B's. Suppose there are b breakpoints in all. Then at least $\lceil \frac{b}{2} \rceil$ translocations and inversions are required to remove these breakpoints, and these are necessarily intergenomic. I.e., $\lceil \frac{b}{2} \rceil \leq n_G$.

4   Hybridization through Interspecific Fertility

Hybrids may be formed by the fertilization event of two distinct though related species, an accident in nature but often feasible in the laboratory, e.g. [4,7]. The parent species A and B may differ from each other by numerous genome rearrangements. The hybrid G' is able to survive and propagate despite the difference between the two haploid components of its diploid genome. Genome rearrangement of the hybrid rapidly ensues, however, first until a normal symmetric diploid configuration G^* is attained, and then while further stabilization of the new genome occurs. This scenario is illustrated in Figure 4. The rapid evolution of the hybrid means that we can often assume the relative stability of the parent genomes A and B if the evolutionary scale is not too lengthy.

Suppose that the rearrangements of the hybrid between G' and G^* are intragenomic. I.e., the two hybrid genomes "conspire" to reorganize internally to a common form, before fixing any intergenomic translocations. Then the inference problem which arises is to find G^*, and the amount of rearrangement which occurred between G' and G^*, and between G^* and the modern genome G.

This is essentially the "median problem" for genomes [6]: Given three genomes A, B and G, find the "median" G^* which minimizes the sum of a genomic distance between G^* and A, G^* and B, and G^* and G. In general, this is a difficult problem, but in one case, namely when the distance is just the sum of the breakpoints between G^* and each of the other three genomes, an algorithm is available

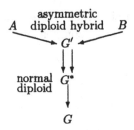

Fig. 4. Rearrangement before and after development of symmetric diploid.

[5], based on a reduction of the median problem to the Traveling Salesman Problem, which functions well for genomes containing a fairly large number of genes.

In the case where the rearrangements between G' and G^* are intergenomic from the start, it is difficult to propose a general model; unrestricted rearrangements in this context allow, for example, two versions of the same chromosomal segment in one haploid component, and zero in the other, meaning that the models of meiosis and mitosis underlying genome rearrangement theory no longer apply.

In the context of hybridization by interspecific fertilization, an additional type of data may be available. Genome typing informs us which chromosomal segments originate in which parental species (cf [7]). This pattern derives from the normal recombination events in the production of gametes. It may or may not be the case that the genomic position of a segment is correlated with that of its homologue in the parental species from which it derives. This illustrates the difference between this mechanism of hybridization and that of Sections 2 and 3, where genome fusion permits the retention of both of a pair of homologous genes, one from each parent.

Discussion

At least four aspects of the study of hybridization and rearrangement play a role in determining the nature of inference problem involved:

– The biological mechanism – genome fusion, interspecific fertilization, or other.

– The kinds of data available – present-day genomes, "ancestral" (i.e. stable or slowly-evolving) genomes, identification of genes in parental species (fusion model), or of segments (fertilization model).

– Assumptions about relative rates of evolution and about types of rearrangement event permitted.

– The entities to be inferred – events, syntenies, gene orders, beginning of intergenomic translocations.

The more detailed the kinds of data, the more detailed the kind of reconstruction that is possible, and the less ambiguity (non-uniqueness) in the results.

For example, the analysis in Section 2 generally reconstructs a large number of optimal solutions, while the one in Section 3 will be less ambiguous.

Each type of problem may require different tools from the inventory of methods developed in recent years for the study of genome rearrangement.

The most obvious domain of application of these methods is in the plant kingdom. The genomes of the cereals are particularly well-mapped and some of these show evidence of hybridization of the genome fusion type. The work of Rieseberg [4,7] illustrates the possibilities of the analysis in Section 4. As more genomic data becomes available, our methods should be more widely applicable.

Acknowledgments

Research supported by grants to DS from the Natural Sciences and Engineering Research Council of Canada and the Canadian Genome Analysis and Technology program. DS is a Fellow, and NEM a Scholar, in the Evolutionary Biology Program of the Canadian Institute for Advanced Research.

References

1. Gaut, B.S., Doebley,J.F.: DNA sequence evidence for the segmental allotetraploid origin of maize. Proceedings of the National Academy of Science (U.S.A.) **94** (1997) 6809–6814.
2. Hannenhalli, S., Pevzner, P.A.: Transforming cabbage into turnip. (polynomial algorithm for sorting signed permutations by reversals). In: Proceedings of the 27th Annual ACM-SIAM Symposium on the Theory of Computing (1995) 178–189.
3. Hannenhalli, S., Pevzner, P.A.: Transforming men into mice (polynomial algorithm for genomic distance problem). In: Proceedings of the IEEE 36th Annual Symposium on Foundations of Computer Science (1995) 581–592.
4. Rieseberg, L.H, Van Fossen, C., Desrochers, S.M.: Hybrid speciation accompanied by genomic reorganization in wild sunflowers. Nature **375** (1995) 313–316.
5. Sankoff, D., Blanchette, M.: The median problem for breakpoints in comparative genomics. In: Computing and Combinatorics, Proceeedings of COCOON '97. (T. Jiang and D.T. Lee, eds) Lecture Notes in Computer Science **1276** Springer Verlag (1997) 251–263.
6. Sankoff, D., Sundaram, G., Kececioglu, J.: Steiner points in the space of genome rearrangements. International Journal of the Foundations of Computer Science **7** (1996) 1–9.
7. Ungerer, M.C., Baird, S.J.E., Pan, J., Rieseberg, L.H.: Rapid hybrid speciation in wild sunflowers. Proceedings of the National Academy of Science (U.S.A.) **95** (1998) 11757–11762.
8. Watterson, G.A., Ewens, W.J., Hall, T.E., Morgan, A.: The chromosome inversion problem. Journal of Theoretical Biology **99** (1982) 1–7.

On the Complexity of Positional
Sequencing by Hybridization

Amir Ben-Dor[1], Itsik Pe'er[2], Ron Shamir[2], and Roded Sharan[2]

[1] Department of Computer Science and Engineering,
University of Washington, Washington, USA.
amirbd@cs.washington.edu.
[2] Department of Computer Science,
Tel-Aviv University, Tel-Aviv, Israel.
{izik,shamir,roded}@math.tau.ac.il.

Abstract. In sequencing by hybridization (SBH), one has to reconstruct a sequence from its k-long substrings. SBH was proposed as a promising alternative to gel-based DNA sequencing approaches, but in its original form the method is not competitive. Positional SBH is a recently proposed enhancement of SBH in which one has additional information about the possible positions of each substring along the target sequence. We give a linear time algorithm for solving the positional SBH problem when each substring has at most two possible positions. On the other hand, we prove that the problem is NP-complete if each substring has at most three possible positions.

1 Introduction

Sequencing by hybridization (SBH) was proposed and patented in the late eighties as an alternative approach to gel-based sequencing [4, 8, 15, 2, 9]. Using DNA chips, cf. [16], one can in principle determine exactly which k-mers (k-tuples) appear as substrings in a target that needs sequencing, and try to infer its sequence. Practical values of k are 8 to 10.

The fundamental computational problem in SBH is the reconstruction of a sequence from its *spectrum* - the list of all k-mers that are included in the sequence (along with their multiplicities). A naive approach to the problem is to look for a Hamiltonian path in a directed graph whose vertices correspond to k-mers in the spectrum, and two vertices are connected if the $(k-1)$-suffix of one equals the $(k-1)$-prefix of the other. This is however a computationally hard problem. Pevzner [10] has shown that the reconstruction problem can be reduced to finding an Eulerian path in another directed graph, an easily solvable problem. In that graph, vertices correspond to $(k-1)$-tuples, and for each k-tuple in the spectrum, an edge connects the vertices corresponding to its $(k-1)$-long prefix and suffix.

The main handicap of SBH is ambiguity of the solution. Alternative solutions are manifested as *branches* in the graph (i.e., two or more edges leaving the same vertex), and unless the number of branches is very small, there is no good way to determine the correct sequence. Theoretical analysis and simulations [17, 11] have shown that the

M. Crochemore, M. Paterson (Eds.): CPM'99, LNCS 1645, pp. 88–100, 1999.
© Springer-Verlag Berlin Heidelberg 1999

average length of a uniquely reconstructible sequence using 8-mer chip is only about two hundred, way below a single read length on a commercial gel-lane machine.

Due to the centrality of the sequencing problem in biotechnology and in the Human Genome Project, and due to its mathematical elegance, SBH continues to draw a lot of attention. Many authors have suggested ways to improve the basic method. Alternative chip designs [4, 7, 12, 13] as well as interactive protocols [14] were suggested. An effective and competitive sequencing solution using SBH has yet to be demonstrated.

Recently, several authors have suggested enhancements of SBH based on adding location information to the spectrum [1, 5, 6]. In *positional sequencing by hybridization* (PSBH), additional information is gathered concerning the position of the k-mers in the target sequence. More precisely, for each k-mer in the spectrum its allowed positions along the target are registered. The reduction to the Eulerian path problem still applies, but for each edge in Pevzner's graph we now have constraints restricting its position in the Eulerian path. Mathematically, this gives rise to the *positional Eulerian path* problem (PEP): Given a directed graph with a list of allowed positions on each edge, decide if there exists an Eulerian path in which each edge appears in one of its allowed positions. Hannenhalli et al. [6] showed that PEP is NP-complete, even if all the lists of allowed positions are intervals of equal length. Note that this leaves open the complexity of PSBH. They also gave a polynomial algorithm for the problem when the length of the intervals is bounded.

In this paper we address the positional sequencing by hybridization problem in the case that the number of allowed positions per k-mer is bounded, and the positions need not be consecutive. We give a linear time algorithm for solving the positional Eulerian path problem, and hence, the PSBH problem, in the case that each edge is allowed at most two positions. On the negative side, we show that the problem of PSBH is NP-complete, even if each k-mer has at most three allowed positions and multiplicity one. We use in our hardness proof a reduction from the positional Eulerian path problem restricted to the case where each edge is allowed at most three positions. The latter problem is shown to be NP-complete as well.

The paper is organized as follows: In Section 2 we define the PSBH and the PEP problems. In Section 3 we describe a linear time algorithm for the PEP problem when each edge has at most two allowed positions. In Section 4 we prove that the PEP problem is NP-complete if each edge has at most three allowed positions. Finally, we show in Section 5 that the PSBH problem is NP-complete when each k-mer is allowed at most three positions. For lack of space, some proofs are omitted.

2 Preliminaries

All graphs in this paper are simple, finite, and directed. Let $D = (V, E)$ be a graph. We denote $m = |E|$ throughout. For a vertex $v \in V$, we define its *in-neighbors* to be the set of all vertices from which there is an edge directed into v. We denote this set by $N_{in}(v) = \{u : (u, v) \in E\}$. We define the *in-degree* of v to be $|N_{in}(v)|$. The *out-neighbors* $N_{out}(v)$ and *out-degree* are similarly defined.

Let $E = \{e_1, \ldots, e_m\}$ and let P be a function mapping each edge of D to a non-empty set of integer labels from $\{1 \ldots m\}$ (its *allowed positions*). We call such a pair

(D, P) a *positional graph*. If for all e, $|P(e)| \leq k$, then (D, P) is called a *k-positional graph*. Let $\pi = \pi(1), \ldots, \pi(m)$ be a permutation of the edges in E. If π defines a path, i.e., for each $1 \leq i < m$, $\pi(i) = (u, v)$ and $\pi(i+1) = (v, w)$, for some $u, v, w \in V$, then we say that π is an *Eulerian path* in D.

An Eulerian path π in D is said to be *compliant* with the positional graph (D, P) if $\pi^{-1}(e) \in P(e)$ for each $e \in E$, that is, each edge in π occupies an allowed position. The k-positional Eulerian path problem is defined as follows:

Problem 1 (k-PEP). **Instance:** A k-positional graph (D, P).
Question: Is there an Eulerian path compliant with (D, P)?

Let $\Sigma = \{A, C, G, T\}$. The *p-spectrum* of a string $X \in \Sigma^*$ is the multi-set of all p-long substrings of X. The problem of sequencing by hybridization is defined as follows:

Problem 2 (SBH). **Instance:** A multi-set S of p-long strings.
Question: Is S the p-spectrum of some string X?

For simplicity, we shall call the input multi-set a spectrum, even if it does not correspond to a sequence. The SBH problem is solvable in polynomial time by a reduction to finding an Eulerian path in Pevzner's graph [11]. More specifically, construct a graph D whose vertices correspond to $(p-1)$-long substrings of strings in S, and edges are directed from $\sigma_1 \ldots \sigma_{p-1}$ to $\sigma_2 \ldots \sigma_p$ for each $\sigma_1 \ldots \sigma_p \in S$. Then every solution to the SBH instance naturally corresponds to an Eulerian path in D.

The *positional SBH* problem is defined as follows:

Problem 3 (PSBH). **Instance:** A multi-set S of p-long strings. For each $s \in S$, a set $P(s) \subseteq \{0, \ldots, |S| - 1\}$.
Question: Is S the p-spectrum of some string X, such that for each $s \in S$ its position along X is in $P(s)$?

If the set of allowed positions for each string is of size at most k, then the corresponding problem is called k-positional SBH, or k-PSBH. k-PSBH is linearly reducible to k-PEP in an obvious manner.

3   A Linear Algorithm for 2-Positional Eulerian Path

In this section we provide a linear time algorithm for solving the 2-positional Eulerian path problem. A key element in our algorithm is reducing the problem to 2-SAT. To this end, the input is preprocessed, discarding unrealizable edge labels (positions).

Let $(D = (V, E), P)$ be the input 2-positional graph. For every $1 \leq t \leq m$ define $\Delta(t)$ to be the set of edges allowed at position t, $\Delta(t) \equiv \{e \in E : t \in P(e)\}$. For every vertex $v \in V$ define $In(v, t)$ as the set of t-labeled edges entering v, $In(v, t) \equiv \{(u, v) : (u, v) \in \Delta(t)\}$. Similarly define $Out(v, t) \equiv \{(v, u) : (v, u) \in \Delta(t)\}$.

The first phase of the algorithm applies the following preprocessing step:

while $\exists t$ such that $\Delta(t) = \{e\}$ ($\Delta(t)$ is a singleton), **do:**
    Suppose $P(e) = \{t, t'\}$.
    Set $\Delta(t') \leftarrow \Delta(t') \setminus \{e\}$.
    Set $P(e) \leftarrow \{t\}$.

Lemma 1. *The preprocessing step does not change the set of Eulerian paths compliant with* (D, P).

When implementing this step, we maintain current $P(e)$ for each e, and $\Delta(t)$ for each t. If at any stage we discover that some set $\Delta(t)$ is empty, then we output *False* and halt, since no edge can be labeled t. The preprocessing phase can be implemented in linear time. We omit further details. In the following we denote by (D, P) the positional graph obtained after the preprocessing phase. The notation Δ refers to the resulting instance as well.

Lemma 2. *In* (D, P) *each position is allowed for at most two edges.*

Proof. The preprocessing ensures that if for some position t, $|\Delta(t)| = 1$, then $e \in \Delta(t)$ satisfies $|P(e)| = 1$. Let R be the set of positions t with $|\Delta(t)| = 1$, and let $r = |R|$. Then there are $m - r$ positions t for which $|\Delta(t)| \geq 2$, and $r' \geq r$ edges e with $|P(e)| = 1$. Thus,

$$2(m - r) \leq \sum_{t \notin R} |\Delta(t)| = \sum_{t} |\Delta(t)| - r = \sum_{e} |P(e)| - r = 2m - r' - r \leq 2(m - r) .$$

Hence, $r = r'$ and each label $t \notin R$ occurs exactly twice, implying that $|\Delta(t)| \in \{1, 2\}$ for all t. ∎

We say that vertex v is *fixed to position* t in (D, P) if $In(v, t) = \Delta(t)$ or $Out(v, t + 1) = \Delta(t+1)$. That is, any Eulerian path compliant with (D, P) must visit v at position t. Define Boolean variables X_e^t for all $t \in P(e)$ ($\sum_e |P(e)| = 2m - r$ variables in total). Define now the following sets of Boolean clauses:

$$X_e^t \quad \text{for every } e \in E \text{ where } P(e) = \{t\} . \tag{1}$$

$$X_{e_1}^t \oplus X_{e_2}^t \quad \text{for every } t \notin R \text{ where } \Delta(t) = \{e_1, e_2\} . \tag{2}$$

$$X_e^{t_1} \oplus X_e^{t_2} \quad \text{for every } e \in E \text{ where } P(e) = \{t_1, t_2\} . \tag{3}$$

$$X_{(a,b)}^t \Leftrightarrow X_{(b,c)}^{t+1} \quad \text{for every } t \in P((a, b)), (t + 1) \in P((b, c)), b \text{ is not fixed to } t . \tag{4}$$

$$\overline{X}_{(u,v)}^t \quad \text{for every } t \in P((u, v)), t < m, \text{ s.t. } Out(v, t + 1) = \emptyset . \tag{5}$$

$$\overline{X}_{(u,v)}^t \quad \text{for every } t \in P((u, v)), t > 1, \text{ s.t. } In(u, t - 1) = \emptyset . \tag{6}$$

Lemma 3. *There is a positional Eulerian path compliant with* (D, P) *iff the set of clauses (1)-(6) is satisfiable.*

Proof. Suppose that a satisfying truth assignment Φ exists. We shall assign an edge e to position t iff $\Phi(X_e^t) = True$. Clauses (1) and (2) guarantee that exactly one edge is assigned to each position. Clauses (1) and (3) guarantee that each edge is assigned to exactly one position, and that this position is allowed to the edge.

It remains to show that the above assignment of edges to positions yields a path in D. Suppose to the contrary that both $X_{(a,b)}^t$ and $X_{(b',c')}^{t+1}$ are assigned *True*, with $b \neq b'$. Then clauses (5) guarantee the existence of an edge $(b, c) \in \Delta(t + 1)$, while clauses (6) guarantee the existence of an edge $(a', b') \in \Delta(t)$. Therefore, b is not fixed to t, and a contradiction follows from clauses (4). Hence, Φ defines an Eulerian path compliant with (D, P).

The converse can be shown in a similar way.∎

Theorem 1. *2-PEP is solvable in linear time.*

Proof. The preprocessing phase is linear. By Lemma 2 the number of clauses (1)-(6) is $O(m)$. Each XOR clause in (2)-(3) and each equivalence clause in (4) can be written as two OR clauses. Moreover, one can generate all clauses in linear time. By Lemma 3 the problem is reduced to an instance of 2-SAT which is solvable in linear time [3].∎

Corollary 1. *2-PSBH is solvable in linear time.*

4   3-Positional Eulerian Path Is NP-Complete

In this section we prove that the 3-PEP problem is NP-complete by reduction from 3-SAT.

Theorem 2. *The 3-PEP problem is NP-complete*

Proof. Membership in NP is trivial. We prove NP-hardness by reduction from *3-SAT*. Let F be a 3-CNF formula with N variables $x_1, \ldots, x_N$, and M clauses $C_1, \ldots, C_M$. We assume, w.l.o.g., that each clause contains three distinct variables, and that all $2N$ literals occur in F. Denote $X_i = \{x_i\} \cup \{\overline{x}_i\}$. For a literal $L \in X_i$, let a_L denote the number of its occurrences in F. For $1 \leq j \leq a_L$ define $L(j) \equiv (L, j)$, thus $L(1), \ldots, L(a_L)$ is an enumeration of indices to the occurrences of L in F. For a clause $C = L \vee L' \vee L''$ introducing the j-th (j', j'') occurrence of L (L', L'', respectively), we write $C = L(j) \vee L'(j') \vee L''(j'')$. We shall construct a directed graph $D = (V, E)$ and a map P from E to integer sets of size at most 3, such that F is satisfiable iff (D, P) has a compliant Eulerian path.

4.1   Outline of the Construction

We now provide a sketch of the main parts of the construction. For each occurrence of a variable in the formula, a *special vertex* is introduced. Special vertices corresponding to the same literal form a *literal path*. Two literal paths of a variable and its negation are connected in parallel to form a *variable subgraph*. For each clause in the formula, the corresponding special vertices are connected by three edges to form a *clause triangle*. Finally, for each special vertex we introduce a triangle incident on it, called its *bypass triangle* (see figure 1).

The sets of allowed positions are chosen so that they force every compliant Eulerian path to visit the literal paths one by one. A compliant Eulerian path corresponds to a satisfying truth assignment. When a special vertex is visited, either its clause triangle, or its bypass triangle are traversed. Traversing the clause triangle while passing through a certain literal's path corresponds to this literal satisfying the clause. We make sure that for one of x_i and $\overline{x}_i$, no clause triangle is visited while passing through its literal path. Eventually, we enable visiting all unvisited bypass triangles.

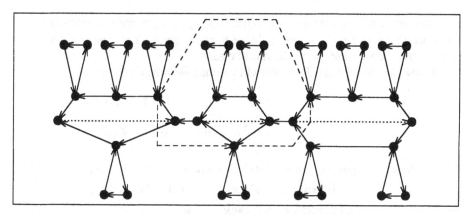

Fig. 1. A schematic sketch of the main elements in our construction. The figure includes three variable subgraphs, with the first variable (whose subgraph is rightmost) having three positive occurrences and two negated occurrences, etc. One of the clause triangles is also drawn, using dashed line.

4.2   Construction in Detail

We introduce the following vertices:

- $u_i, \hat{u}_i$ for each variable x_i, $1 \leq i \leq N$.
- $v_{L(j)}, \hat{v}_{L(j)}$ for each occurrence $L(j)$ of the literal L. $v_{L(j)}$ is called *special*. For $L \in X_i$, we shall denote u_i also by $v_{L(0)}$, and $\hat{u}_i$ also by $v_{L(a_L+1)}$.
- $r(C_c)$ for each clause C_c, $1 \leq c \leq M$, identifying $\hat{u}_N$ as $r(C_0)$ and u_1 as $r(C_{M+1})$.

We introduce the following edges:

- For each clause $C = L(j) \vee L'(j') \vee L''(j'')$, a clause triangle with the edges $\{(v_{L(j)}, v_{L'(j')}), (v_{L'(j')}, v_{L''(j'')}), (v_{L''(j'')}, v_{L(j)})\}$.
- For each occurrence $L(j)$ of the literal L in the clause C, a bypass triangle with the edges $\{(v_{L(j)}, \hat{v}_{L(j)}), (\hat{v}_{L(j)}, r(C)), (r(C), v_{L(j)})\}$.
- A literal path $lpath(L)$: $(u_i, v_{L(1)}), (v_{L(1)}, v_{L(2)}), (v_{L(2)}, v_{L(3)}), \ldots, (v_{L(a_L)}, \hat{u}_i)$, for each literal $L \in X_i$.
- For $i = 1, \ldots, N$, *back* edges $(\hat{u}_i, u_i)$; for $i = 1, \ldots, N - 1$, *forward* edges $(\hat{u}_i, u_{i+1})$.
- A *finishing path* $(\hat{u}_N, r(C_1)), (r(C_1), r(C_2)), (r(C_2), r(C_3)), \ldots, (r(C_M), u_1)$.

Figure 2 shows an example of this constructed graph. The motivation for this construction is the following: Using the position sets, we intend to force the literal paths of the different variables to be traversed in the natural order, where the only degree of freedom is switching order between $lpath(x_i)$ and $lpath(\bar{x}_i)$. This switch will correspond to a truth assignment for variable x_i, by assigning *True* to the literal in X_i whose $lpath$ was visited first. After visiting a special vertex along this first path, we either visit its clause triangle, or its bypass triangle. Along the other path (the one of the literal assigned *False*) only a bypass triangle can be visited.

Eventually, the finishing path is traversed. Its vertices are visited in the natural order. Upon visiting a vertex $r(C)$, we visit only one bypass triangle - the yet unvisited triangle among those corresponding to the literals of clause C.

We now describe the sets $P(e)$. We will use the following notation:

$$b_i = a_{x_i} + a_{\overline{x}_i} \quad \text{for } i = 1, \dots, N .$$

$$B_i = \sum_{j=1}^{i} b_j \quad \text{for } i = 0, \dots, N \ (B_0 = 0) .$$

$$Base_L = Base_{\overline{L}} = Base_i = 4B_{i-1} + 4(i-1) \quad \text{for } L \in X_i .$$

$$Alternate_L = Base_i + 4a_{\overline{L}} + 2 \quad \text{for } L \in X_i .$$

$$ClauseBase_c = Base_{N+1} + 4c .$$

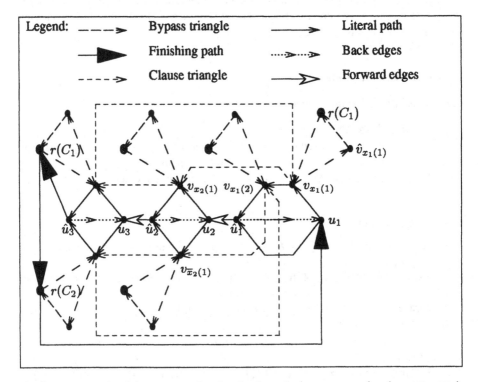

Fig. 2. An example of the construction for the formula $(x_1 \vee x_2 \vee x_3) \wedge (x_1 \vee \overline{x}_2 \vee \overline{x}_3)$. All large grey (black) vertices are actually the same vertex $r(C_1)$ $(r(C_2))$.

- For each forward edge $e = (\hat{u}_{i-1}, u_i)$, $2 \le i \le N$, we set $P(e) = \{Base_i\}$. This is intended to ensure that the literal paths are traversed in a constrained order: $lpath(x_i)$ and $lpath(\overline{x}_i)$ are allocated a time interval $[Base_i + 1, Base_{i+1} - 1]$ of length $4b_i + 3$, during which they must be traversed.

- For each back edge $e = (\hat{u}_i, u_i)$ we set $P(e) = \{Alternate_{x_i}, Alternate_{\overline{x}_i}\}$. This enables either visiting $lpath(x_i)$ first, then e and $lpath(\overline{x}_i)$, or visiting $lpath(\overline{x}_i)$ first, followed by e and $lpath(x_i)$.
- For each literal path edge $e = (v_{L(j)}, v_{L(j+1)})$, with $L \in X_i$, $0 \le j \le a_L$, we set $P(e) = \{Base_i + 4j + 1, Alternate_L + 4j + 1\}$. Consecutive edges in a literal path are thus positioned 4 time units apart (allowing a triangle in-between).
- For each clause $C = L_1(j_1) \lor L_2(j_2) \lor L_3(j_3)$ with the clause triangle $\{e_1 = (v_{L_1(j_1)}, v_{L_2(j_2)}), e_2 = (v_{L_2(j_2)}, v_{L_3(j_3)}), e_3 = (v_{L_3(j_3)}, v_{L_1(j_1)})\}$ such that $L_k \in X_{i_k}$, define $t_k = Base_{i_k} + 4j_k - 2$ and set

$$P(e_1) = \{t_1, t_3 + 1, t_2 + 2\},$$
$$P(e_2) = \{t_2, t_1 + 1, t_3 + 2\},$$
$$P(e_3) = \{t_3, t_2 + 1, t_1 + 2\}.$$

This means that the edges of a clause triangle must be visited consecutively during the traversal of $lpath(L_k)$, for some k. Furthermore, note that this may happen only if $lpath(L_k)$ is traversed immediately after time $Base_{L_k}$, that is, only if it precedes $lpath(\overline{L}_k)$ (see figure 3).

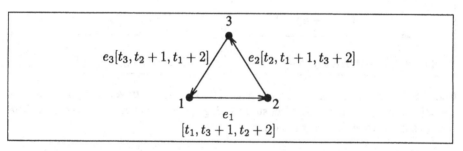

Fig. 3. A clause triangle, with vertex $v_{L_k(j_k)}$ denoted by k. The allowed positions for each edge appear in brackets.

- For each finishing edge $e = (r(C_c), r(C_{c+1}))$, $0 \le c \le M$, we set $P(e) = \{ClauseBase_c\}$, thus determining the order in which the vertices of the finishing path are visited, allowing a time slot $[ClauseBase_c + 1, ClauseBase_{c+1} - 1]$ of length 3, for the bypass triangle visited while traversing $r(C_c)$.
- For a bypass triangle with the edges $\{e = (v_{L(j)}, \hat{v}_{L(j)}), e' = (\hat{v}_{L(j)}, r(C_c)), e'' = (r(C_c), v_{L(j)})\}$, we set:

$$P(e) = \{Base_L + 4j - 2, Alternate_L + 4j - 2, ClauseBase_c + 2\},$$
$$P(e') = \{Base_L + 4j - 1, Alternate_L + 4j - 1, ClauseBase_c + 3\},$$
$$P(e'') = \{Base_L + 4j, Alternate_L + 4j, ClauseBase_c + 1\}.$$

This means that the bypass triangle edges must be visited consecutively, and there are three possible time slots for that:

- While traversing $lpath(L)$, before traversing $lpath(\overline{L})$.
- While traversing $lpath(L)$, after traversing $lpath(\overline{L})$.
- While traversing $r(C_c)$ along the finishing path.

The reduction is obviously polynomial. We now prove validity of the construction.

$\Leftarrow$ Suppose that F is satisfiable. We will show that (D, P) is a "yes" instance of the 3-positional Eulerian path problem. Let ϕ be a truth assignment satisfying F. For each clause C_c, let $L_c(j_c)$ be a specific literal occurrence satisfying C_c.
We describe an Eulerian path π in D. Set $\pi(ClauseBase_c) = (r(C_c), r(C_{c+1}))$, for $c = 0, \ldots, M$. Set $\pi(Base_i) = (\hat{u}_{i-1}, u_i)$, for $i = 2, \ldots, N$. For all i, if $\phi(x_i) =$ *True*, set $\pi(Alternate_{\overline{x}_i}) = (\hat{u}_i, u_i)$. Otherwise, set $\pi(Alternate_{x_i}) = (\hat{u}_i, u_i)$.
For each literal $L \in X_i$:

- If $\phi(L) =$ *True*: For each $0 \le j \le a_L$, set $\pi(Base_i + 4j + 1) = (v_{L(j)}, v_{L(j+1)})$ (see figure 4, top).

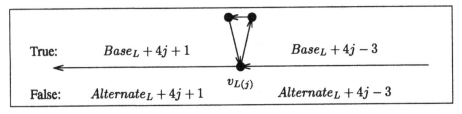

True:	$Base_L + 4j + 1$		$Base_L + 4j - 3$
		$v_{L(j)}$	
False:	$Alternate_L + 4j + 1$		$Alternate_L + 4j - 3$

Fig. 4. Either a clause triangle or a bypass triangle must be traversed upon visiting a special vertex $v_{L(j)}$, due to time constraints. Edge positions in case L is assigned *True* (*False*) are shown at the top (bottom).

We further distinguish between two cases:

* If $L(j) = L_c(j_c)$ for the clause $C_c = L(j) \vee L'(j') \vee L''(j'')$ in which $L(j)$ occurs, then set π to visit the edges of the clause triangle of C_c as follows:

$$\pi(Base_L + 4j - 2) = (v_{L(j)}, v_{L'(j')}) ,$$
$$\pi(Base_L + 4j - 1) = (v_{L'(j')}, v_{L''(j'')}) ,$$
$$\pi(Base_L + 4j) = (v_{L''(j'')}, v_{L(j)}) .$$

Furthermore, in this case we set π to visit the edges of the bypass triangle of $L(j)$ as follows:

$$\pi(ClauseBase_c + 1) = (r(C_c), v_{L(j)}) ,$$
$$\pi(ClauseBase_c + 2) = (v_{L(j)}, \hat{v}_{L(j)}) ,$$
$$\pi(ClauseBase_c + 3) = (\hat{v}_{L(j)}, r(C_c)) .$$

* Otherwise, $L(j) \neq L_c(j_c)$ for the clause C_c in which $L(j)$ occurs. In this case we set π to visit the edges of the bypass triangle of $L(j)$ as follows:

$$\pi(Base_L + 4j - 2) = (v_{L(j)}, \hat{v}_{L(j)}),$$
$$\pi(Base_L + 4j - 1) = (\hat{v}_{L(j)}, r(C_c)),$$
$$\pi(Base_L + 4j) = (r(C_c), v_{L(j)}).$$

• If $\phi(L)$ =False: For each $0 \leq j \leq a_L$, set $\pi(Alternate_L + 4j + 1) = (v_{L(j)}, v_{L(j+1)})$. Furthermore, in this case we set π to visit the edges of the bypass triangle of $L(j)$ as follows (see figure 4, bottom):

$$\pi(Alternate_L + 4j - 2) = (v_{L(j)}, \hat{v}_{L(j)}),$$
$$\pi(Alternate_L + 4j - 1) = (\hat{v}_{L(j)}, r(C_c)),$$
$$\pi(Alternate_L + 4j) = (r(C_c), v_{L(j)}).$$

Examining all the cases shows that π is a permutation of the edges, and if $\pi(k) = (u, v), \pi(k + 1) = (u', v')$ then $v = u'$. Hence, π is an Eulerian path. Furthermore, by our construction π is compliant with (D, P), proving that (D, P) is a "yes" instance.

⇒ Let π be an Eulerian path compliant with (D, P). We shall construct an assignment ϕ satisfying F. In order to determine $\phi(x_i)$ we consider the edge $\pi(Base_i + 1)$. By construction, $\pi(Base_i + 1) = (u_i, v_{L(1)})$ for $L \in X_i$. We therefore set $\phi(L)$ =True (and of course $\phi(\overline{L})$ =False). We observe that for any other edge $e' = (v_{L(j)}, v_{L(j+1)})$ along $lpath(L)$, we must have $\pi(Base_i + 4j + 1) = e'$ iff $\phi(L)$ =True.

We now prove that ϕ satisfies each clause $C_c = L_1(j_1) \vee L_2(j_2) \vee L_3(j_3)$ in F. Consider the clause triangle of C_c: $\{e_1 = (v_{L_1(j_1)}, v_{L_2(j_2)}), e_2 = (v_{L_2(j_2)}, v_{L_3(j_3)}), e_3 = (v_{L_3(j_3)}, v_{L_1(j_1)})\}$. Denote $t_k = Base_{L_k} + 4j_k - 2$. Suppose that $\pi^{-1}(e_k) \neq t_k$, for some $1 \leq k \leq 3$, then by the positional constraints the edge visited prior to e_k must be in the clause triangle. Therefore, there exists some $1 \leq k \leq 3$ for which $\pi(t_k) = e_k$. Furthermore, the edge e preceding e_k in π must have $t_k - 1 = Base_{L_k} + 4(j_k - 1) + 1 \in P(e)$. The only such edge entering $v_{L_k(j_k)}$ is the literal path edge $(v_{L_k(j_k-1)}, v_{L_k(j_k)})$. Therefore, $\phi(L_k)$ =True, satisfying C_c.

This proves that F is satisfiable iff (D, P) is a "yes" instance, completing the proof of Theorem 2.■

Observe that the graph constructed in the proof of Theorem 2 has in-degree and out-degree bounded by 4, giving rise to the following result:

Corollary 2. *3-PEP is NP-complete, even when restricted to graphs with in-degree and out-degree bounded by 4 .*

Henceforth, we call this restricted problem *(3,4)-PEP*. We comment that a slight modification of the construction results in a graph whose in-degree and out-degree are bounded by 2.

5   3-Positional SBH Is NP-Complete

We show in this section that the problem of sequencing by hybridization with at most 3 positions per spectrum element is NP-complete, even if each element in the spectrum is unique. The proof is by reduction from (3,4)-PEP.

Theorem 3. *The 3-PSBH problem is NP-complete, even if all spectrum elements are of multiplicity one.*

Proof. It is easy to see that the problem is in NP. We reduce (3,4)-PEP to 3-PSBH. Let $(D = (V, E), P)$ be an instance of (3,4)-PEP. Let $k = \lceil \log_4 |V| \rceil + 2, p = 3k + 1$ and $c = p + 1$. In order to construct an instance of 3-PSBH we first encode the edges and vertices of D. In the following, we denote string concatenation by $|$. We let $\sigma_1 =$'A', $\sigma_2 =$'C', $\sigma_3 =$'G' and $\sigma_4 =$'T'.

For each $v \in V$ we assign a unique string in Σ^{k-2}. We add a leading 'T' symbol and a trailing 'T' symbol to this string, and call the resulting k-long sequence the *name* of v. We also assign the (unique) sequence 'A...A' of length k to encode a *space*. Each vertex is encoded by a $3k$-long sequence containing two copies of its name separated by a space. We denote the encoding of v by $en(v)$. Each edge $(u, v) \in E$ is encoded by two symbols chosen as follows: Let $N_{out}(u) = \{v_1, \ldots, v_l\}$, where $v = v_i$ for some i, and $l \leq 4$. Let $N_{in}(v) = \{u_1, \ldots, u_r\}$, where $u = u_j$ for some j, and $r \leq 4$. Then (u, v) is encoded by $\sigma_i | \sigma_j$, and we denote its encoding by $en(u, v)$. We let $EN((u, v)) = en(u)|en(u, v)|en(v)$ (see figure 5).

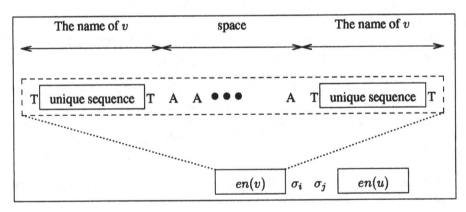

Fig. 5. The encoding of vertices and edges into strings.

We now construct a 3-PSBH instance, i.e., a set S with position constraints T, as follows: For every edge $(u, v) \in E$ the set S contains all p-long substrings of the $2p$-long sequence $EN((u, v))$ (c substrings in total). Let $s^i_{(u,v)}$ denote the i'th substring, $i = 0, \ldots, p$. Let $P((u, v)) = \{t_1, \ldots, t_l\}, 1 \leq l \leq 3$, be the set of allowed positions for (u, v). Then we set $T(s^i_{(u,v)}) = \{c(t_1 - 1) + i, \ldots, c(t_l - 1) + i\}$ for all i (substring positions are 0-up).

Lemma 4. *Each of the p-long substrings defined above is unique.*

Proof. Suppose that $s = s^i_{(u,v)} = s^j_{(w,z)}$. We first claim that $i = j$. There are two cases to examine: If s contains a space at positions $r, \ldots, r + k - 1, 0 \leq r \leq 2k + 1$, then $i = j = (c + k - r) \pmod{c}$. Otherwise, s begins with a run of 'A' symbols of length $0 \leq r' \leq k - 1$. This run belongs to a space in $en(u)$ and $en(w)$, and must be followed by the symbol 'T'. In this case $i = j = 2k - r'$.

By construction, s contains a name of a vertex plus a unique symbol identifying an edge entering or leaving that vertex, implying that $(u, v) = (w, z)$. ∎

We now show the validity of the reduction.

⇐ Suppose that $\pi = (v_0, v_1), (v_1, v_2), \ldots, (v_{m-1}, v_m)$ is a solution of the (3,4)-PEP instance. We claim that $X = en(v_0)|en((v_0, v_1))|en(v_1)|en((v_1, v_2))|\ldots|en(v_m)$ is a solution of the 3-PSBH instance. By Lemma 4 each p-long substring of X occurs exactly once in X. As π visits all edges in D, we have that S is the p-spectrum of X. The fact that position constraints are obeyed follows directly from the construction.

⇒ Let X be a solution of the 3-PSBH instance. Consider the m substrings of length p, whose starting positions are integer multiples of c. By the position constraints, the r-th such substring is an encoding of some vertex v_r, followed by a symbol σ_{i_r}. Denote by w_r the i_r-th out-neighbor of v_r. We prove that $\pi = (v_1, w_1), \ldots, (v_m, w_m)$ is an Eulerian path compliant with (D, P).

Since each string in the p-spectrum of X is unique, π is a permutation of the edges in D. To prove that π is a path in D we have to show that $w_r = v_{r+1}$ for $r = 1, \ldots, m - 1$. Let x be the p-long substring of X starting at position $rc + 2k$. We observe that x must begin with the last k symbols of $en(v_r)$, which compose $name(v_r)$, followed by σ_{i_r}, some symbol, and the first $2k - 1$ symbols of $en(v_{r+1})$, which contain $name(v_{r+1})$. The uniqueness of $name(v_r)$, $name(v_{r+1})$ and the index i_r among the out-neighbors of v_r, implies that $w_r = v_{r+1}$. The claim now follows, since position constraints are trivially satisfied by π. ∎

Acknowledgments

The first author was supported by the program for mathematics and molecular biology. The second author was supported by the Clore foundation scholarship. The third author was supported in part by a grant from the Ministry of Science, Israel. The fourth author was supported by Eshkol scholarship from the Ministry of Science, Israel.

References

[1] L. M. Adleman. Location sensitive sequencing of DNA. Technical report, University of Southern California, 1998.

[2] R. Drmanac amd R. Crkvenjakov, 1987. Yugoslav Patent Application 570.

[3] B. Apsvall, M. F. Plass, and R.E. Tarjan. A linear-time algorithm for testing the truth of certain quantified boolean formulas. *Information Processing Letters*, 8(3):121–123, 1979.

[4] W. Bains and G. C. Smith. A novel method for nucleic acid sequence determination. *J. Theor. Biology*, 135:303–307, 1988.

[5] S. D. Broude, T. Sano, C. S. Smith, and C. R. Cantor. Enhanced DNA sequencing by hybridization. *Proc. Nat. Acad. Sci. USA*, 91:3072–3076, 1994.

[6] S. Hannenhalli, P. Pevzner, H. Lewis, and S. Skiena. Positional sequencing by hybridization. *Computer Applications in the Biosciences*, 12:19–24, 1996.

[7] K. R. Khrapko, Yu. P. Lysov, A. A. Khorlyn, V. V. Shick, V. L. Florentiev, and A. D. Mirzabekov. An oligonucleotide hybridization approach to DNA sequencing. *FEBS letters*, 256:118–122, 1989.

[8] Y. Lysov, V. Floretiev, A. Khorlyn, K. Khrapko, V. Shick, and A. Mirzabekov. DNA sequencing by hybridization with oligonucleotides. *Dokl. Acad. Sci. USSR*, 303:1508–1511, 1988.

[9] S. C. Macevics, 1989. International Patent Application PS US89 04741.

[10] P. A. Pevzner. 1-tuple DNA sequencing: computer analysis. *J. Biomol. Struct. Dyn.*, 7:63–73, 1989.

[11] P. A. Pevzner and R. J. Lipshutz. Towards DNA sequencing chips. In *Symposium on Mathematical Foundations of Computer Science*, pages 143–158. Springer, 1994. LNCS vol. 841.

[12] P. A. Pevzner, Yu. P. Lysov, K. R. Khrapko, A. V. Belyavsky, V. L. Florentiev, and A. D. Mirzabekov. Improved chips for sequencing by hybridization. *J. Biomol. Struct. Dyn.*, 9:399–410, 1991.

[13] F. Preparata, A. Frieze, and Upfal E. On the power of universal bases in sequencing by hybridization. In *Proceedings of the Third Annual International Conference on Computational Molecular Biology (RECOMB '99)*, pages 295–301, 1999.

[14] S. S. Skiena and G. Sundaram. Reconstructing strings from substrings. *J. Comput. Biol.*, 2:333–353, 1995.

[15] E. Southern, 1988. UK Patent Application GB8810400.

[16] E. M. Southern. DNA chips: analysing sequence by hybridization to oligonucleotides on a large scale. *Trends in Genetics*, 12:110–115, 1996.

[17] E. M. Southern, U. Maskos, and J. K. Elder. Analyzing and comparing nucleic acid sequences by hybridization to arrays of oligonucleotides: evaluation using experimental models. *Genomics*, 13:1008–1017, 1992.

GESTALT: Genomic Steiner Alignments

Giuseppe Lancia[*1] and R. Ravi[**2]

[1] Dipartimento Elettronica ed Informatica, University of Padova,
lancia@dei.unipd.it
[2] GSIA, Carnegie Mellon University,
ravi@cmu.edu

Abstract. We describe GESTALT (GEnomic sequences STeiner ALignmenT), a public–domain suite of programs for generating multiple alignments of a set of biosequences. We allow the use of either of the two popular objectives, Tree Alignment or Sum-of-Pairs. The main distinguishing feature of our method is that the alignment is obtained via a tree in which the internal nodes (ancestors) are labeled by Steiner sequences for triples of the input sequences. Given lists of candidate labels for the ancestral sequences, we use dynamic programming to choose an optimal labeling under either objective function. Finally, the fully labeled tree of sequences is turned into into a multiple alignment. Enhancements in our implementation include the traditional space-saving ideas of Hirschberg as well as new data-packing techniques. The running-time bottleneck of computing exact Steiner sequences is handled by a highly effective but much faster heuristic alternative. Finally, other modules in the suite allow automatic generation of linear-program input files that can be used to compute new lower bounds on the optimal values. We also report on some preliminary computational experiments with GESTALT.

1 Introduction

Comparing genomic sequences drawn from individuals of the same or different species is one of the fundamental problems in computational molecular biology. These comparisons can (i) lead to the identification of highly conserved (and therefore presumably functionally relevant) genomic regions, (ii) spot fatal mutations, (iii) suggest evolutionary relationships, (iv) help in correcting sequencing errors etc. Therefore, the mathematical formulation and solution of the *Multiple Sequence Alignment* problem has been and remains a fundamental challenge for computational molecular biologists.

Aligning a set of sequences consists in arranging them in a matrix having each sequence in a row. This is obtained by possibly inserting spaces (gaps) in each sequence so that they all have the same length. The following is a simple example of an alignment of the sequences **ATTCGAC**, **TTCCGTC** and **ATCGTC**.

---

[*] Most of this work was done when this author was visiting CMU during Summer '98, under a grant from the CMU Faculty Development Fund.

[**] Supported in part by an NSF CAREER grant CCR-9625297

M. Crochemore, M. Paterson (Eds.): CPM'99, LNCS 1645, pp. 101–114, 1999.
© Springer-Verlag Berlin Heidelberg 1999

```
A T T - C G A - C
- T T C C G - T C
A - T - C G - T C
```

There are many popular formulations of the alignment problem. The choice of the objective function for multiple alignments depends mainly on the presence or absence of extra input information in the form of a phylogenetic tree relating the sequences to their unknown ancestors. In fact, when such tree is given, knowledge of the ancestral sequences would imply the possibility of aligning the given sequences by progressively aligning each sequence to its ancestor in the tree all the way to the root and chaining these pairwise alignments together [6]. Hence when a phylogeny is given, the *tree alignment (TA)* objective consists in finding the best ancestral sequences to label this tree and deriving the induced alignment. Guided by parsimony, the best labeling is taken to be one minimizing the total evolutionary change represented in the tree, namely, the total distance of all the edges in the tree. When the phylogenetic tree is not available, a popular multiple alignment objective is the *Sum-of-pairs (SP)* objective, which attempts to minimize the average distance between a pair of sequences in the multiple alignment. This objective results naturally by extending the alignment objective for pairs of sequences, namely, that of minimizing the edit-distance between the pair, to more than two sequences. The SP objective has been popular in the literature and several heuristic implementations addressing it proceed by first finding a heuristic tree spanning the sequences and aligning them progressively as mentioned earlier to obtain the final alignment.

Historically, the SP objective is the one to which more attention has been devoted by computational biologists, and correspondingly a set of programs have been developed which are now widely in use. Among them, the only program that computes optimal SP alignments is MSA by Lipman, Altschul, Kececioglu, Gupta and Schaeffer [2,8]. A variety of other multiple sequence alignment programs implicitly use the SP objective in guiding heuristic construction of the multi-alignments: An example is CLUSTAL V [11] (see also the various methods described in the surveys [16,5] for other examples). As for tree alignment, the only implementation that addresses this problem directly that we are aware of is the recent TAAR by Jiang and Liu [13]. This program implements some of the ideas from the approximation algorithms of Jiang, Lawler and Wang [27] to heuristically compute tree alignments, phylogenies and generalized tree alignments.

In this paper we introduce and describe a new public–domain suite of programs for multiple sequence alignment that produce heuristic alignments under both the TA and SP alignment objectives. Like TAAR, Our methods are based on ideas used in an approximation algorithm for tree alignment due to Ravi and Kececioglu [17]. However, unlike the methods of Jiang, Lawler and Wang [27] on which TAAR is based, whose refined heuristics require very high running times, the ideas of Ravi and Kececioglu are based on mainly computing and using Steiner sequences as candidates for the unlabeled ancestral sequences in the tree. Intuitively, a Steiner sequence for a given set of sequences is a "central" sequence to them, one whose sum of distances to all these sequences is minimized.

Once these Steiner sequences for appropriate subsets of the input sequences have been computed, dynamic programming can be used to efficiently pick one such sequence for each ancestral node so as to minimize the total resulting distance in the tree, as in [27]. Thus, this method is adaptable for efficient implementation giving us the freedom to specify the subsets of sequences for which the Steiner sequences must be computed. Further, we can effectively adapt this general idea by modifying the dynamic program to provide an efficient heuristic even for the SP objective using the postulated Steiner ancestors.

Further refinements in our implementation include incorporating the traditional space-saving ideas of Hirschberg [12] as well as some new data-packing techniques to reduce the space overhead; The running-time bottleneck in our method of computing exact Steiner sequences is effectively handled by a much faster heuristic alternative that has never shown more than two percent degradation in quality in our extensive preliminary testing. Finally, other programs in the suite allow automatic generation of linear-programming models as files that can be input to the popular commercial CPLEX package. The solution of these programs give lower bounds on the minimum TA and SP alignment values for the given set of sequences, thus providing the deviations from optimality on a case-by-case basis.

We formally describe the various objectives and methods in the remainder of this section. In Sect. 2 we give a high–level description of the algorithms in GESTALT, together with an analysis of the individual steps. In Sect. 3 we report on some experimental results on real data.

1.1   Edit Distance

At the heart of any alignment algorithm lies the procedure for optimally comparing two given sequences. This problem is called pairwise alignment, and is formulated as follows. Given symmetric costs $c(a, b)$ for replacing a symbol a with a symbol b and costs $c(a, -)$ for deleting (inserting) symbol a, find a minimum–cost set of symbol operations that turn a sequence S' into a sequence S''. It is well known that this problem can be solved by dynamic programming in time and space $O(l^2)$, where l is the length of the sequences. The value of an optimal solution is called the *edit distance* of S' and S'' and denoted by $d(S', S'')$.

An *alignment* $\mathcal{A}$ of two (or more) sequences is a way of inserting "$-$" characters (gaps) in the sequences so that the resulting sequences have the same length. For two sequences S' and S'', the value $d_{\mathcal{A}}(S', S'')$ of their alignment is obtained by adding up the costs for the pairs of characters in corresponding positions. It is immediate that $d(S', S'') = \min_{\mathcal{A}} d_{\mathcal{A}}(S', S'')$.

1.2   The Sum–of–Pairs Alignment Problem

The SP score is the generalization to many sequences of the pairwise alignment objective, in which the cost of the alignment is obtained by adding the costs of the symbols matched up at the same positions. Analogously, in a multiple alignment the cost is obtained by adding up the matching characters, over all the positions and for all the pairs of sequences.

Minimizing SP is NP-hard [26]. In [9] Gusfield showed that a tree-based progressive alignment method due to Feng and Doolittle (described below) using the minimum cost star gives a 2–approximation. In the program described in this paper we push this idea further, by considering also trees that are not only stars and also employing alignments with sequences which are not in the original set, but are derived from it as Steiner sequences of some of the original ones.

1.3   The Tree Alignment Problem

In the tree alignment problem, we are given n sequences related by an evolutionary tree T. The sequences label the leaves of the tree, while the internal nodes correspond to the unknown ancestral sequences from which the others have evolved. The problem consists in finding the sequences at the internal nodes which minimize the cost of the tree, defined as $\sum_{(S_i,S_j)\in T} d(S_i, S_j)$. When T is a star, the problem is called a *Steiner problem*, and the optimal sequence for the center is called the *Steiner sequence* for the leaves.

The first exact algorithm for tree alignment was proposed by Sankoff in [18], and is based on dynamic programming. Later Altschul and Lipman [1] introduced some bounding rules to reduce the size of the dynamic programming lattice. Due to the prohibitive worst case complexity of exact methods, approximation algorithms for this problem were devised, by Jiang, Lawler and Wang [27] first, and improved by Wang and Gusfield [25] later. In [27] a 2–approximation method is described, based on what are called *lifted alignments*. In lifted alignments, the internal nodes can only be labeled by sequences occurring at the leaves. The running time of their algorithm is $O(n^2l^2 + n^3)$ for a tree of n leaves of length l. For trees of bounded degree d, they also provided the first PTAS for the problem. For any t, their approximation scheme guarantees a solution within a factor $1 + \frac{3}{t}$ of optimal, in time $O(n^{2+d^{t-1}} l^{d^{t-1}-1/d-1})$.

For regular d–ary trees on n sequences, Ravi and Kececioglu gave in [17] a $\frac{d+1}{d-1}$–approximation algorithm with running time roughly $(O(2kn)^d)$ – the main ideas of their algorithm are briefly described in Sect. 2. The program GESTALT described in this paper is the first implementation of the ideas in [17].

1.4   A Tree-Based Progressive Alignment Method

A reasonable requirement on the cost function is that $c(a, a) = 0 \; \forall a$, and it obeys triangle inequality. In this case, the edit distance induces a metric over the space of all sequences and, given n sequences, we can talk of graphs having the sequences as vertices and for which an edge is weighted by the edit distance between the endpoints. In this setting, graph theoretical concepts such as spanning trees, stars and Steiner points, have been widely used in the design and analysis of effective alignment algorithms. In particular, a folklore approach to multiple alignments is due to Feng and Doolittle [6] and shows how we can use any tree to align a set of n sequences. The appeal of the approach is that for $n - 1$ out of $n(n - 1)/2$ pairs, the pairwise alignment induced is in fact optimal.

Proposition 1. *For any tree T over a set of sequences, there exists a multiple alignment $\mathcal{A}(T)$ of the sequences such that $d_{\mathcal{A}(T)}(S', S'') = d(S', S'')$ for all the pairs of sequences (S', S'') connected by an edge of the tree.*

Feng and Doolittle's method can be used to turn the solution of the tree alignment problem, namely a labeling of the internal nodes of the given tree, into a multiple alignment of the leaves. Moreover, it is straightforward to upper bound the distance in this alignment of pairs that are not endpoints of a tree edge. In fact, denote by $d(S', S'', T)$ the length of the path in T between two sequences S' and S''. Then, by triangular inequality we have that $d_{\mathcal{A}(T)}(S', S'') \leq d(S', S'', T)$. This inequality suggests that, given a tree with sequences at the leaves for which we want to minimize average pairwise distance in the resulting multiple alignment, a good labeling for the internal nodes is one which minimizes the total inter-leaf distance in the tree. This strategy is adopted in this work to obtain alignments of small SP value, as described in 2.3.

1.5 GESTALT Program Suite

In this paper we describe the program GESTALT (GEnomic sequences STeiner ALignmenT), which can be used for both TA and SP multiple alignments. GESTALT is in fact a program suite, including modules for computing LP-based lower bounds for TA and SP, and optimal alignments of two or three sequences.

The main program takes as input a set $\mathcal{L} = \{S_1, \ldots, S_n\}$ of n sequences and possibly a tree T of which $\mathcal{L}$ are the leaves. If the phylogenetic tree is not available, the algorithm internally computes one, which is then used to find an alignment of small SP value. If the tree is given, then the TA objective is optimized[1]. The output of the algorithm consists of a multiple alignment of the input sequences, plus some extra information, such as the Steiner sequences computed at the internal nodes of the phylogenetic tree.

GESTALT is based on the ideas introduced by Ravi and Kececioglu in [17] of using Steiner sequences of the leaves to label the internal nodes of the tree. While in their paper Ravi and Kececioglu show that if the tree is d–ary the method gives a $\frac{d+1}{d-1}$ approximation for TA, in our work we do not restrict the degree of each node to a constant. Therefore we do not have the same approximation guarantee. However, among all the labelings considered is included the best lifted labeling of [27] and therefore we still have a performance guarantee of 2 for the TA objective. As is typically the case, this bound turns out to be largely pessimistic and our computational results show that the algorithm performs much better in practice.

The 2–approximation guarantee holds also for the SP alignments we output. Recall that we include, among all the labelings considered, one in which the

---

[1] The choice of the objective in the presence or absence of the tree can also be user-specified

internal nodes of the tree are all labeled with any leaf S. For this particular labeling, the resulting tree is equivalent to a star centered at S, and as remarked before [9], the best star centered at a leaf gives a 2–approximation.

2   Procedure Overview

Our program is largely based on a heuristic procedure by Ravi and Kececioglu ([17]) for solving the tree alignment problem. Their algorithm relies on labeling the internal nodes with Steiner sequences for subsets of p leaves, where p is a parameter. The procedure is divided in two phases. In the first phase a Steiner sequence is computed for every subset of $q \le p$ leaves, obtaining a set $\mathcal{F}$ of all such Steiner sequences. In the second phase, dynamic programming is used to compute the best labeling of the internal nodes among those in which only labels from $\mathcal{F}$ are allowed.

In this work, we have decided to solve the TA problem by employing Ravi and Kececioglu's algorithm, with the following variants: (i) Because computing exact Steiner sequences is expensive, we have limited the size of the subsets for which a Steiner problem is solved to $p = 3$. (ii) In addition to Sankoff's exact algorithm for Steiner sequences, with complexity $O(l^3)$, we also use a heuristic algorithm, with average (empirical) complexity $O(l^2)$. (iii) We do not necessarily compute the Steiner sequences for all the $\binom{n}{3}$ possible triples of leaves, but provide alternate, heuristic methods of sampling significant triples. (iv) We also perform a final re-optimization step, as introduced by Sankoff et al ([20]).

Our program can be used also to optimize SP. In this case, we first compute a tree having the given sequences for leaves and then assign tentative labels to the internal nodes by using Steiner sequences, as for the TA objective. In choosing the best label at each node, however, we use dynamic programming to minimize the total leaf–to–leaf distance in the tree, which is an upper bound on the final SP score. A final reoptimization phase can be run to improve the alignment.

The outline of our multiple alignment heuristic procedure is given below.

1. *Tree computation.*
   - TA: none (the tree is given).
   - SP: We compute a phylogenetic tree having the given sequences as leaves
     - this is derived from a MST on the sequence graph.
2. *Solution of Steiner problems.* We tentatively assign to each of the internal nodes of the phylogenetic tree a set of labels, given by the Steiner sequences of some subsets of the leaves.
3. *Optimal labeling by Dynamic Programming.* We find for each internal node the best sequence among those in its set of possible labels.
   - TA: The objective is to minimize the total tree-length.
   - SP: The objective is to minimize the total leaf–to–leaf distance in the tree.

4. *Local re-optimization.*
   - TA: At each node of degree three we replace the current sequence by the Steiner sequence of its neighbors. We iterate as long as there are improvements.
   - SP: (after step 5.) We iteratively break up the alignment into two subalignments that are then realigned optimally. The subalignments chosen have a large average difference in the current value versus the edit distance.
5. *Final alignment by Feng and Doolittle.* We compute a multiple alignment of all the resulting sequences (both leaves and internal nodes) by the progressive alignment method of Feng and Doolittle.

We elaborate on some of these steps next.

2.1   Tree Computation

In order to derive a phylogenetic tree T relating a set of sequences when one is not input, we use a simple greedy approach. We start with T being a minimum cost spanning tree of the edit distance graph. Let (u, v) be the largest cost edge of T. Break up T by deleting edge (u, v) into two trees T_u containing u and T_v containing v. Recursively, apply the same procedure to T_u and T_v, obtaining two new trees, $T_{u'}$ and $T_{v'}$ rooted at new nodes u' and v' respectively. Finally, join these two subtrees by means of edges (u', w) and (v', w) to a new root node w, thus obtaining the final phylogenetic tree.

2.2   Solution of Steiner Problems

Choice of Steiner Sequences Given a set of possible sequences (labels) for each internal node of the tree, choosing the best label is done by dynamic programming (described in 2.3) and is very fast in practice. On the other hand, computing the labels is very expensive. Therefore once some labels have been computed, it is convenient to store them at every internal node, i.e. all the nodes will have the same set $\mathcal{G}$ of labels. As previously noted, the labels allowed at the internal nodes will only be Steiner sequences for some subsets of $q \leq 3$ leaves. When $q = 1$ or 2, a Steiner sequence is simply a leaf, so that it will always be $\mathcal{G} = \mathcal{L} \cup \mathcal{G}'$, where $\mathcal{G}'$ is a set of Steiner sequences for some triples of leaves. Let us denote by $Y(S_i, S_j, S_k)$ a Steiner sequences for the triple (S_i, S_j, S_k). We allow three possibilities for $\mathcal{G}'$:

- $\mathcal{G}' = \emptyset$. In this case the internal nodes are labeled with leaves sequences only. This option results in the fastest running time, but may produce poor final alignments, especially when the given sequences are very dissimilar. Note that among the alignments based on these labels are included all lifted alignments [27] for TA. Similarly, these labels contain also all star alignments for SP.

- $\mathcal{G}' = \{Y(S_i, S_j, S_k) : i < j < k\}$. This is computationally the most expensive option, since it requires the solution of $\binom{n}{3}$ Steiner problems. On the other hand, the larger set of possible labels at the internal nodes guarantees a better value of the final alignment.
- Let $S_1, S_2, \ldots, S_n$ be the sequence of leaves as encountered by performing a depth–first visit of the tree. Then, $\mathcal{G}' = \{Y(S_j, S_k, S_h) : h = k + 1 = j + 2 \text{ or } h = k + \Delta = j + 2\Delta\}$ where $\Delta = \lfloor \frac{n}{3} \rfloor$. The intention is to heuristically obtain a uniform sampling by selecting triples of leaves from different positions in an Euler tour of the tree. This option is quick –there are only $O(n)$ such triples– but ensures that each sequence is included in some triples, and that all the sequences are given the same representation in the samples.

Exact Steiner Sequences Assume we are interested in finding a Steiner sequence for three sequences U_1, U_2 and U_3. The dynamic programming procedure computes the optimal alignment of the variable Steiner sequence and U_1, U_2 and U_3. This is done backwards from the final column of the alignment, which will be of the form $(x_1, x_2, x_3, y)'$, where each x_i is either the last letter of the sequence U_i or a blank (but at least one x_i must be nonblank), and y is any nonblank letter of the alphabet Σ (representing the metter in the Steiner sequence being constructed). For any letter x, define $1 \cdot x = x$ and $0 \cdot x = -$. Let $B^+ = \{0,1\}^3 \setminus (0,0,0)$ be the set of nonnull binary 3–vectors and let $V(l_1, l_2, l_3)$ be the cost of an optimal Steiner sequence for the the the first l_1, l_2 and l_3 characters respectively of U_1, U_2 and U_3. The recursive dynamic programming relation is then

$$V(l_1, l_2, l_3) = \min_{b \in B^+} \left\{ V(l_1 - b_1, l_2 - b_2, l_3 - b_3) + \min_{y \in \Sigma} \sum_{i=1}^{3} c(b_i \cdot U_i[l_i], y) \right\}$$

The Steiner sequence is given, as customary in dynamic programming, by backtracking through the values $V(l_1, l_2, l_3)$ along the path for an optimal solution and listing the letters $\widehat{y} = \arg\min \sum_{i=1}^{3} c(b_i \cdot S_i[l_i], y)$ which achieve the minimum in the above expression. Note that the above recurrence requires time and space complexity of $O(7l^3)$, provided that for all $(x_1, x_2, x_3) \in \Sigma^3$, the values $C(x_1, x_2, x_3) := \min_{y \in \Sigma} \sum_{i=1}^{3} c(x_i, y)$ have been computed in a preliminary step and stored in a look-up table. In our implementation we have reduced the space complexity to $O(l^2)$ for the matrix $V(i, j, k)$ using ideas from [12].

Heuristic Steiner Sequences Computing exact Steiner sequences is very time consuming. For instance, the solution of a problem on sequences of about 200 letters each takes roughly half minute on a Pentium PC. Considering that for aligning 10 sequences we may have to solve $\binom{10}{3} = 120$ such problems, we see that speeding up the computation of Steiner sequences would be greatly beneficial. Therefore, we have devised an alternative, heuristic way of computing Steiner sequences which is extremely fast and turns out to be almost–optimal after extensive testing (see Sect. 3).

The idea is to first find all optimal alignments of two of the three sequences, say S_1 and S_2. They correspond to all the shortest paths from $(0,0)$ to $(|S_1|, |S_2|)$ in the $|S_1| \times |S_2|$ dynamic programming lattice used for the pairwise alignment, and can be represented in a compact form as the subgraph of the lattice of all the edges on some optimal path. Note that this subgraph is typically much smaller than the whole lattice (empirically, $O(l)$ versus $O(l^2)$). Then, we perform a graph–to–sequence alignment, i.e. we find the best completion of an optimal alignment of S_1 and S_2 with S_3. In this case, "best" is taken with respect to the Steiner objective.

The value of the final solution may depend on the ordering of the sequences, since S_3 is clearly used differently than S_1 and S_2. We have observed in our experiments that choosing S_1 and S_2 to be the two closest sequences results in the best Steiner sequences over the three possible choices. However, since the algorithm is very fast, we compute all three possibilities of first aligning together two sequences and then versus the third, and return the best solution found. We conclude this section by remarking that the computation of heuristic Steiner sequences takes on the average one second for sequences of length 200, while returning a solution whose value was never more than 2% larger than the optimum in our extensive testing.

2.3 Optimal Labeling by Dynamic Programming

In this section we consider the problem of optimally assigning a sequence from a given set $\mathcal{G}$ to each internal node of the tree. Denote by $w_1, \ldots, w_t$ the nodes which are immediate descendants of a node i. Let $V(i, S)$ be the optimal value for the subtree rooted at i when node i is labeled with a sequence $S \in \mathcal{G}$. We have the following dynamic programming recurrence:

$$V(i,S) = \begin{cases} 0 \text{ if } i \text{ is a leaf} \\ \min_{L_1,\ldots,L_t \in \mathcal{G}} \sum_{j=1}^{t} \left(\lambda(i, w_j)d(S, L_j) + V(w_j, L_j) \right) \text{ otherwise} \end{cases}$$

The coefficients $\lambda(i, w_j)$ allow us to distinguish between the two objective functions - TA and SP. For the TA objective, $V(i, S)$ represents the minimum total length of the subtree, among the labelings that assigns S to i. This is obtained by setting all the λ equal to 1. For the SP objective, we want to find the labels which minimize the total leaf–to–leaf distance. For any edge (u, v) of T, we set $\lambda(u, v)$ to be the number of pairs of leaves whose connecting path in the tree goes through (u, v). This value, called the load of the edge, is equal to $k(n - k)$, where k is the number of leaves on one shore of the cut identified by (u, v). By using the loads, the total leaf–to–leaf distance can be rewritten as $\sum_{S_i, S_j} d(S_i, S_j, T) = \sum_{(u,v) \in T} \lambda(u, v)d(L(u), L(v))$, where $L(u)$ and $L(v)$ are the sequences labeling nodes u and v.

Using the above relation, first the value of each label at each node is computed bottom–up, and later, proceeding top–down from the root, it is determined which label to pick at each node for obtaining an optimal solution. The overall complexity is $O(n|\mathcal{G}|^2)$, i.e. a very fast procedure.

2.4 Reoptimization

The reoptimization for TA objective is the same as in Sankoff et al [20]. For SP, however, we use a new approach. As in other works (e.g. [7]) we repeatedly break up the alignment into two pieces that are then realigned optimally via the basic dynamic program for edit distance. The new idea relies in how these alignments are chosen. Since for each pair of sequences in the same subalignment the distance remains the same, the only improvement can be for sequences that are in different subalignments. Let $\delta(S, S') = d_A(S, S') - d(S, S')$. If A_1 and A_2 are the subalignments, $\delta(A_1, A_2) = \sum_{S \in A_1, S' \in A_2} \delta(S, S')$ is the δ–value of the cut (A_1, A_2) in the graph of all sequences, and $\delta(A_1, A_2)/|A_1||A_2|$ is a per-sequence measure of how bad the alignment currently is versus the lower bound given by the edit distance. Hence we want to reoptimize some cuts of high (per-sequence) value, which we find through standard greedy heuristics. We have different settings on how far the reoptimization phase can be pushed. In the most expensive setting, for each pair (S, S') of sequences we find a large–value cut separating them and relign it. We iterate as long as there are improvements.

3   Computational Experiences

For our preliminary tests, we used two popular data sets. First, we obtained the sets of protein sequences of Mc Clure [16], used extensively to benchmark programs guided by the SP objective. For the Tree Alignment problem, we have used a famous instance by Sankoff et al [20], used as a benchmark in [10,13].

As for the cost matrix, in our experiments we have used a distance matrix due to Taylor [23] for amino acid sequences, and the matrix in Sankoff [20] for DNA sequences. Our program also works with all the common score matrices (e.g. PAM, BLOSUM, etc).

1. **Lower Bounds.** A unique feature of the GESTALT suite is a procedure to generate linear programming (LP) based lower bounds on the TA and SP objective values of the given instance by using the Steiner sequences for triples computed so far. We describe the LP for the TA problem. We use a nonnegative variable for the length of every edge of the tree, and the objective is to minimize the sum of lengths of all tree edges. A distance of d between a pair of leaves S_i and S_j allows us to add the constraint that the sum of the values of the edge lengths on the path between S_i and S_j in the tree must be at least D. Similarly, given a value of $TA(i, j, k)$ for the minimum sum of the distances from an optimal Steiner sequence for the triple (S_i, S_j, S_k) to the three sequences S_i, S_j and S_k, we add the constraint that the sum of the lengths of all the edges in the tree induced by the three leaves S_i, S_j and S_k must be at least $TA(i, j, k)$. The objective function in the LP is to minimize the sum of the values of the edge variables. The set of constraints for distances between pairs of leaves was

Table 1. Heuristic vs exact Steiner sequences. Times in seconds, Pentium 133Mhz

instance	tot seqs	tot triples	relative error			time exact		time heuristic	
			avg	min	max	min	max	min	max
sank	9	84	0.003	0	0.02	15.8	41.0	0.6	1.9
mc582x6	6	20	0.004	0	0.01	52.3	75.6	0.5	3.0
mc586x6	6	20	0.007	0	0.017	17.8	42.5	0.6	2.1
mc587x6	6	20	0.01	0.003	0.019	29.2	71.9	0.8	2.7

experimented with in [10], while the strengthening to triples gives better bounds as reported below.

For the SP objective for multiple alignment, a simple averaging argument using the usage of Steiner triples yields a simple lower bound of $\sum_{i,j,k} SP(i,j,k)/(n-2)$ for n sequences, where $SP(i,j,k)$ denotes the optimal sum-of-pair value for the triple S_i, S_j and S_k. This may be further extended to a LP lower bound with one nonnegative variable for the distance between every pair of sequences in the multiple alignment. The constraints now require that for every triple S_i, S_j, S_k of distinct sequences, the sum of the values of the three variables involving the three pairs from this triple must be at least $SP(i,j,k)$. The objective is to minimize the sum of all the variables over all pairs of sequences.

2. Steiner Sequences. First, we determined the quality of heuristic vs exact Steiner sequences. The results are reported in Table 1. For these tests, we used four data sets, i.e. the sequences from Sankoff and three sets of sequences from McClure. These sequences have between one hundred and two hundred letters each. For each set, we have computed for each triple the exact and heuristic Steiner sequences, and compared the relative errors. It should be noted that on these sequences, the heuristic is roughly thirty times faster than the exact procedure, while the average error is less than one percent. A striking result was that in 41 out of 84 triples for the **sank** instance, the heuristic solution was in fact optimal.

3. Tree Alignment. A second experiment was performed to access the quality of the solution to the Tree Alignment problem, and the relative performance with different settings of the program. We have run GESTALT on Sankoff's problem with all possible combinations of user choices. The results are reported in Table 2. Again, it should be noted that using heuristic Steiner sequences is greatly beneficial to the computing time, and, since the whole procedure is heuristic in nature, can even lead to better solutions than the exact option. This is indeed the case here.

In order to evaluate the quality of the results, we have computed the lower bound on the problem by using our LP module. The LP lower bound based on all the Steiner sequences of triples for the TA objective is 266.375 improving over the best bound of 253.5 previously known [10]. The optimal lifted alignment finds a value of 364, as also reported in [10]. Using heuristic Steiner sequences,

Table 2. TA results on the instance **sank**. Times in seconds, Pentium PC

Triples	Steiner	Reopt	Value	Time
ALL	HEUR	EXACT	302	592
ALL	HEUR	HEUR	302.25	424
ALL	EXACT	EXACT	303.25	2802
SOME	EXACT	EXACT	304	493
ALL	EXACT	HEUR	304.25	2599
SOME	EXACT	HEUR	304.5	267
SOME	HEUR	EXACT	314	201
SOME	HEUR	HEUR	315.75	23
NONE	-	EXACT	320	152
NONE	-	HEUR	320.5	6
ALL	HEUR	NONE	322.25	298
ALL	EXACT	NONE	322.5	2387
SOME	EXACT	NONE	333.5	258
SOME	HEUR	NONE	333.75	15
NONE	-	NONE	364	1

we find a solution of value about 302 in about 7 minutes. Contrast this with the best upper bound of 295.5 by Sankoff et al. [20]. Our improved lower bound shows that Sankoff's solution is within 11% of optimal.

4. **Sum of Pairs.** For the SP objective, we report some results for the McClure data sets (Table 3). For each problem, we have computed the trivial lower bound given by the sum of edit distances, and two lower bounds based on the optimal SP alignment of triples of sequences - one uses a simple averaging argument (LB triples) and the other the solution to an LP relaxation (LB lp). We ran GESTALT with heuristic Steiner sequences, sampling all triples. Our solutions are in an interval of 2 to 9 percent from the lower bound. The table shows also the effectiveness of local reoptimization. For comparison, we also report the SP value of the star alignment (Gusfield, [9]).

Table 3. SP lower and upper bounds for McClure data sets

Instance	LB pairs	LB triples	LB lp	Star align.	GESTALT	Err %	GESTALT+reop	Err %
mc582x6	25411	26056	26100	28444	27647	0.06	26963	0.03
mc586x6	25191	25979	26029	29307	28605	0.10	27498	0.05
mc587x6	29914	30802	30864	34085	34152	0.11	32664	0.05
mc582x10	70718	72274	72757	82011	77676	0.07	75131	0.03
mc586x10	81745	84211	84662	99140	97725	0.15	91754	0.08
mc587x10	95002	97889	98349	115918	110463	0.12	105806	0.07
mc582x12	98810	100720	101464	113328	105674	0.04	103803	0.02
mc586x12	116889	120409	121130	143792	139398	0.15	131980	0.08
mc587x12	140679	145043	145804	174270	164883	0.13	160256	0.09

References

1. S. Altschul and D. Lipman, Trees, Stars and Multiple Sequence Alignment, *SIAM J. Appl. Math.* **49** (1989) 197–209
2. S. Altschul, D. Lipman and J. D. Kececioglu, A tool for multiple sequence alignment. *Proc. Natl. Acad. Sci. USA* **86** (1989) 4412–4415
3. V. Bafna, E. L. Lawler and P. Pevzner. Approximation Algorithms for Multiple Sequence Alignment. *Proceedings of the 5th Combinatorial Pattern Matching conference* LNCS **807** (1994) 43–53
4. H. Carrillo and D. Lipman. The multiple sequence alignment problem in biology. *SIAM J. Appl. Math.* **49:1** (1989) 197–209
5. S. C. Chan, A. K. C. Wong and D. K. Y. Chiu, "A survey of multiple sequence comparison methods," *Bull. Math. Biol.* **54** (1992) 563–598
6. D. Feng and R. Doolittle. Progressive sequence alignment as a prerequisite to correct phylogenetic trees. *J. Molec. Evol.* **25** (1987) 351–360
7. O. Gotoh, Optimal alignment between groups of sequences and its application to multiple sequence alignment, *CABIOS* **9:3** (1993) 361–370
8. S. K. Gupta, J. Kececioglu, and A. A. Schaffer, Making the Shortest-Paths Approach to Sum-of-Pairs Multiple Sequence Alignment More Space Efficient in Practice, (extended abstract) *Proceedings of the 6th Combinatorial Pattern Matching conference* (1995)
9. D. Gusfield, Efficient methods for multiple sequence alignment with guaranteed error bounds, *Bulletin of Mathematical Biology* **55** (1993) 141–154
10. D. Gusfield and L. Wang, New Uses for Uniform Lifted Alignments, Submitted for publication (1996)
11. D. G. Higgins, A. J. Bleasby and R. Fuchs, Clustal V: Improved software for multiple sequence alignment, *CABIOS* **8** (1992) 189–191
12. D. Hirschberg, A linear space algorithm for computing maximal common subsequences, *Communications of the ACM* **18** (1975) 341–343
13. T. Jiang and F. Liu, Tree Alignment And Reconstruction application software, Version 1.0, February 1998. Available from http://www.dcss.mcmaster.ca/~fliu.
14. D. Lipman, S. Altschul and J. D. Kececioglu, A tool for multiple sequence alignment. *Proc. Natl. Acad. Sci. USA* **86** (1989) 4412–4415
15. S. B. Needleman and C. D. Wunsch. A general method applicable to search the similarities in the amino acid sequences of two proteins. *J. Mol. Biol.*, **48** (1970) 444
16. M. A. McClure, T. K. Vasi and W. M. Fitch. Comparative analysis of multiple protein–sequence alignment methods, *Mol. Biol. Evol.* **11** (1994) 571–592
17. R. Ravi and J. Kececioglu. Approximation algorithms for multiple sequence alignment under a fixed evolutionary tree, *Proceedings of the 6th Combinatorial Pattern Matching conference* (1995) 330–339
18. D. Sankoff, Minimal mutation trees of sequences, *SIAM J. Applied Math.* **28(1)** (1975) 35–42
19. D. Sankoff and R. Cedergren, Simultaneous comparison of three or more sequences related by a tree, in D. Sankoff and J. Kruskal editors, *Time warps, string edits and macromolecules: the theory and practice of sequence comparison,* Addison Wesley (1983) 253–264
20. D. Sankoff, R. Cedergren and G. Laplame, Frequency of insertion-deletion, transversion, and transition in the evolution of the 5s ribosomal rna, *J. Mol. Evol.* **7** (1976) 133–149

21. D. Sankoff, Analytical approaches to genomic evolution, *Biochimie* **75** (1993) 409–413

22. T. F. Smith and M. S. Waterman. Comparison of Biosequences. *Adv. Appl. Math.* (1981) 482–489

23. W. R. Taylor and D. T. Jones. Deriving an Amino Acid Distance Matrix, *J. Theor. Biol.* **164** (1993) 65–83

24. M. Vingron and P. Argos. A fast and sensitive multiple sequence alignment algorithm. *Comput. Appl. Biosci.* **5** (1989) 115–121

25. L. Wang and D. Gusfield. Improved Approximation Algorithms for Tree Alignment, *Proceedings of the 7th Combinatorial Pattern Matching conference* (1996) 220–233

26. L. Wang and T. Jiang. On the complexity of multiple sequence alignment, *J. Comp. Biol.* **1** (1994) 337–348

27. L. Wang, T. Jiang and E. L. Lawler. Aligning sequences via an evolutionary tree: complexity and approximation, *Algorithmica*, to appear. Also presented at the *26th ACM Symp. on Theory of Computing* (1994)

Bounds on the Number of String Subsequences

Daniel S. Hirschberg

Department of Information and Computer Science
University of California at Irvine
Irvine CA 92697-3425
dan@ics.uci.edu
http://www.ics.uci.edu/~dan/

Abstract. The problem considered is that of determining the number of subsequences obtainable by deleting t symbols from a string of length n over an alphabet of size s. Recurrences are proven and solved for the maximum and average case values, and bounds on these values are exhibited.

1 Problem Definition

We prove bounds on the number of subsequences of a given length that a string on a fixed-size alphabet can have. Such bounds have been the basis for an efficient algorithm that reconstructs a binary string from knowledge of a sufficient number of its subsequences [6]. This research area is linked to applications of Levenshtein distance whose usage "plays the central role" in "the study of block codes capable of correcting substitution and synchronization errors" [2].

An *L-string* is a string over alphabet L, where $|L| = s$; L_n denotes the set of all length n L-strings. A *series* is a maximal run of identical symbols and $\tau(X)$ denotes the number of series in string X. A *subsequence* Y of string X is a string obtained by deleting 0 or more symbols from X, and X is said to be a *supersequence* of Y. $D_t(X)$ denotes the set of subsequences of X that can be obtained by deleting exactly t symbols from X.

Calabi ([1], as cited in [2]) proved that a particular string form attains the maximum value of $|D_t(X)|$ and found an expression for the generating function of that maximum value. We present a direct alternative proof of the upper bound and prove a simple underlying recurrence.

Levenshtein [3] proved that, for any binary string X, $\binom{\tau(X)-t+1}{t} \le |D_t(X)| \le \binom{\tau(X)+t-1}{t}$. These bounds can be generalized to L-strings [5]. However, while the upper bound is tight, the lower bound is not. We prove a tight lower bound.

Assuming that X is equally likely any string in L_n, we derive and solve a recurrence on the average value of $|D_t(X)|$.

2 Upper Bounds for $|D_t(X)|$

We determine an upper bound on the number of subsequences obtainable by deleting t symbols from a string of length n over an alphabet of size s.

M. Crochemore, M. Paterson (Eds.): CPM'99, LNCS 1645, pp. 115–122, 1999.
© Springer-Verlag Berlin Heidelberg 1999

Let $L = \{\sigma_1, \ldots, \sigma_s\}$, where the $\{\sigma_i\}$ are listed in some order. Let $C_n = c_1, \ldots, c_n$ be a string in L_n, where $c_i = \sigma_{1+(i-1 \bmod s)}$. Thus, C_n has the symbols of L in circular order, cycling as many times as needed.

Let $d_s(t, n)$ denote $|D_t(C_n)|$, the number of subsequences obtainable by deleting t symbols from C_n, where L has cardinality s.

Calabi ([1], as cited in [2]) proved that, for all X in L_n, $|D_t(X)| \leq d_s(t, n)$, and that $d_s(t, n)$ is the coefficient of x^n in the generating function: $\phi(x) = (\sum_{j=1}^{s} x^j)^{n-t}(\sum_{j=0}^{\infty} x^j)$. Apparently, "the proof is rather involved" and was not published. In our Theorem 1, we present a direct alternative proof of the upper bound. In our Theorem 2, we prove a simple recurrence on $d_s(t, n)$.

We use Q^a to denote the subset of strings in set Q that begin with symbol a. For example, $D_t^b(X)$ denotes a set of subsequences of X that start with symbol b. If Q and R are sets then we use $Q + R$ to denote $Q \cup R$ with the assertion that Q and R are disjoint and, thus, $|Q + R| = |Q| + |R|$.

Lemma 1. *For any L-string X, $D_t(X) = \sum_{a \in L} D_t^a(X)$.*

Proof. The set of strings is partitioned into subsets organized by each string's first symbol.   □

Lemma 2. *For $s \geq 1$, $d_s(t-1, n-1) \leq d_s(t, n)$.*

Proof. $d_s(t-1, n-1)$ counts subsequences of length $n - t$ as does $d_s(t, n)$, but of a smaller string.   □

If Q is a set of L-strings and $\sigma \in L$ is a symbol then σQ denotes the set of L-strings $\{\sigma q | q \in Q\}$.

Lemma 3. *For $s \geq 1$, $d_s(t, n) = \sum_{i=1}^{s} d_s(t+1-i, n-i)$.*

Proof. Let $C_n^{(j)} = c_1, \ldots, c_n$ be a string in L_n, where $c_i = \sigma_{j+(i-1 \bmod s)}$. Thus, $C_n^{(j)}$ has the symbols of L in circular order, beginning with σ_j. Using Lemma 1, we see that $D_t(C_n) = \sum_{i=1}^{s} D_t^{\sigma_i}(C_n^{(1)}) = \sum_{i=1}^{s} \sigma_i D_{t+1-i}(C_{n-i}^{(i+1)})$. The statement of the lemma follows directly.   □

Theorem 1. *For $s \geq 1$ and for any $X \in L_n$, $|D_t(X)| \leq d_s(t, n)$.*

Proof. By induction on n and $n - t$. The theorem is trivially true for $n \leq 1$ and $n - t \leq 1$. Let $X = x_1 \ldots x_n$. Let f_i be the smallest index j such that $x_j = \sigma_i$ (and f_i is $n + 1$ if σ_i does not appear in X), where the elements of L, $\{\sigma_i\}$, are ordered by their first appearance in X, thereby ordering f_i smallest to largest. Consequentially, $f_i \geq i$. We use $X[i : j]$ to denote the substring $x_i \ldots x_j$ of X.

Using Lemma 1, we have

$$D_t(X) = \sum_{i=1}^{s} D_t^{\sigma_i}(X) = \sum_{i=1}^{s} \sigma_i D_{t+1-f_i}(X[f_i + 1 : n]) .$$

Therefore,

$$|D_t(X)| = \sum_{i=1}^{s} |D_{t+1-f_i}(X[f_i + 1 : n])|$$

$$\leq \sum_{i=1}^{s} |D_{t+1-f_i}(C_{n-f_i})|, \text{ using the inductive hypothesis,}$$

$$\leq \sum_{i=1}^{s} |D_{t+1-i}(C_{n-i})|, \text{ because } f_i \geq i \text{ and by applying Lemma 2,}$$

$$= d_s(t, n), \text{ by applying Lemma 3.}$$

□

Theorem 2. *For $0 \leq t \leq n$ and $s \geq 2$, $d_s(t, n) = \sum_{i=0}^{t} \binom{n-t}{i} d_{s-1}(t - i, t)$.*

Proof. By induction on n and $n - t$.

For the basis when $n = 0$, t must be zero and it suffices to see that $d_s(0, 0) = \binom{0}{0} d_{s-1}(0, 0) = 1$.

For the basis when $n - t = 0$, since $\binom{0}{i}$ is zero unless $i = 0$, it suffices to see that $d_s(n, n) = \sum_{i=0}^{n} \binom{0}{i} d_{s-1}(n - i, n) = d_{s-1}(n, n) = 1$.

For the induction, using the recurrence of Lemma 3,

$$d_s(t, n) = d_s(t, n - 1) + \sum_{k=2}^{s} d_s(t + 1 - k, n - k) .$$

Let $r = \sum_{k=2}^{s} d_s(t + 1 - k, n - k)$. Then

$$r = \sum_{k=2}^{s} \sum_{i=0}^{t+1-k} \binom{n-t-1}{i} d_{s-1}(t + 1 - k - i, t + 1 - k), \text{ using the inductive hyp.,}$$

$$= \sum_{j=1}^{s-1} \sum_{i=0}^{t-j} \binom{n-t-1}{i} d_{s-1}(t - j - i, t - j), \text{ by letting } j = k - 1,$$

$$= \sum_{j=1}^{s-1} \sum_{i=0}^{t} \binom{n-t-1}{i} d_{s-1}(t - j - i, t - j), \text{ as } d_{s-1}(t - j - i, t - j) = 0 \text{ if } i > t - j,$$

$$= \sum_{i=0}^{t} \binom{n-t-1}{i} \sum_{j=1}^{s-1} d_{s-1}(t - i - j, t - j)$$

$$= \sum_{i=0}^{t} \binom{n-t-1}{i} d_{s-1}(t - i - 1, t), \text{ by using Lemma 3,}$$

$$= \sum_{j=1}^{t+1} \binom{n-t-1}{j-1} d_{s-1}(t - j, t), \text{ by letting } j = i + 1,$$

$$= \sum_{j=0}^{t} \binom{n-t-1}{j-1} d_{s-1}(t-j,t), \text{ changing } j\text{'s range by noting that,}$$

when j is 0, $\binom{n-t-1}{j-1} = 0$, and that, when j is $t+1$, $d_{s-1}(t-j,t) = 0$.

Therefore,

$d_s(t,n) = d_s(t,n-1) + r$, using the recurrence of Lemma 3,

$$= \sum_{i=0}^{t} [\binom{n-t-1}{i} + \binom{n-t-1}{i-1}] d_{s-1}(t-i,t), \text{ using the inductive hypothesis,}$$

$$= \sum_{i=0}^{t} \binom{n-t}{i} d_{s-1}(t-i,t), \text{ using the binomial recurrence.}$$

$\square$

Corollary 1. *For* $0 \le t \le n, d_2(t,n) = \sum_{i=0}^{t} \binom{n-t}{i}$; *for* $0 \le t \le n, d_3(t,n) = \sum_{i=0}^{t} \binom{n-t}{i} \sum_{j=0}^{t-i} \binom{i}{j}$.

Proof. This follows immediately from Theorem 2 and the fact that $d_1(t,n) = 1$ for $0 \le t \le n$. $\square$

Observations. By evaluating $d_s(t,n)$ and expressing its difference from $\binom{n}{t}$ as a power series, one can see that, for $t \ge s, d_s(t,n) = \binom{n}{t} - n^{t+1-s}/(t-s)! + O(n^{t-s})$.

We note that the problem of calculating the number, $i_s(t,n)$, of supersequences obtainable by inserting t symbols in a length n string X on an alphabet of size s is much simpler, and is invariant over X. It is known [4] that, using a binary alphabet, $i_2(t,n) = \sum_{j=0}^{t} \binom{n+t}{j}$. This can be generalized to $s \ge 2$ [5]. It is easy to see that $i_s(t,n) = i_s(t,n-1) + (s-1)i_s(t-1,n)$, with boundary conditions $i_s(0,n) = 1$ and $i_s(t,0) = s^t$. (Let $X \in L_{n-1}, Y \in L_{n+t-1}$, and $a, b \in L$. Then bY is a supersequence of aX if and only if either (1) $a = b$ and Y is a supersequence of X, or (2) $a \ne b$ and Y is a supersequence of aX.) This recurrence is solved by $i_s(t,n) = \sum_{j=0}^{t}(s-1)^j \binom{n+t}{j}$.

As we have just seen, the recurrence for supersequences is very simple and intuitive. The recurrence of Theorem 2 for subsequences is simple but currently lacks an intuitive explanation.

3   A Lower Bound for $|D_t(X)|$

It was stated [6,3] that, for any binary string X, $|D_t(X)| \ge \binom{\tau(X)-t+1}{t}$. We note that this bound is the same as $\binom{\tau(X)-t}{t} + \binom{\tau(X)-t}{t-1}$. We will improve and generalize this bound. We first need a few lemmas.

Lemma 4. *For any L-strings* U, V *and any* $\sigma \in L, |D_t(UV)| \le |D_t(U\sigma V)|$.

Proof. For all subsequences u of U and v of V, $uv \in D_t(UV) \rightarrow u\sigma v \in D_t(U\sigma V)$ and $uv \neq u'v' \rightarrow u\sigma v \neq u'\sigma v'$.   □

Lemma 5. *If X is an L-string such that $\tau(X) = n$ then there exists a string $Y \in L_n$, with $\tau(Y) = n$, such that $|D_t(Y)| \leq |D_t(X)|$.*

Proof. Let Y be the length n L-string consisting of one symbol from each of the series in X. String X can be obtained from Y by a sequence of symbol insertions. The statement of the lemma then follows from repeated applications of Lemma 4.   □

Lemma 6. *If X is a string in L_n such that $\tau(X) = n$ then $|D_t(X)| \geq d_2(t, n)$.*

Proof. By induction on n and $n - t$. The lemma is trivially true for the base cases, when $n \leq 2$ or $n - t \leq 2$. For the induction step, let $X = abY$, where $a \neq b$ because each series in X has length 1. Then,

$$D_t(X) = D_t^a(X) + D_t^b(X) + \sum_{\sigma \neq a, b} D_t^\sigma(X)$$
$$\supseteq D_t^a(X) + D_t^b(X) = aD_t(bY) + bD_{t-1}(Y) .$$

Using the inductive hypothesis and Lemma 3, we obtain

$$|D_t(X)| \geq |D_t(bY)| + |D_{t-1}(Y)|$$
$$\geq d_2(t, n-1) + d_2(t-1, n-2) = d_2(t, n) .$$

□

Theorem 3. *For any L-string X, $|D_t(X)| \geq \sum_{i=0}^{t} \binom{\tau(X)-t}{i}$ and this bound is tight.*

Proof. Follows directly from Corollary 1 and Lemmas 5 and 6.   □

4   The Average Number of Subsequences

Under the assumption that all length n L-strings are equiprobable, the average number of subsequences obtainable by deleting one symbol has been shown to be $(n(s-1)+1)/s$ [2]. We develop and solve a recurrence on the average number of subsequences obtainable by deleting t symbols.

Let $G_t(n) = \sum_{X \in L_n} |D_t(X)|$ be the sum, over all strings in L_n, of the number of subsequences obtainable by deleting t symbols. Similarly, $G_t^a(n) = \sum_{X \in L_n} |D_t^a(X)|$ is the sum when the subsequences are restricted to begin with symbol a.

We see that, for $0 < t < n$,

$$G_t^a(n) = \sum_{X \in L_n} |D_t^a(X)| = \sum_{b \in L} \sum_{X \in L_n^b} |D_t^a(X)| . \tag{1}$$

If $b = a$ then $\{X \in L_n^a\} = \{aY | Y \in L_{n-1}\}$. We note that the count of subsequences of aY that start with a and have length $n - t$ is the same as the count of subsequences of Y that have length $n - 1 - t$ because of a simple bijection between those two sets of subsequences. As a result, we see that $\sum_{X \in L_n^a} |D_t^a(X)| = \sum_{Y \in L_{n-1}} |D_t^a(aY)| = \sum_{Y \in L_{n-1}} |D_t(Y)|$.

If $b \neq a$ then $\{X \in L_n^b\} = \{bY | Y \in L_{n-1}\}$. We note that the count of subsequences of bY that start with a and have length $n - t$ is the same as the count of subsequences of Y that start with a and have length $n - t$ because the leading b of bY can just be discarded. As a result, we see that $\sum_{X \in L_n^b} |D_t^a(X)| = \sum_{Y \in L_{n-1}} |D_t^a(bY)| = \sum_{Y \in L_{n-1}} |D_{t-1}(Y)|$.

Therefore,

$$G_t^a(n) = \sum_{X \in L_{n-1}} |D_t(X)| + (s - 1) \sum_{X \in L_{n-1}} |D_{t-1}^a(X)| . \qquad (2)$$

We then see that

$$G_t^a(n) = G_t(n - 1) + (s - 1)G_{t-1}^a(n - 1) \qquad (3)$$

follows immediately from (1), (2) and the definitions.

From the fact that $G_t(n) = \sum_{a \in L} G_t^a(n)$, using (3) s times, once for each symbol in L, we obtain

$$G_t(n) = sG_t(n - 1) + (s - 1)G_{t-1}(n - 1) . \qquad (4)$$

Boundary conditions, $G_0(n) = G_n(n) = s^n$, hold because there is only one string obtainable by deleting none or all of the symbols in each of the s^n strings in L_n.

Let $E_t(n)$ be the average (or expected) value of $|D_t(X)|$, where X can equally likely be any string in L_n, and let $\lambda = 1 - 1/s$.

Theorem 4. *For $0 < t < n$, $E_t(n) = E_t(n - 1) + \lambda E_{t-1}(n - 1)$, and $E_0(n) = E_t(t) = 1$.*

Proof. This follows from the recurrence (4) and boundary conditions on G and the fact that $E_t(n) = G_t(n)/s^n$. $\qquad \square$

Theorem 5. $E_t(n) = \sum_{i=0}^{t} \binom{n-1-t+i}{i} \lambda^i$.

Proof. We use a generating function (see, for example, Liu [7]).

Let $F_n(x) = \sum_{t=0}^{\infty} E_t(n)x^t$. We note that $E_t(n) = 0$ if $t > n$, and that $E_t(n) = 1$ if $t = n$ or $t = 0$. Then,

$$F_n(x) - 1 - x^n = \sum_{t=1}^{n-1} E_t(n)x^t$$

$$= \sum_{t=1}^{n-1} E_t(n - 1)x^t + \lambda \sum_{t=1}^{n-1} E_{t-1}(n - 1)x^t$$

$$= \sum_{t=1}^{n-1} E_t(n-1)x^t + \lambda x \sum_{t=0}^{n-2} E_t(n-1)x^t$$

$$= [F_{n-1}(x) - E_0(n-1)] + \lambda[xF_{n-1}(x) - x^n E_{n-1}(n-1)] .$$

Therefore,

$$F_n(x) = (1 + \lambda x)F_{n-1}(x) + x^n(1 - \lambda) . \tag{5}$$

By iterated expansion of (5), we obtain

$$F_n(x) = (1 + \lambda x)^n + \sum_{i=0}^{n-1} (1 + \lambda x)^i (1 - \lambda)x^{n-i}$$

$$= \sum_{i=0}^{n} \binom{n}{i}\lambda^i x^i + \sum_{i=0}^{n-1} \sum_{j=0}^{i} \binom{i}{j}\lambda^j (1 - \lambda)x^{n+j-i} . \tag{6}$$

In the expression (6) of $F_n(x)$, the coefficient of x^t is $E_t(n) = \binom{n}{t}\lambda^t + \sum_{i=0}^{n-1} \binom{i}{j}\lambda^j(1 - \lambda)$, where $j = i - (n - t) \geq 0$. Therefore we can restrict the summation to $i \geq n - t$ and, letting $k = n - i$, we get

$$E_t(n) = \binom{n}{t}\lambda^t + \sum_{k=1}^{t} \binom{n-k}{t-k}\lambda^{t-k}(1 - \lambda) .$$

Noting that $\binom{n-k}{t-k} - \binom{n-k-1}{t-k-1} = \binom{n-k-1}{t-k}$, we finally obtain

$$E_t(n) = \binom{n}{t}\lambda^t + \sum_{k=1}^{t-1} \binom{n-k-1}{t-k}\lambda^{t-k} - \binom{n-1}{t-1}\lambda^t + \binom{n-t}{0}$$

$$= \binom{n-1}{t}\lambda^t + \sum_{i=1}^{t-1} \binom{n-1-t+i}{i}\lambda^i + 1 .$$

$\square$

Observation. For $t < n\lambda/2$, $E_t(n) \geq \sum_{j=0}^{3} \binom{n-1-j}{t-j}\lambda^{t-j} \geq .93E_t(n)$.

References

1. L. Calabi, "On the computation of Levenshtein's distances," TM-9-0030, Parke Math. Labs., Inc., Carlisle, Mass., 1967.
2. L. Calabi and W.E. Hartnett, "Some general results of coding theory with applications to the study of codes for the correction of synchronization errors," *Information and Control 15*,3 (Sept 1969) 235-249. Reprinted in: W. E. Hartnett (ed), *Foundations of Coding Theory* (1974) Chapter 7, pp.107-121.
3. V. I. Levenshtein, "Binary codes capable of correcting deletions, insertions and reversals," *Soviet Phys. Dokl. 10* (1966) 707-710.
4. V. I. Levenshtein, "Elements of the coding theory," in: *Discrete Math. and Math. Probl. of Cybern.*, Nauka, Moscow (1974) 207-235 (in Russian).

5. V. I. Levenshtein, "On perfect codes in deletion and insertion metric," *Discrete Math. Appl.* 2,3 (1992) 241-258. Originally published in *Diskretnaya Matematika* 3,1 (1991) 3-20 (in Russian).
6. V. I. Levenshtein, "Reconstructing binary sequences by the minimum number of their subsequences or supersequences of a given length," *Proc. 5th Int. Wkshp on Alg. & Comb. Coding Theory*, Sozopol, Bulgaria (1996) 176-183.
7. C. L. Liu, *Introduction to Combinatorial Mathematics*, McGraw-Hill, 1986, pp. 80-86.

Approximate Periods of Strings

Jeong Seop Sim[1] *, Costas S. Iliopoulos[2,4] **, Kunsoo Park[1] *, and
William F. Smyth[3,4] * * *

[1] Department of Computer Engineering, Seoul National University
{jssim, kpark}@theory.snu.ac.kr
[2] Department of Computer Science, King's College London
csi@dcs.kcl.ac.uk
[3] Department of Computing & Software, McMaster University
smyth@mcmaster.ca
[4] School of Computing, Curtin University

Abstract. The study of approximately periodic strings is relevant to
diverse applications such as molecular biology, data compression, and
computer-assisted music analysis. Here we study different forms of approximate periodicity under a variety of distance rules. We consider three
related problems, for two of which we derive polynomial-time algorithms;
we then show that the third problem is NP-complete.

1 Introduction

Repetitive or periodic strings have been studied in such diverse fields as molecular biology, data compression, and computer-assisted music analysis. In response
to requirements arising out of a variety of applications, interest has arisen in algorithms for finding *regularities* in strings; that is, periodicities of an approximate
nature. Some important regularities that have been studied in the literature are
the following:

- **Periods:** A string p is called a *period* of a string x if x can be written as
  $x = p^k p'$ where $k \geq 1$ and p' is a prefix of p. The shortest period of x is
  called *the period* of x. For example, if $x = abcabcab$, then abc, $abcabc$, and x
  are periods of x, while abc is *the* period of x. If x has a period p such that
  $|p| \leq |x|/2$, then x is said to be *periodic*. Further, if setting $x = p^k$ implies
  $k = 1$, x is said to be *primitive*; if $k \geq 2$, p^k is called a *repetition*.
- **Covers:** A string w is called a *cover* of x if x can be constructed by concatenations and superpositions of w. For example, if $x = ababaaba$, then aba and
  x are the covers of x. If x has a cover $w \neq x$, x is said to be *quasiperiodic*;
  otherwise, x is *superprimitive*.
- **Seeds:** A substring w of x is called a *seed* of x if it is a cover of any superstring
  of x. For example, aba and $ababa$ are some seeds of $x = ababaab$.

* Supported by KOSEF Grant 981-0925-128-2.
** Supported in part by the CCSLAAR Royal Society Researh Grant.
* * * Supported by NSERC Grant No. A8180.

- **Repetitions:** A substring w of x that is a repetition is called a *repetition* or *tandem repeat in* x. For example, if $x = aababab$, then aa and $ababab$ are repetitions in x; in particular, $a^2 = aa$ is called a *square* and $(ab)^3 = ababab$ is called a *cube*.

The notions *cover* and *seed* are generalizations of periods in the sense that superpositions as well as concatenations are used to define them. A significant amount of research has been done on each of these four notions:

- **Periods:** The preprocessing of the Knuth-Morris-Pratt algorithm [19] finds all periods of x in linear time — in fact, all periods of every prefix of x. In parallel computation, Apostolico, Breslauer and Galil [2] gave an optimal $O(\log \log n)$ time algorithm for finding all periods, where n is the length of x.
- **Covers:** Apostolico, Farach and Iliopoulos [4] introduced the notion of covers and described a linear-time algorithm to test whether x is superprimitive or not (see also [7,8,17]). Moore and Smyth [26] and recently Li and Smyth [22] gave linear-time algorithms for finding all covers of x. In parallel computation, Iliopoulos and Park [16] obtained an optimal $O(\log \log n)$ time algorithm for finding all covers of x. Apostolico and Ehrenfeucht [3] and Iliopoulos and Mouchard [15] considered the problem of finding maximal quasiperiodic substrings of x. A two-dimensional variant of the covering problem was studied in [11,13], and minimum covering by substrings of given length in [18].
- **Seeds:** Iliopoulos, Moore and Park [14] introduced the notion of seeds and gave an $O(n \log n)$ time algorithm for computing all seeds of x. For the same problem Berkman, Iliopoulos and Park [6] presented a parallel algorithm that requires $O(\log n)$ time and $O(n \log n)$ work.
- **Repetitions:** There are several $O(n \log n)$ time algorithms for finding all the repetitions in a string [10,5,24]. In parallel computation, Apostolico and Breslauer [1] gave an optimal $O(\log \log n)$ time algorithm (i.e., total work is $O(n \log n)$) for finding all the repetitions.

A natural extension of the repetition problems is to allow errors. Approximate repetitions are common in applications such as molecular biology and computer-assisted music analysis [9,12]. Among the four notions above, only approximate repetitions have been studied. If $x = uww'v$ where w and w' are similar, ww' is called an *approximate square* or *approximate tandem repeat*. When there is a nonempty string y between w and w', we say that w and w' are an *approximate nontandem repeat*. In [21], Landau and Schmidt gave an $O(kn \log k \log n)$ time algorithm for finding repeated patterns whose edit distance is at most k in a text of length n. Schmidt also gave an $O(n^2 \log n)$ algorithm for finding approximate tandem or nontandem repeats in [28] which uses an arbitrary score for similarity of repeated strings.

In this paper, we introduce the notion of *approximate periods* which can be considered as an approximate version of three notions *periods*, *covers*, and *seeds.* Here we study different forms of approximate periodicity under a variety of distance rules. We consider three related problems, for two of which we derive polynomial-time algorithms; we then show that the third problem is NP-complete.

2   Preliminaries

A *string* is a sequence of zero or more characters from an alphabet Σ. The set of all strings over the alphabet Σ is denoted by Σ^*. The empty string is denoted by ϵ. The ith character of a string x is denoted by $x[i]$. A substring of x that starts at position i and ends at position j is denoted by $x[i..j]$.

A string w is a *prefix* of x if $x = wu$ for $u \in \Sigma^*$. Similarly, w is a *suffix* of x if $x = uw$ for $u \in \Sigma^*$. A string w is a *subsequence* of x (or x is a *supersequence* of w) if w is obtained by deleting zero or more characters (at any positions) from x. For example, *ace* is a subsequence of *aabcdef*.

2.1   Measures

ABSOLUTE MEASURES. To measure the similarity (or distance) between two strings, the Hamming distance and the edit distance are widely used. The *Hamming distance* between two strings x and y is defined to be the smallest number of *change* operations to convert x to y. The *edit distance* is defined to be the smallest number of *change*, *insert*, and *delete* operations to convert x to y. In more general cases, especially in molecular biology, a penalty matrix is used. A penalty matrix specifies the substitution cost for each pair of characters and the insertion/deletion cost for each character. An arbitrary penalty matrix can also be used as a relative measure because it can contain both positive and negative costs [28]. It is common to assume that a penalty matrix satisfies the triangle inequality [30].

RELATIVE MEASURES. When we want to compare the similarity between x and y and the similarity between x' and y', we need relative measures (rather than absolute measures) because the lengths of the strings x, y, x', y' may be different. There are two ways to define relative measures between x and y:
- First, we can fix one of the two strings and define a relative measure with respect to the fixed string. The *error ratio with respect to x* is defined to be $t/|x|$, where t is an absolute measure between x and y.
- Second, we can define a relative measure symmetrically. The *symmetric error ratio* is defined to be t/l, where t is an absolute measure between x and y, and $l = (|x| + |y|)/2$ [29]. Note that we may take $l = |x| + |y|$ (then everything is the same except that the ratio is multiplied by 2).

3   Problem Definitions

Given two strings x and p, we define approximate periods as follows. If there exists a partition of x into disjoint blocks of substrings, i.e., $x = p_1 p_2 \cdots p_r$

($p_i \neq \epsilon$) such that the distance between p and p_i for every $1 \leq i \leq r$ is less than or equal to t, we say that p is a *t-approximate period* of x (or p is an approximate period of x with distance t). Each p_i, $1 \leq i \leq r$, will be called a partition block of x. Note that there can be several versions of approximate periods according to the definition of *distance*. This definition of approximate periods can be considered as an approximate version of the three notions *periods*, *covers*, and *seeds* discussed above, because

(i) superpositions in defining covers and seeds and
(ii) extra characters at the ends of a given string in defining periods and seeds

can be accounted for in some degree when we use edit distances for the measure. Of course, if we allow overlaps between p_i's, then we could extend the definition of an approximate period. But this will merely increase the complexity of problems of finding approximate periods.

We consider the following problems related to approximate periods.

Problem 1. Given x and p, find the minimum t such that p is a t-approximate period of x.

Since p is fixed in this case, it makes no difference whether we use the absolute Hamming (or edit) distance or the error ratio with respect to p. We can also use a penalty matrix for the measure. If a threshold k on the edit distance is given as input in Problem 1, the problem asks whether p is a k-approximate period of x or not.

Problem 2. Given a string x, find a substring p of x that is an approximate period of x with the minimum distance.

Since the length of p is not (a priori) fixed in this problem, we need to use relative measures (i.e., error ratios or penalty matrices) rather than absolute measures.

Problem 3. Given a string x, find a string p that is an approximate period of x with the minimum distance.

This problem is harder than Problem 2 because p can be any string, not necessarily a substring of x.

4   Algorithms and NP-Completeness

Basically we will use arbitrary penalty matrices for the measure of similarity in each problem. Recall that a penalty matrix defines the substitution cost for each pair of characters and the insertion or deletion cost for each character.

4.1   Problem 1

Our algorithm for Problem 1 consists of two steps. Let $n = |x|$ and $m = |p|$.

1. Compute the distance between p and every substring of x.
2. Compute the minimum t such that p is a t-approximate period of x. We use dynamic programming to compute t. Let w_{ij} be the distance between p and $x[i..j]$. These values of w_{ij} are obtained from the first step. Let t_i be the minimum value such that p is a t_i-approximate period of $x[1..i]$. Let $t_0 = 0$. For $i = 1$ to n, we compute t_i by the following formula:

$$t_i = \min_{0 \le h < i} \left(\max(t_h, w_{h+1,i}) \right).$$

The value t_n is the minimum t such that p is a t-approximate period of x.

To compute the distances in step 1, we use the dynamic programming table called *the D table*. To compute the distance between two strings x and y, a D table of size $(|x|+1) \times (|y|+1)$ is used. Each entry $D[i, j]$ $(0 \le i \le |x|, 0 \le j \le |y|)$ stores the minimum cost of transforming $x[1..i]$ to $y[1..j]$. Initially, $D[0,0] = 0$, $D[i,0] = D[i-1,0] + \delta(x[i], \Delta)$, and $D[0,j] = D[0,j-1] + \delta(\Delta, y[j])$. Then we can compute all the entries of the D table in $O(|x||y|)$ time by the following recurrence:

$$D[i,j] = \min \begin{cases} D[i-1,j] & + \delta(x[i], \Delta) \\ D[i,j-1] & + \delta(\Delta, y[j]) \\ D[i-1,j-1] & + \delta(x[i], y[j]) \end{cases}$$

where $\delta(a, b)$ is the cost of transforming the character a to b. (Δ is a space, so $\delta(a, \Delta)$ means the deletion cost of a and $\delta(\Delta, a)$ means the insertion cost of a.)

Theorem 1. *Problem 1 can be solved in $O(mn^2)$ time when an arbitrary penalty matrix is used for the measure of similarity. If the edit distance (resp. the Hamming distance) is used for the measure, it can be solved in $O(mn)$ time (resp. in $O(n)$ time).*

Proof. For an arbitrary penalty matrix, step 1 takes $O(mn^2)$ time since we make a D table of size $m \times (n-i+1)$ for each position i of x. In step 2, we can compute the minimum t in $O(n^2)$ time since we compare $O(n)$ values at each position of x. Thus, the total time complexity is $O(mn^2)$.

When the edit distance is used for the measure of similarity, this algorithm for Problem 1 can be improved. In this case, $\delta(a, b)$ is always 1 if $a \ne b$; $\delta(a, b) = 0$, otherwise. Now it is not necessary to compute the edit distances between p and the substrings of x whose lengths are larger than $2m$ because their edit distances with p will exceed m. (It is trivially true that p is an m-approximate period of x.) Step 1 now takes $O(m^2n)$ time since we make a D table of size $m \times 2m$ for each position of x. Also, step 2 can be done in $O(mn)$ time since we compare $O(m)$ values at each position of x. Thus the time complexity is reduced to $O(m^2n)$.

However, we can do better. Step 1 can be solved in $O(mn)$ time by the algorithm due to Landau, Myers, and Schmidt [20]. Given two strings x and y and a forward (resp. backward) solution for the comparison between x and y, the algorithm in [20] incrementally computes a solution for x and by (resp. yb) in $O(k)$ time, where b is an additional character and k is a threshold on the edit distance. This can be done due to the relationship between the solution for x and y and the solution for x and by. When $k = m$ (i.e., the threshold is not given), we can compute all the edit distances between p and every substring of x whose length is at most $2m$ in $O(mn)$ time using this algorithm. Therefore, we can solve Problem 1 in $O(mn)$ time if the edit distance is used for the measure of similarity.

If we use the Hamming distance for the measure, it takes trivially $O(n)$ time since x must be partitioned into blocks of size m. $\qquad\qquad\qquad\qquad\square$

When the threshold k on the edit distance is given as input for Problem 1, it can be solved in $O(kn)$ time because each step of the above algorithm takes $O(kn)$ time.

4.2   Problem 2

Let p be a candidate string for the approximate period of x. If the Hamming (or edit) distance is used for Problem 2, we need to use relative measures because the length of p varies. (If the absolute Hamming or edit distance is used, every substring of x of length 1 is a 1-approximate period of x.) We can use the *error ratio* t/l for the measure of similarity, where t is the Hamming (or edit) distance between the two strings and l is either the average length of the two strings (symmetric error ratio) or the length of p (error ratio with respect to p).

When the relative edit distance is used for the measure of similarity, Problem 2 can be solved in $O(n^4)$ time by our algorithm for Problem 1. If we take each substring of x as p and apply the $O(mn)$ algorithm for Problem 1 (that uses the algorithm in [20]), it takes $O(|p|n)$ time for each p. Since there are $O(n^2)$ substrings of x, the overall time is $O(n^4)$.

Without using the somewhat complicated algorithm in [20], however, we can solve Problem 2 in $O(n^4)$ time by the following simple algorithm for arbitrary penalty matrices.

Let R be the minimum distance so far. Initially, $R = \infty$. For $i = 1$ to n, we do the following. For each i, we process the $n - i + 1$ substrings that start at position i. Let m be the length of a chosen substring of x as p. Let $m = 1$.

1. Take $x[i..i + m - 1]$ as p and compute the distance between p and every substring of x. This can be done by making n D tables with p and each of n suffixes of x. By adding just one row to each of previous D tables (i.e., n D tables when $p = x[i..i + m - 2]$), we can compute these new D tables in $O(n^2)$ time. (Note that when $m = 1$, we create new D tables.)
2. Compute the minimum distance t such that p is a t-approximate period of x. This step is similar to the second step of the algorithm for Problem 1. Let

w_{hj} be the distance between p and $x[h..j]$ which is obtained from step 1. Let t_j be the minimum value such that p is a t_j-approximate period of $x[1..j]$ and let $t_0 = 0$. For $j = 1$ to n, we compute t_j by the following formula:

$$t_j = \min_{0 \le h < j} (\max(t_h, w_{h+1,j})).$$

The value t_n is the minimum t such that p is a t-approximate period of x. If t is smaller than R, we update R with t. If $m < n - i + 1$, increase m by 1 and go to step 1.

When all the steps are completed, the final value of R is the minimum distance and a substring that is an R-approximate period of x is an answer to Problem 2.

Theorem 2. *Problem 2 can be solved in $O(n^4)$ time when an arbitrary penalty matrix is used for the measure of similarity. If the Hamming distance is used for the measure, it can be solved in $O(n^3)$ time.*

Proof. For an arbitrary penalty matrix, we make n D tables in $O(n^2)$ time in step 1 and compute the minimum distance in $O(n^2)$ time in step 2. For $m = 1$ to $n - i + 1$, we repeat the two steps. Therefore, it takes $O(n^3)$ time for each i and the total time complexity of this algorithm is $O(n^4)$. If the relative edit distance is used, this algorithm can be slightly simplified as in Problem 1, but it still takes time $O(n^4)$.

   If the relative Hamming distance is used for the measure, Problem 2 can be solved in $O(n^3)$ time because there are $O(n^2)$ candidates for p and $O(n)$ time is required for each candidate.   □

4.3   Problem 3

Given a set of strings, the *shortest common supersequence* (SCS) problem is to find a shortest common supersequence of all strings in the set. The SCS problem is NP-complete [23,27]. We will show that Problem 3 is NP-complete by a reduction from the SCS problem. In this section we will call Problem 3 *the AP problem* (abbreviation of the approximate period problem).

   The decision versions of the SCS and AP problems are as follows:

Definition 1. *Given a positive integer m and a finite set S of strings from Σ^* where Σ is a finite alphabet, the SCS problem is to decide if there exists a string w with $|w| \le m$ such that w is a supersequence of each string in S.*

Definition 2. *Given a number t, a string x from $(\Sigma')^*$ where Σ' is a finite alphabet, and a penalty matrix, the AP problem is to decide if there exists a string u such that u is a t-approximate period of x.*

Now we transform an instance of the SCS problem to an instance of the AP problem. We can assume that $\Sigma = \{0, 1\}$ since the SCS problem is NP-complete even

	0	1	a	b	$*_1$	$*_2$	#	$	Δ
0	0	2	1	2	2	2			1
1	2	0	2	1	2	2			1
a	1	2	0	2	1	1			1
b	2	1	2	0	1	1			1
$*_1$	2	2	1	1	0	2			2
$*_2$	2	2	1	1	2	0			2
#							0		
$								0	
Δ	1	1	1	1	2	2			0

Fig. 1. The penalty matrix M

if $\Sigma = \{0,1\}$ [25,27]. First, we set $\Sigma' = \Sigma \cup \{a, b, \#, \$, *_1, *_2, \Delta\}$. Assume that there are n strings $s_1, \ldots, s_n$ in S. Let $x = \#s_1\$\#s_2\$ \cdots \#s_n\$\#*_1{}^m\$\#*_2{}^m\$$. Then, set $t = m$ and define the penalty matrix as in Figure 1, where a shaded entry can be any value greater than m. It is easy to see that this transformation can be done in polynomial time. Note that the penalty matrix M is a metric.

Lemma 1. *Assume that x is constructed as above. If u is an m-approximate period of x, then u is of the form $\#\alpha\$$ where $\alpha \in \{a, b\}^m$.*

Proof. We first show that u must have one # and one $.

1. Suppose that u has no # (resp. $). Clearly, there exists a partition block of x which has at least one # (resp. $), and the distance between u and the block is greater than m. Therefore, u must have at least one # and at least one $.

2. Suppose that u has more than one # (or $). Assume that u has two #'s. (The other cases are similar.) Then u must also have two $'s because unless the number of #'s equals that of $'s in u, at least one partition block of x cannot have the same numbers of #'s and $'s to those of u. Consider the last partition block of x. Since the last block must have two #'s and two $'s as u, it contains $\#*_1{}^m\$\#*_2{}^m\$$. For the distance between u and the last block of x to be at most m, u must have at least m characters from $\{*_1, *_2\}$. In such cases, however, the distance between u and any other partition block of x will exceed m.

It remains to show that $u = \#\alpha\$$ where $\alpha \in \{a, b\}^m$. Since u has one # and one $, x must be partitioned just after every occurrence of $. Let u be of the form $\beta\#\alpha\$\gamma$, where $\beta, \alpha, \gamma \in \{0, 1, a, b, *_1, *_2, \Delta\}^*$. Consider the last two blocks

$\#*_1{}^m\$$ and $\#*_2{}^m\$$ of x. If α contains i $*_1$'s for $i \geq 1$, α must also have i $*_2$'s and the remaining $m - 2i$ characters in α must be from $\{a, b\}$ so that the distances between u and the last two blocks of x do not exceed m. However, this makes the distance between u and any other partition block of x exceed m due to $*_1$'s and $*_2$'s in α. Hence α cannot have $*_1$ or $*_2$. Also, α cannot have any character from $\{0, 1, \Delta\}$ since $0, 1$ and Δ have cost 2 with $*_1$ and $*_2$ in the last two blocks of x. For the distances between u and the last two blocks of x to be at most m, β and γ must be empty and α must be of the form $\{a, b\}^m$.   □

Theorem 3. *The AP problem is NP-complete.*

Proof. It is easy to see that the AP problem is in NP. To show that the AP problem is NP-complete, we need to show that S has a common supersequence w such that $|w| \leq m$ if and only if there exists a string u such that u is an m-approximate period of x.

(if) By Lemma 1, $u = \#\alpha\$$ where $\alpha \in \{a, b\}^m$. Since u is an m-approximate period of x, the distance between u and each partition block $\#s_i\$$ is at most m. (The distances between u and the last two blocks $\#*_1{}^m\$$ and $\#*_2{}^m\$$ are always m.) Since $|\alpha| = m$ and the distance between α and s_i is at most m, each 0 (resp. 1) in s_i must be aligned with a (resp. b) in α. That is, each a (resp. b) in α must be aligned with 0 (resp. 1) or Δ in s_i. If we substitute 0 for a and 1 for b in α, we obtain a common supersequence w of $s_1, \ldots, s_n$ such that $|w| = m$. (Note that if a or b in α is aligned with Δ for all s_i, we can delete the character in α and we can obtain a common supersequence which is shorter than m.) A similar alignment was used by Wang and Jiang [30].

(only if) Let s be a common supersequence of S such that $|s| \leq m$. Let α be the string constructed by substituting a for 0 and b for 1 in s. Partition x just after every occurrence of $\$$. The distance between each partition block of x and $\#\alpha\$$ is at most m since each a (resp. b) in α can be aligned with 0 (resp. 1), Δ, $*_1$, or $*_2$ in each partition block. Therefore, $\#\alpha\$$ is an m-approximate period of x.   □

References

1. A. Apostolico and D. Breslauer, An optimal $O(\log \log N)$-time parallel algorithm for detecting all squares in a string, *SIAM Journal on Computing* 25, 6 (1996), 1318-1331.
2. A. Apostolico, D. Breslauer and Z. Galil, Optimal parallel algorithms for periods, palindromes and squares, *Proc. 19th Int. Colloq. Automata Languages and Programming*, Lecture Notes in Computer Science 623 (1992), 296-307.
3. A. Apostolico and A. Ehrenfeucht, Efficient detection of quasiperiodicities in strings, *Theoretical Computer Science* 119, 2(1993), 247-265.
4. A. Apostolico, M. Farach and C.S. Iliopoulos, Optimal superprimitivity testing for strings, *Information Processing Letters* 39, 1 (1991), 17-20.
5. A. Apostolico, F.P. Preparata, Optimal off-line detection of repetitions in a string, *Theoretical Computer Science* 22, (1983), 297-315.

6. O. Berkman, C.S. Iliopoulos and K. Park, The subtree max gap problem with application to parallel string covering, *Information and Computation* 123, 1 (1995), 127-137.

7. D. Breslauer, An on-line string superprimitivity test, *Information Processing Letters* 44 (1992), 345-347.

8. D. Breslauer, Testing string superprimitivity in parallel, *Information Processing Letters* 49, 5 (1994), 235-241.

9. T. Crawford, C.S. Iliopoulos and R. Raman, String Matching Techniques for Musical Similarity and Melodic Recognition, *Computing in Musicology*, 11 (1998), 73-100.

10. M. Crochemore, An optimal algorithm for computing the repetitions in a word, *Information Processing Letters* 12, 5 (1981), 244-250.

11. M. Crochemore, C.S. Iliopoulos and M. Korda, Two-dimensional Prefix String Matching and Covering on Square Matrices, *Algorithmica* 20 (1998), 353-373.

12. M. Crochemore, C.S. Iliopoulos and H. Yu, Algorithms for computing evolutionary chains in molecular and musical sequences, *Proc. 9th Australasian Workshop on Combinatorial Algorithms* (1998), 172-185.

13. C.S. Iliopoulos and M. Korda, Optimal parallel superprimitivity testing on square arrays, *Parallel Processing Letters* 6, 3 (1996), 299-308.

14. C.S. Iliopoulos, D.W.G. Moore and K. Park, Covering a string, *Algorithmica* 16 (1996), 288-297.

15. C.S. Iliopoulos and L. Mouchard, An $O(n \log n)$ algorithm for computing all maximal quasiperiodicities in strings, to appear in the *Proceedings of CATS'99: "Computing: Australasian Theory Symposium"*, Auckland, New Zealand, Lecture Notes in Computer Science (1999), 262–272.

16. C.S. Iliopoulos and K. Park, A work-time optimal algorithm for computing all string covers, *Theoretical Computer Science* 164 (1996), 299-310.

17. C.S. Iliopoulos and K. Park, An optimal $O(\log \log n)$-time algorithm for parallel superprimitivity testing, *J. Korea Inform. Sci. Soc.* 21 (1994), 1400-1404.

18. C.S. Iliopoulos and W.F. Smyth, On-line algorithms for k-covering, *Proc. 9th Australasian Workshop on Combinatorial Algorithms* (1998), 97-106.

19. D.E. Knuth, J.H. Morris and V.R. Pratt, Fast pattern matching in strings, *SIAM Journal on Computing* 6, 1 (1977), 323-350.

20. G.M. Landau, E.W. Myers and J.P. Schmidt, Incremental string comparison, *SIAM Journal on Computing* 27, 2 (1998), 557-582.

21. G.M. Landau and J.P. Schmidt, An algorithm for approximate tandem repeats, *Proc. 4th Symp. Combinatorial Pattern Matching*, Lecture Notes in Computer Science 648 (1993), 120-133.

22. Y. Li and W.F. Smyth, An optimal on-line algorithm to compute all the covers of a string, preprint.

23. D. Maier, The complexity of some problems on subsequences and supersequences, *J. Assoc. Comput. Mach.* 25 (1978), 322-336.

24. M.G. Main and R.J. Lorentz, An algorithm for finding all repetitions in a string, *Journal of Algorithms* 5 (1984), 422-432.

25. M. Middendorf, More on the complexity of common superstring and supersequence problems, *Theoretical Computer Science* 125, 2 (1994), 205-228.

26. D. Moore and W.F. Smyth, A correction to "An optimal algorithm to compute all the covers of a string.", *Information Processing Letters* 54, 2 (1995), 101-103.

27. K.J. Räihä and E. Ukkonen, The shortest common supersequence problem over binary alphabet is NP-complete. *Theoretical Computer Science* 16 (1981), 187-198.

28. J.P. Schmidt, All highest scoring paths in weighted grid graphs and its application to finding all approximate repeats in strings, *SIAM Journal on Computing* 27, 4 (1998), 972-992.

29. P.H. Sellers, Pattern recognition genetic sequences by mismatch density, *Bulletin of Mathematical Biology* 46, 4 (1984), 501-514.

30. L. Wang and T. Jiang, On the complexity of multiple sequence alignment, *J. Comp. Biol.* 1 (1994), 337-348.

Finding Maximal Pairs with Bounded Gap

Gerth Stølting Brodal[1]*, Rune B. Lyngsø[1] *, Christian N.S. Pedersen[1] *, and
Jens Stoye[2]

[1] Basic Research in Computer Science (BRICS), Centre of the Danish National
Research Foundation, Department of Computer Science,
University of Aarhus, Ny Munkegade, 8000 Århus C, Denmark.
{gerth,rlyngsoe,cstorm}@brics.dk
[2] Deutsches Krebsforschungszentrum (DKFZ), Theoretische Bioinformatik,
Im Neuenheimer Feld 280, 69120 Heidelberg, Germany.
j.stoye@dkfz-heidelberg.de

Abstract. A pair in a string is the occurrence of the same substring
twice. A pair is maximal if the two occurrences of the substring cannot
be extended to the left and right without making them different. The gap
of a pair is the number of characters between the two occurrences of the
substring. In this paper we present methods for finding all maximal pairs
under various constraints on the gap. In a string of length n we can find
all maximal pairs with gap in an upper and lower bounded interval in
time $O(n \log n + z)$ where z is the number of reported pairs. If the upper
bound is removed the time reduces to $O(n + z)$. Since a tandem repeat is
a pair where the gap is zero, our methods can be seen as a generalization
of finding tandem repeats. The running time of our methods equals the
running time of well known methods for finding tandem repeats.

1 Introduction

A pair in a string is the occurrence of the same substring twice. A pair is left-
maximal (right-maximal) if the characters to the immediate left (right) of the
two occurrences of the substring are different. A pair is maximal if it is both
left- and right-maximal. The gap of a pair is the number of characters between
the two occurrences of the substring. For example, the two occurrences of the
substring *ma* in the string *maximal* form a maximal pair of *ma* with gap two.

Gusfield [9, Sect. 7.12.3] describes how to report all maximal pairs in a string
using the suffix tree of the string in time $O(n + z)$ and space $O(n)$, where n
is the length of the string and z is the number of reported pairs. Since there
is no restriction on the gap of the maximal pairs reported by this algorithm,
many of them probably describe occurrences of substrings that are overlapping
or far apart in the string. In many applications in computational biology this
is unfortunate, so several papers address the problem of finding occurrences of
similar substrings not too far apart [13,17,23].

---

* Supported by the ESPRIT Long Term Research Programme of the EU under project
number 20244 (ALCOM-IT).

M. Crochemore, M. Paterson (Eds.): CPM'99, LNCS 1645, pp. 134–149, 1999.

In this paper we will describe how to find all maximal pairs in a string with gap in an upper and lower bounded interval in time $O(n \log n + z)$ and space $O(n)$. The interval of allowed gaps can be chosen such that we report a maximal pair only if the gap is between constants c_1 and c_2, but more generally, it can be chosen such that we report a maximal pair of α only if the gap is between $g_1(|\alpha|)$ and $g_2(|\alpha|)$, where g_1 and g_2 are functions that can be computed in constant time. This, for example, makes it possible to find all maximal pairs with gap between zero and some fraction of the length of the repeated substring. If we remove the upper bound on allowed gaps, and only require the gap of a reported pair of α to be at least $g_1(|\alpha|)$, then the running time reduces to $O(n + z)$. The methods we present all use the suffix tree as the fundamental data structure combined with efficient methods for merging search trees and heap-ordered trees.

The problem of finding occurrences of repeated substrings in a string is well studied. Most of the work has been concerned with efficient methods for finding occurrences of contiguously repeated substrings. An occurrence of a substring of the form $\alpha\alpha$ is called an occurrence of a square or a tandem repeat. Most well-known methods for finding the occurrences of all tandem repeats in a string require time $O(n \log n + z)$, where n is the length of the string and z is the number of reported occurrences of tandem repeats [5,2,18,15,24]. Work has also been done on just detecting whether or not a string contains a tandem repeat [19,6]. Recently, extending on the idea presented in [6], two methods have been presented that find a compact representation of all tandem repeats in a string in time $O(n)$ [14,10]. Other papers consider the problem of finding occurrences of contiguous repeats of substrings that are within some Hamming- or edit-distance of each other [16].

In biological sequence analysis searching for tandem repeats is used to reveal structural and functional information [9, pp. 139–142], but searching for exact tandem repeats can be too restrictive because of sequencing and other experimental errors. By searching for maximal pairs with small gaps (maybe depending on the length of the substring) it could be possible to compensate for these errors. On the other hand, finding maximal pairs with a gap within an interval can be seen as a generalization of finding occurrences of tandem repeats. Stoye and Gusfield [24] say that an occurrence of the tandem repeat $\alpha\alpha$ is a branching occurrence of the tandem repeat $\alpha\alpha$ if and only if the characters to the immediate right of the two occurrences of α are different, and they explain how to deduce the occurrence of all tandem repeats in a string from the occurrences of branching tandem repeats in time proportional to the number of tandem repeats. Since a branching occurrence of a tandem repeat is just a right-maximal pair with gap zero, the methods presented in this paper can be used to find all tandem repeats in time $O(n \log n + z)$. This matches the time bounds of previous published methods for this problem [5,2,18,15,24].

The rest of this paper is organized as follows. In Sect. 2 we define pairs and suffix trees and describe how in general to find pairs using the suffix tree. In Sect. 3 we present facts about efficient merging of search trees, and use them to formulate methods for finding all maximal pairs in a string with gap in an upper and lower bounded interval. In Sect. 4 we briefly discuss how to find all maximal pairs in a string with gap in a lower bounded interval. Finally, in Sect. 5 we summarize our work and discuss open problems.

2   Preliminaries

Throughout this paper S will denote a string of length n over a finite alphabet Σ. We will use $S[i]$, for $i = 1, 2, \ldots, n$, to denote the ith character of S, and use $S[i..j]$ as notation for the substring $S[i]S[i+1]\cdots S[j]$ of S. To be able to refer to the characters to the left and right of every character in S without worrying about the first and last character, we define $S[0]$ and $S[n+1]$ to be two distinct characters not appearing anywhere else in S.

In order to formulate methods for finding repetitive structures in S, we need a proper definition of such structures. An obvious definition is to find all pairs of identical substrings in S. This, however, leads to a lot of redundant output, e.g. in the string that consists of n identical characters there are $\Theta(n^3)$ such pairs. To limit the redundancy without sacrificing any meaningful structures Gusfield [9] defines maximal pairs.

Definition 1 (Pair). *We say that $(i, j, |\alpha|)$ is a pair of α in S formed by i and j if and only if $1 \leq i < j \leq n - |\alpha| + 1$ and $\alpha = S[i..i+|\alpha|-1] = S[j..j+|\alpha|-1]$. The pair is left-maximal (right-maximal) if the characters to the immediate left (right) of two occurrences of α are different, i.e. left-maximal if $S[i-1] \neq S[j-1]$ and right-maximal if $S[i+|\alpha|] \neq S[j+|\alpha|]$. The pair is maximal if it is right- and left-maximal. The gap of a pair $(i, j, |\alpha|)$ is the number of characters $j - i - |\alpha|$ between the two occurrences of α in S.*

It follows from the definition that a string of length n in the worst case contains $\Theta(n^2)$ right-maximal pairs. The string a^n contains the worst case number of right-maximal pairs but only $\Theta(n)$ maximal pairs. The string $(aab)^{n/3}$ however contains $\Theta(n^2)$ maximal pairs. This shows that the worst case number of maximal pairs and right-maximal pairs in a string are asymptotically equal.

Figure 1 illustrates the occurrence of a pair. In some applications it might be interesting only to find pairs that obey certain restrictions on the gap, e.g. to filter out pairs of substrings that are overlapping or far apart and thus to reduce the number of pairs to report. Using the "smaller-half trick", see Sect. 3.1, and Lemma 3 it is easy to prove that a string of length n in the worst case contains $\Theta(n \log n)$ right-maximal pairs with gap in an interval of constant size.

In this paper we present methods for finding all right-maximal and maximal pairs $(i, j, |\alpha|)$ in S with gap in a bounded interval. These methods all use the suffix tree of S as the fundamental data structure. We briefly review the suffix tree and refer to [9] for a more comprehensive treatment.

Definition 2 (Suffix tree). *The suffix tree $T(S)$ of the string S is the compressed trie of all suffixes of S. Each leaf in $T(S)$ represents a suffix $S[i..n]$ of S and is annotated with the index i. We refer to the set of indices stored at the*

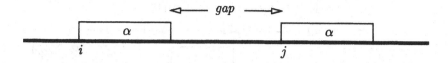

Fig. 1. An occurrence of a pair $(i, j, |\alpha|)$ with gap $j - i - |\alpha|$.

leaves in the subtree rooted at node v as the leaf-list of v and denote it $LL(v)$. Each edge in $T(S)$ is labelled with a nonempty substring of S such that the path from the root to the leaf annotated with index i spells the suffix $S[i..n]$. We refer to the substring of S spelled by the path from the root to node v as the path-label of v and denote it $L(v)$.

The suffix tree $T(S)$ can be constructed in time $O(n)$ [26,20,25,7]. It follows from the definition that all internal nodes in $T(S)$ have out-degree between two and $|\Sigma|$. We can turn the suffix tree $T(S)$ into the binary suffix tree $T_B(S)$ by replacing every node v in $T(S)$ with out-degree $d > 2$ by a binary tree with $d-1$ internal nodes and $d-2$ internal edges in which the d leaves are the d children of node v. We label each new internal edge with the empty string such that the $d-1$ nodes replacing node v all have the same path-label as node v has in $T(S)$. Since $T(S)$ has n leaves, constructing the binary suffix tree $T_B(S)$ requires adding at most $n-2$ new nodes. Since each new node can be added in constant time, the binary suffix tree $T_B(S)$ can be constructed in time $O(n)$.

The binary suffix tree is an essential component of our methods. Definition 2 implies that there is a node v in $T(S)$ with path-label α if and only if α is the longest common prefix of $S[i..n]$ and $S[j..n]$ for some $1 \le i < j \le n$. In other words, there is a node v with path-label α if and only if $(i, j, |\alpha|)$ is a right-maximal pair in S. Since $S[i + |\alpha|] \ne S[j + |\alpha|]$ the indices i and j cannot be elements in the leaf-list of the same child of v. Using the binary suffix tree $T_B(S)$ we can thus formulate the following lemma.

Lemma 3. *There is a right-maximal pair $(i, j, |\alpha|)$ in S if and only if there is a node v in the binary suffix tree $T_B(S)$ with path-label α and distinct children w_1 and w_2 where $i \in LL(w_1)$ and $j \in LL(w_2)$.*

Lemma 3 gives an approach to find all right-maximal pairs in S; for every internal node v in the binary suffix tree $T_B(S)$ consider the leaf-lists at its two children w_1 and w_2, and for every element (i, j) in $LL(w_1) \times LL(w_2)$ report a right-maximal pair $(i, j, |\alpha|)$ if $i < j$ and $(j, i, |\alpha|)$ if $j < i$. To find all maximal pairs in S the problem remains to filter out all right-maximal pairs that are not left-maximal.

3   Pairs with Upper and Lower Bounded Gap

We want to find all maximal pairs $(i, j, |\alpha|)$ in S with gap between $g_1(|\alpha|)$ and $g_2(|\alpha|)$, i.e. $g_1(|\alpha|) \le j - i - |\alpha| \le g_2(|\alpha|)$, where g_1 and g_2 are functions

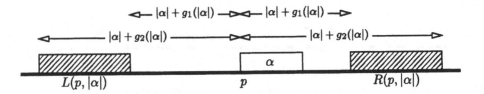

Fig. 2. If $(p, q, |\alpha|)$ (respectively $(q, p, |\alpha|)$) is a pair with gap between $g_1(|\alpha|)$ and $g_2(|\alpha|)$, then one occurrence of α is at position p and the other occurrence is at a position q in the interval $R(p, |\alpha|)$ (respectively $L(p, |\alpha|)$) of positions.

that can be computed in constant time. An obvious approach is to generate all maximal pairs in S and only report those with gap between $g_1(|\alpha|)$ and $g_2(|\alpha|)$, but as shown above there might be asymptotically fewer maximal pairs in S with gap between $g_1(|\alpha|)$ and $g_2(|\alpha|)$ than maximal pairs in S in total. We therefore want to find all maximal pairs $(i, j, |\alpha|)$ in S with gap between $g_1(|\alpha|)$ and $g_2(|\alpha|)$ *without* generating and considering all maximal pairs in S. A step towards finding all maximal pairs with gap between $g_1(|\alpha|)$ and $g_2(|\alpha|)$ is to find all right-maximal pairs with gap between $g_1(|\alpha|)$ and $g_2(|\alpha|)$.

Figure 2 shows that if one occurrence of α in a pair with gap between $g_1(|\alpha|)$ and $g_2(|\alpha|)$ is at position p, then the other occurrence of α must be at a position q in one of the two intervals $L(p, |\alpha|) = [p - |\alpha| - g_2(|\alpha|) \mathinner{\ldotp\ldotp} p - |\alpha| - g_1(|\alpha|)]$ or $R(p, |\alpha|) = [p + |\alpha| + g_1(|\alpha|) \mathinner{\ldotp\ldotp} p + |\alpha| + g_2(|\alpha|)]$. Together with Lemma 3 this gives an approach to find all right-maximal pairs in S with gap between $g_1(|\alpha|)$ and $g_2(|\alpha|)$; from every internal node v in the binary suffix tree $T_B(S)$ with path-label α and children w_1 and w_2, we report for every p in $LL(w_1)$ the pairs $(p, q, |\alpha|)$ for all q in $LL(w_2) \cap R(p, |\alpha|)$ and the pairs $(q, p, |\alpha|)$ for all q in $LL(w_2) \cap L(p, |\alpha|)$.

To report right-maximal pairs efficiently using this procedure, we must be able to find for every p in $LL(w_1)$, without looking at all the elements in $LL(w_2)$, the proper elements q in $LL(w_2)$ to report it against. It turns out that search trees make this possible. In this paper we use AVL trees, but other types of search trees, e.g. (a, b)-trees [11] or red-black trees [8], can also be used as long as they obey Lemmas 4 and 5 stated below. Before we can formulate algorithms we review some useful facts about AVL trees.

3.1   Data Structures

An AVL tree T is a balanced search tree that stores an ordered set of elements. AVL trees were introduced in [1], but are explained in almost every textbook on data structures. We say that an element e is in T, or $e \in T$, if it is stored at a node in T. For short notation we use e to denote both the element and the node at which it is stored in T. We can keep links between the nodes in T in such a way that we in constant time from the node e can find the nodes $next(e)$ and $prev(e)$ storing the next and previous element in increasing order. We use $|T|$ to denote the size of T, i.e. the number of elements stored in T.

Efficient merging of two AVL trees is essential to our methods. Hwang and Lin [12] show how to merge two sorted lists using the optimal number of comparisons. Brown and Tarjan [4] show how to implement merging of two height-balanced search trees, e.g. AVL trees, in time proportional to the optimal number of comparisons. Their result is summarized in Lemma 4, which immediately implies Lemma 5.

Lemma 4. *Two AVL trees of size at most n and m can be merged in time $O(\log \binom{n+m}{n})$.*

Lemma 5. *Given a sorted list of elements $e_1, e_2, \ldots, e_n$ and an AVL tree T of size at most m, $m \geq n$, we can find $q_i = \min\{x \in T \mid x \geq e_i\}$ for all $i = 1, 2, \ldots, n$ in time $O(\log \binom{n+m}{n})$.*

Proof. Construct the AVL tree of the elements $e_1, e_2, \ldots, e_n$ in time $O(n)$. Merge this AVL tree with T according to Lemma 4, except that whenever the merge-algorithm would insert one of the elements $e_1, e_2, \ldots, e_n$ into T, we change the merge-algorithm to report the neighbor of the element in T instead. This modification does not increase the running time.   □

The "smaller-half trick" is essential to several methods for finding tandem repeats [5,2,24]. It says that the sum over all nodes v in an arbitrary binary tree of size n of terms that are $O(n_1)$, where $n_1 \leq n_2$ are the numbers of leaves in the subtrees rooted at the two children of v, is $O(n \log n)$. Our methods rely on a stronger version of the "smaller-half trick" hinted at in [21, Ex. 35] and used in [22, Chap. 5, p. 84]; we summarize it in the following lemma.

Lemma 6. *Let T be an arbitrary binary tree with n leaves. The sum over all internal nodes v in T of terms that are $O(\log \binom{n_1+n_2}{n_1})$, where n_1 and n_2 are the numbers of leaves in the subtrees rooted at the two children of v, is $O(n \log n)$.*

Proof. As the terms are $O(\log \binom{n_1+n_2}{n_1})$ we can find constants, a and b, such that the terms are upper bounded by $a + b \log \binom{n_1+n_2}{n_1}$. We will by induction in the number of leaves of the binary tree prove that the sum is upper bounded by $(2n-1)a + b \log n!$. As $\log n! = O(n \log n)$ the lemma follows.

If T is a leaf then the upper bound holds vacuously. Now assume inductively that the upper bound holds for all trees with at most $n-1$ leaves. Let T be a tree with n leaves where the number of leaves in the subtrees rooted at the two children of the root are $n_1 < n$ and $n_2 < n$. According to the induction hypothesis the sum over all nodes in these two subtrees, i.e. the sum over all nodes of T except the root, is bounded by $(2n_1 - 1)a + b \log n_1! + (2n_2 - 1)a + b \log n_2!$ and thus the entire sum is bounded by

$$(2n_1 - 1)a + b\log n_1! + (2n_2 - 1)a + b\log n_2! + a + b\log \binom{n_1 + n_2}{n_1}$$

$$= (2(n_1 + n_2) - 1)a + b\log n_1! + b\log n_2! +$$
$$b\log(n_1 + n_2)! - b\log n_1! - b\log n_2!$$
$$= (2n - 1)a + b\log n!$$

which proves the lemma. □

3.2   Algorithms

We first describe an algorithm that finds all right-maximal pairs in S with bounded gap using AVL trees to keep track of the elements in the leaf-lists during a traversal of the binary suffix tree $T_B(S)$. We then extend it to find all maximal pairs in S with bounded gap using an additional AVL tree to filter out efficiently all right-maximal pairs that are not left-maximal. Both algorithms run in time $O(n\log n + z)$ and space $O(n)$, where z is the number of reported pairs. In the following we assume, unless stated otherwise, that v is a node in the binary suffix tree $T_B(S)$ with path-label α and children w_1 and w_2 named such that $|LL(w_1)| \leq |LL(w_2)|$. We say that w_1 is the small child of v and that w_2 is the big child of v.

Right-Maximal Pairs with Upper and Lower Bounded Gap   To find all right-maximal pairs in S with gap between $g_1(|\alpha|)$ and $g_2(|\alpha|)$ we consider every node v in the binary suffix tree $T_B(S)$ in a bottom-up fashion, e.g. during a depth-first traversal. At every node v we use AVL trees storing the leaf-lists $LL(w_1)$ and $LL(w_2)$ at its two children to report the proper right-maximal pairs of its path-label α. The details are given in Algorithm 1 and explained below.

At every node v in $T_B(S)$ we construct an AVL tree, the leaf-list tree T, that stores the elements in $LL(v)$. If v is a leaf then we construct T directly in Step 1. If v is an internal node then $LL(v)$ is the union of the disjoint leaf-lists $LL(w_1)$ and $LL(w_2)$ which by assumption are stored in the already constructed T_1 and T_2, so we construct T by merging T_1 and T_2, $|T_1| \leq |T_2|$, using Lemma 4. Before constructing T in Step 2c we use T_1 and T_2 to report right-maximal pairs from node v by reporting every p in $LL(w_1)$ against every q in $LL(w_2) \cap L(p, |\alpha|)$ and $LL(w_2) \cap R(p, |\alpha|)$. This is done in two steps. In Step 2a we find for every p in $LL(w_1)$ the minimum element $q_r(p)$ in $LL(w_2) \cap R(p, |\alpha|)$ and the minimum element $q_l(p)$ in $LL(w_2) \cap L(p, |\alpha|)$ by searching in T_2 using Lemma 5. In Step 2b we report pairs $(p, q, |\alpha|)$ and $(q, p, |\alpha|)$ for every p in $LL(w_1)$ and increasing q's in $LL(w_2)$ starting with $q_r(p)$ and $q_l(p)$ respectively, until the gap violates the upper or lower bound.

To argue that Algorithm 1 finds all right-maximal pairs with gap between $g_1(|\alpha|)$ and $g_2(|\alpha|)$ it is enough to argue that we for every p in $LL(w_1)$ report all right-maximal pairs $(p, q, |\alpha|)$ and $(q, p, |\alpha|)$ with gap between $g_1(|\alpha|)$ and $g_2(|\alpha|)$. The rest follows because we at every node v in $T_B(S)$ consider every p in $LL(w_1)$. Consider the call $\mathsf{Report}(q_r(p), p + |\alpha| + g_2(|\alpha|))$ in Step 2b.

Algorithm 1 Find all right-maximal pairs in string S with bounded gap.

---

1. *Initializing:* Build the binary suffix tree $T_B(S)$ and create at each leaf an AVL tree of size one that stores the index at the leaf.

2. *Reporting and merging:* When the AVL trees T_1 and T_2, $|T_1| \leq |T_2|$, at the two children w_1 and w_2 of node v with path-label α are available, we do the following:

   (a) Let $\{p_1, p_2, \ldots, p_s\}$ be the elements in T_1 in sorted order. For each element p in T_1 we find

   $$q_r(p) = \min\{x \in T_2 \mid x \geq p + |\alpha| + g_1(|\alpha|)\}$$
   $$q_l(p) = \min\{x \in T_2 \mid x \geq p - |\alpha| - g_2(|\alpha|)\}$$

   by searching in T_2 with the sorted lists $\{p_i + |\alpha| + g_1(|\alpha|) \mid i = 1, 2, \ldots, s\}$ and $\{p_i - |\alpha| - g_2(|\alpha|) \mid i = 1, 2, \ldots, s\}$ using Lemma 5.

   (b) For each element p in T_1 we do $\mathsf{Report}(q_r(p), p + |\alpha| + g_2(|\alpha|))$ and $\mathsf{Report}(q_l(p), p - |\alpha| - g_1(|\alpha|))$ where Report is the following procedure.

   def Report(*from*, *to*) :
       $q = from$
       while $q \leq to$:
           report pair $(p, q, |\alpha|)$ if $p < q$, and $(q, p, |\alpha|)$ otherwise
           $q = next(q)$

   (c) Build the leaf-list tree T at node v by merging T_1 and T_2 using Lemma 4.

From the implementation of **Report** follows that this call reports p against every q in $LL(w_2) \cap [q_r(p) .. p + |\alpha| + g_2(|\alpha|)]$. By construction of $q_r(p)$ and definition of $R(p, |\alpha|)$ follows that $LL(w_2) \cap [q_r(p) .. p + |\alpha| + g_2(|\alpha|)]$ is equal to $LL(w_2) \cap R(p, |\alpha|)$, so the call reports all pairs $(p, q, |\alpha|)$ with gap between $g_1(|\alpha|)$ and $g_2(|\alpha|)$. Similarly we can argue that the call $\mathsf{Report}(q_l(p), p - |\alpha| - g_1(|\alpha|))$ reports all pairs $(q, p, |\alpha|)$ with gap between $g_1(|\alpha|)$ and $g_2(|\alpha|)$.

Now consider the running time of Algorithm 1. Building the binary suffix tree $T_B(S)$ and creating an AVL tree of size one at each leaf in Step 1 takes time $O(n)$. At every internal node in $T_B(S)$ we do Step 2. Since $|T_1| \leq |T_2|$ searching in Step 2a and merging in Step 2c takes time $O(\log \binom{|T_1|+|T_2|}{|T_1|})$ by Lemmas 5 and 4 respectively. Reporting of pairs in Step 2b takes time proportional to $|T_1|$, because we consider every p in $LL(w_1)$, plus the number of reported pairs. Summing this over all nodes gives by Lemma 6 that the total running time is $O(n \log n + z)$, where z is the number of reported pairs. Since constructing and keeping $T_B(S)$ requires space $O(n)$, and since no element at any time is in more than one leaf-list tree, Algorithm 1 requires space $O(n)$.

Theorem 7. *Algorithm 1 finds all right-maximal pairs $(i, j, |\alpha|)$ in a string S with gap between $g_1(|\alpha|)$ and $g_2(|\alpha|)$ in space $O(n)$ and time $O(n \log n + z)$, where z is the number of reported pairs and n is the length of S.*

Maximal Pairs with Upper and Lower Bounded Gap We now turn towards finding all maximal pairs in S with gap between $g_1(|\alpha|)$ and $g_2(|\alpha|)$. Our approach to find all maximal pairs in S with gap between $g_1(|\alpha|)$ and $g_2(|\alpha|)$ is to extend Algorithm 1 to filter out all right-maximal pairs that are not left-maximal. A simple solution is to extend the procedure Report to check if $S[p-1] \neq S[q-1]$ before reporting the pair $(p, q, |\alpha|)$ or $(q, p, |\alpha|)$ in Step 2b. This solution takes time proportional to the number of inspected right-maximal pairs, and not time proportional to the number of reported maximal pairs. Even though the maximum number of right-maximal pairs and maximal pairs in strings of a given length are asymptotically equal, many strings contain significantly fewer maximal pairs than right-maximal pairs. We therefore want to filter out all right-maximal pairs that are not left-maximal *without* inspecting all right-maximal pairs. In the remainder of this section we describe one way to do this.

Consider the reporting step in Algorithm 1 and assume that we are about to report from a node v with children w_1 and w_2. The leaf-list trees T_1 and T_2, $|T_1| \leq |T_2|$, are available and they make it possible to access the elements in $LL(w_1) = \{p_1, p_2, \ldots, p_s\}$ and $LL(w_2) = \{q_1, q_2, \ldots, q_t\}$ in sorted order. We divide the sorted leaf-list $LL(w_2)$ into blocks of contiguous elements such that the elements q_{i-1} and q_i are in the same block if and only if $S[q_{i-1}-1] = S[q_i-1]$. We say that we divide the sorted leaf-list into blocks of elements with equal left-characters. To filter out all right-maximal pairs that are not left-maximal we must avoid to report p in $LL(w_1)$ against any element q in $LL(w_2)$ in a block of elements with left-character $S[p-1]$. This gives the overall idea of the extended algorithm; we extend the reporting step in Algorithm 1 such that whenever we are about to report p in $LL(w_1)$ against q in $LL(w_2)$ where $S[p-1] = S[q-1]$ we skip all elements in the current block containing q and continue reporting p against the first element q' in the following block, which by the definition of blocks satisfies that $S[p-1] \neq S[q'-1]$.

To implement this extended reporting step efficiently we must be able to skip all elements in a block without inspecting each of them. We achieve this by constructing an additional AVL tree, the block-start tree, that keeps track of the blocks in the leaf-list. At each node v during the traversal of $T_B(S)$ we thus construct two AVL trees; the leaf-list tree T that stores the elements in $LL(v)$, and the block-start tree B that keeps track of the blocks in the sorted leaf-list by storing all the elements in $LL(v)$ that start a block. We keep links from the block-start tree to the leaf-list tree such that we in constant time can go from an element in the block-start tree to the corresponding element in the leaf-list tree. Figure 3 illustrates the leaf-list tree, the block-start tree and the links between them. Before we present the extended algorithm and explain how to use the block-start tree to efficiently skip all elements in a block, we first describe how to construct the leaf-list tree T and block-start tree B at node v from the leaf-list trees, T_1 and T_2, and block-start trees, B_1 and B_2, at its two children w_1 and w_2.

Since $LL(v)$ is the union of the disjoint leaf-lists $LL(w_1)$ and $LL(w_2)$ stored in T_1 and T_2 respectively, we can construct the leaf-list tree T by merging T_1 and T_2 using Lemma 4. It is more involved to construct the block-start tree B.

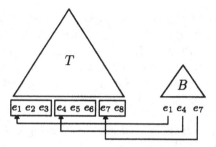

Fig. 3. The data structure constructed at each node v in $T_B(S)$. The leaf-list tree T stores all elements in $LL(v)$. The block-start tree B stores all elements in $LL(v)$ that start a block in the sorted leaf-list. We keep links from the elements in the block-start tree to the corresponding elements in the leaf-list tree.

The reason is that an element p_i that starts a block in $LL(w_1)$ or an element q_j that starts a block in $LL(w_2)$ does not necessarily start a block in $LL(v)$ and vice versa, so we cannot construct B by merging B_1 and B_2. Let $\{e_1, e_2, \ldots, e_{s+t}\}$ be the elements in $LL(v)$ in sorted order. By definition the block-start tree B contains all elements e_k in $LL(v)$ where $S[e_{k-1} - 1] \neq S[e_k - 1]$. We construct B by modifying B_2. We choose to modify B_2, and not B_1, because $|LL(w_1)| \leq |LL(w_2)|$, which by the "smaller-half trick" allows us to consider all elements in $LL(w_1)$ without spending too much time in total. To modify B_2 to become B we must identify all the elements that are in B but not in B_2 and vice versa.

Lemma 8. *If e_k is in B but not in B_2 then $e_k \in LL(w_1)$ or $e_{k-1} \in LL(w_1)$.*

Proof. Assume that e_k is in B and that e_k and e_{k-1} both are in $LL(w_2)$. In $LL(w_2)$ the elements e_k and e_{k-1} are neighboring elements q_j and q_{j-1}. Since e_k starts a block in $LL(v)$ then $S[q_j - 1] = S[e_k - 1] \neq S[e_{k-1} - 1] = S[q_{j-1} - 1]$. This shows that $q_j = e_k$ is in B_2 and the lemma follows. $\square$

Let NEW be the set of elements e_k in B where e_k or e_{k-1} are in $LL(w_1)$. It follows from Lemma 8 that this set contains at least all elements in B that are not in B_2. It is easy to see that we can construct NEW in sorted order while merging T_1 and T_2; whenever an element e_k from T_1, i.e. $LL(w_1)$, is placed in T, i.e. $LL(v)$, we include it, and/or the next element e_{k+1} placed in T, in NEW if they start a block in $LL(v)$.

If we insert the elements in NEW we are halfway done modifying B_2 to become B. We still need to identify and remove the elements that should be removed from B_2, that is, the elements that are in B_2 but not in B.

Lemma 9. *An element q_j in B_2 is not in B if and only if the largest element e_k in NEW smaller than q_j in B_2 has the same left-character as q_j.*

Proof. If q_j is in B_2 but does not start a block in $LL(v)$, then it must be in a block by some element e_k with the same left-character as q_j. This block cannot contain q_{j-1} because q_j being in B_2 implies that $S[q_j - 1] \neq S[q_{j-1} - 1]$. We thus have the ordering $q_{j-1} < e_k < q_j$. This implies that e_k is the largest element in NEW smaller than q_j. If e_k is the largest element in NEW smaller than q_j, then no block starts in $LL(v)$ between e_k and q_j, i.e. all elements e in $LL(v)$ where $e_k < e < q_j$ satisfy that $S[e-1] = S[e_k-1]$, so if $S[e_k-1] = S[q_j-1]$ then q_j does not start a block in $LL(v)$.   □

By searching in B_2 with the sorted list NEW using Lemma 5 it is straightforward to find all pairs of elements (e_k, q_j), where e_k is the largest element in NEW smaller than q_j in B_2. If the left-characters of e_k and q_j in such a pair are equal, i.e. $S[e_k - 1] = S[q_j - 1]$, then by Lemma 9 the element q_j is not in B and must therefore be removed from B_2. It follows from the proof of Lemma 9 that if this is the case then $q_{j-1} < e_k < q_j$, so we can, without destroying the order among the nodes in B_2, remove q_j from B_2 and insert e_k instead, simply by replacing the element q_j with the element e_k at the node storing q_j in B_2.

We can now summarize the three steps it takes to modify B_2 to become B. In Step 1 we construct the sorted set NEW that contains all elements in B that are not in B_2. This is done while merging T_1 and T_2 using Lemma 4. In Step 2 we remove the elements from B_2 that are not in B. The elements in B_2 being removed and the elements from NEW replacing them are identified using Lemmas 5 and 9. In Step 3 we merge the remaining elements in NEW into the modified B_2 using Lemma 4. Adding links from the new elements in B to the corresponding elements in T can be done while replacing and merging in Steps 2 and 3. Since $|NEW| \leq 2|T_1|$ and $|B_2| \leq |T_2|$, the time it takes to construct B is dominated by the the time it takes merge a sorted list of size $2|T_1|$ into an AVL tree of size $|T_2|$. By Lemma 4 this is within a constant factor of the time it takes to merge T_1 and T_2, so the time is takes to construct B is dominated by the time it takes to construct the leaf-list tree T.

Now that we know how to construct the leaf-list tree T and block-start tree B at node v from the leaf-list trees, T_1 and T_2, and block-start trees, B_1 and B_2, at its two children w_1 and w_2, we can proceed with the implementation of the extended reporting step. The details are shown in Algorithm 2. This algorithm is similar to Algorithm 1 except that we at every node v in $T_B(S)$ construct two AVL trees; the leaf-list tree T that stores the elements in $LL(v)$, and the block-start tree B that keeps track of the blocks in $LL(v)$ by storing the subset of elements that start a block. If v is a leaf, we construct T and B directly. If v is an internal node, we construct T by merging the leaf-list trees T_1 and T_2 at its two children w_1 and w_2, and we construct B by modifying the block-start tree B_2 as explained above.

Before constructing T and B we report all maximal pairs from node v with gap between $g_1(|\alpha|)$ and $g_2(|\alpha|)$ by reporting every p in $LL(w_1)$ against every q in $LL(w_2) \cap L(p, |\alpha|)$ and $LL(w_2) \cap R(p, |\alpha|)$ where $S[p-1] \neq S[q-1]$. This is done in two steps. In Step 2a we find for every p in $LL(w_1)$ the minimum elements $q_l(p)$ and $q_r(p)$, as well as the minimum elements $b_l(p)$ and $b_r(p)$ that start a block, in $LL(w_2) \cap L(p, |\alpha|)$ and $LL(w_2) \cap R(p, |\alpha|)$ respectively. This is done by searching in T_2 and B_2 using Lemma 5. In Step 2b we report pairs $(p, q, |\alpha|)$ and $(q, p, |\alpha|)$

Algorithm 2 Find all maximal pairs in string S with bounded gap.

1. *Initializing:* Build the binary suffix tree $T_B(S)$ and create at each leaf two AVL trees of size one, the leaf-list and the block-start tree, both storing the index at the leaf.

2. *Reporting and merging:* When the leaf-list trees T_1 and T_2, $|T_1| \leq |T_2|$, and the block-start trees B_1 and B_2 at the two children w_1 and w_2 of node v with path-label α are available, we do the following:

   (a) Let $\{p_1, p_2, \ldots, p_s\}$ be the elements in T_1 in sorted order. For each element p in T_1 we find

   $$q_r(p) = \min\{x \in T_2 \mid x \geq p + |\alpha| + g_1(|\alpha|)\}$$
   $$q_l(p) = \min\{x \in T_2 \mid x \geq p - |\alpha| - g_2(|\alpha|)\}$$
   $$b_r(p) = \min\{x \in B_2 \mid x \geq p + |\alpha| + g_1(|\alpha|)\}$$
   $$b_l(p) = \min\{x \in B_2 \mid x \geq p - |\alpha| - g_2(|\alpha|)\}$$

   by searching in T_2 and B_2 with the sorted lists $\{p_i + |\alpha| + g_1(|\alpha|) \mid i = 1, 2, \ldots, s\}$ and $\{p_i - |\alpha| - g_2(|\alpha|) \mid i = 1, 2, \ldots, s\}$ using Lemma 5.

   (b) For each element p in T_1 we do ReportMax$(q_r(p), b_r(p), p + |\alpha| + g_2(|\alpha|))$ and ReportMax$(q_l(p), b_l(p), p - |\alpha| - g_1(|\alpha|))$ where ReportMax is the following procedure.

   ```
   def ReportMax(from_T, from_B, to):
       q = from_T
       b = from_B
       while q ≤ to:
           if S[q − 1] ≠ S[p − 1]:
               report pair (p, q, |α|) if p < q, and (q, p, |α|) otherwise
               q = next(q)
           else:
               while b ≤ q:
                   b = next(b)
               q = b
   ```

   (c) Build the leaf-list tree T at node v by merging T_1 and T_2 using Lemma 4. Build the block-start tree B at node v by modifying B_2 as described in the text.

---

for every p in $LL(w_1)$ and increasing q's in $LL(w_2)$ starting with $q_r(p)$ and $q_l(p)$ respectively, until the gap violates the upper or lower bound. Whenever we are about to report p against q where $S[p-1] = S[q-1]$, we instead use the block-start tree B_2 to skip all elements in the block containing q and continue with reporting p against the first element in the following block.

To argue that Algorithm 2 finds all maximal pairs with gap between $g_1(|\alpha|)$ and $g_2(|\alpha|)$ it is enough to argue that we for every p in $LL(w_1)$ report all maximal pairs $(p, q, |\alpha|)$ and $(q, p, |\alpha|)$ with gap between $g_1(|\alpha|)$ and $g_2(|\alpha|)$. The rest follows because we at every node in $T_B(S)$ consider every p in $LL(w_1)$. Consider the call ReportMax($q_r(p), b_r(p), p + |\alpha| + g_2(|\alpha|)$) in Step 2b. From the implementation of ReportMax follows that unless we skip elements by increasing b then we consider every q in $LL(w_2) \cap R(p, |\alpha|)$. The test $S[q - 1] \neq S[p - 1]$ before reporting a pair ensures that we only report maximal pairs and whenever $S[q - 1] = S[p - 1]$ we increase b until $b = \min\{x \in B_2 \mid x > q\}$. This is, by construction of B_2 and $b_r(p)$, the element that starts the block following the block containing q, so all elements q', $q < q' < b$, we skip by setting q to b satisfy that $S[p - 1] = S[q - 1] = S[q' - 1]$. We thus conclude that ReportMax($q_r(p), b_r(p), p + |\alpha| + g_2(|\alpha|)$) reports p against exactly those q in $LL(w_2) \cap R(p, |\alpha|)$ where $S[p - 1] \neq S[q - 1]$, i.e. it reports all maximal pairs $(p, q, |\alpha|)$ at node v with gap between $g_1(|\alpha|)$ and $g_2(|\alpha|)$. Similarly, the call ReportMax($q_l(p), b_l(p), p - |\alpha| - g_1(|\alpha|)$) reports all maximal pairs $(q, p, |\alpha|)$ with gap between $g_1(|\alpha|)$ and $g_2(|\alpha|)$.

Now consider the running time of Algorithm 2. We first argue that the call ReportMax($q_r(p), b_r(p), p + |\alpha| + g_2(|\alpha|)$) takes constant time plus time proportional to the number of reported pairs $(p, q, |\alpha|)$. To do this all we have to show is that the time used to skip blocks, i.e. the number of times we increase b, is proportional to the number of reported pairs. By construction $b_r(p) \geq q_r(p)$, so the number of times we increase b is bounded by the number of blocks in $LL(w_2) \cap R(p, |\alpha|)$. Since neighboring blocks contain elements with different left-characters, we report p against an element from at least every second block in $LL(w_2) \cap R(p, |\alpha|)$. The number of times we increase b is thus proportional to the number of reported pairs. The call ReportMax($q_l(p), b_l(p), p - |\alpha| - g_1(|\alpha|)$) also takes constant time plus time proportional to the number of reported pairs $(q, p, |\alpha|)$. We thus have that Step 2b takes time proportional to $|T_1|$ plus the number of reported pairs. Everything else we do at node v, i.e. searching in T_2 and B_2 and constructing the leaf-list tree T and block-start tree B, takes time $O(\log \binom{|T_1| + |T_2|}{|T_1|})$. Summing this over all nodes gives by Lemma 6 that the total running time of the algorithm is $O(n \log n + z)$ where z is the number of reported pairs. Since constructing and keeping $T_B(S)$ requires space $O(n)$, and since no element at any time is in more than one leaf-list tree, and maybe one block-start tree, Algorithm 2 requires space $O(n)$.

Theorem 10. *Algorithm 2 finds all maximal pairs $(i, j, |\alpha|)$ in a string S with gap between $g_1(|\alpha|)$ and $g_2(|\alpha|)$ in space $O(n)$ and time $O(n \log n + z)$, where z is the number of reported pairs and n is the length of S.*

We observe that Algorithm 2 never uses the block-start tree B_1 at the small child w_1. This observation can be used to ensure that only one block-start tree exists during the execution of the algorithm. If we implement the traversal of $T_B(S)$ as a depth-first traversal in which we at each node v first recursively traverse the subtree rooted at the small child w_1, then we do not need to store the block-start tree returned by this recursive traversal while recursively traversing the subtree rooted at the big child w_2. This implies that only one block-start tree exists at all times during the recursive traversal of $T_B(S)$. The drawback is that we at each node v need to know in advance which child is the small child, but this

knowledge can be obtained in linear time by annotating each node with the size
of the subtree it roots.

4   Pairs with Lower Bounded Gap

If we relax the constraint on the gap and only want to find all maximal pairs
in S with gap at least $g(|\alpha|)$, where g is a function that can be computed
in constant time, then a straightforward solution is to use Algorithm 2 with
$g_1(|\alpha|) = g(|\alpha|)$ and $g_2(|\alpha|) = n$. This obviously finds all maximal pairs with
gap at least $g_1(|\alpha|) = g(|\alpha|)$ in time $O(n \log n + z)$. However, the missing upper
bound on the gap, i.e. the trivial upper bound $g_2(|\alpha|) = n$, makes it possible to
reduce the running time to $O(n + z)$ since reporting from each node during the
traversal of the binary suffix tree is simplified.

The reporting of pairs from node v with children w_1 and w_2 is simplified,
because the lack of an upper bound on the gap implies that we do not have
to search $LL(w_2)$ for the first element to report against the current element
in $LL(w_1)$. Instead we can start by reporting the current element in $LL(w_1)$
against the biggest (and smallest) element in $LL(w_2)$ and then continue reporting it against decreasing (and increasing) elements from $LL(w_2)$ until the gap
becomes smaller than $g(|\alpha|)$. Unfortunately this simplification alone does not re-
duce the asymptotic running time because inspecting every element in $LL(w_1)$
and keeping track of the leaf-lists in AVL trees alone requires time $\Theta(n \log n)$. To
reduce the running time we must thus avoid to inspect every element in $LL(w_1)$
and find another way to store the leaf-lists.

We achieve this by using a data structure based on heap-ordered trees to
store the leaf-lists during the traversal of the binary suffix tree. The key feature
of the data structure is that it allows us to merge two trees in amortized constant
time. The details of the data structure and the methods using it to find pairs
with gap at least $g(|\alpha|)$ is given in [3, Sect. 4]. Here we just summarize the result.

Theorem 11. *All maximal pairs $(i, j, |\alpha|)$ in a string S with gap at least $g(|\alpha|)$
can be found in space $O(n)$ and time $O(n + z)$, where z is the number of reported
pairs and n is the length of S.*

5   Conclusion

We have presented efficient and flexible methods to find all maximal pairs
$(i, j, |\alpha|)$ in a string under various constraints on the gap $j - i - |\alpha|$. If the gap
is required to be between $g_1(|\alpha|)$ and $g_2(|\alpha|)$, the running time is $O(n \log n + z)$
where n is the length of the string and z is the number of reported pairs. If the
gap is only required to be at least $g_1(|\alpha|)$, the running time reduces to $O(n + z)$.
In both cases we use space $O(n)$.

In some cases it might be interesting only to find maximal pairs $(i, j, |\alpha|)$
fulfilling additional requirements on $|\alpha|$, e.g. to filter out pairs of short substrings.
This is straightforward to do using our methods by only reporting from the nodes

in the binary suffix tree whose path-label α fulfills the requirements on $|\alpha|$. In other cases it might be of interest just to find the vocabulary of substrings that occur in maximal pairs. This is also straightforward to do using our methods by just reporting the path-label α of a node if we can report one or more maximal pairs from the node.

Instead of just looking for maximal pairs, it could be interesting to look for an array of occurrences of the same substring in which the gap between consecutive occurrences is bounded by some constants. This problem requires a suitable definition of a maximal array. One definition and approach is presented in [23]. Another definition inspired by the definition of a maximal pair could be to require that every pair of occurrences in the array is a maximal pair. This definition seems very restrictive. A more relaxed definition could be to only require that we cannot extend all the occurrences in the array to the left or to the right without destroying at least one pair of occurrences in the array.

Acknowledgments This work was initiated while Christian N. S. Pedersen and Jens Stoye were visiting Dan Gusfield at UC Davis. We would like to thank Dan Gusfield, as well as Rob Irwing, for listening to some preliminary results.

References

1. G. M. Adel'son-Vel'skii and Y. M. Landis. An algorithm for the organization of information. *Doklady Akademii Nauk SSSR*, 146:263–266, 1962. English translation in *Soviet Math. Dokl.*, 3:1259–1262.
2. A. Apostolico and F. P. Preparata. Optimal off-line detection of repetitions in a string. *Theoretical Computer Science*, 22:297–315, 1983.
3. G. S. Brodal, R. B. Lyngsø, C. N. S. Pedersen, and J. Stoye. Finding maximal pairs with bounded gap. Technical Report RS-99-12, BRICS, April 1999.
4. M. R. Brown and R. E. Tarjan. A fast merging algorithm. *Journal of the ACM*, 26(2):211–226, 1979.
5. M. Crochemore. An optimal algorithm for computing the repetitions in a word. *Information Processing Letters*, 12(5):244–250, 1981.
6. M. Crochemore. Tranducers and repetitions. *Theoretical Computer Science*, 45:63–86, 1986.
7. M. Farach. Optimal suffix tree construction with large alphabets. In *Proceedings of the 38th Annual Symposium on Foundations of Computer Science (FOCS)*, pages 137–143, 1997.
8. L. J. Guibas and R. Sedgewick. A dichromatic framework for balanced trees. In *Proceedings of the 19th Annual Symposium on Foundations of Computer Science (FOCS)*, pages 8–21, 1978.
9. D. Gusfield. *Algorithms on Strings, Trees and Sequences: Computer Science and Computational Biology*. Cambridge University Press, 1997.
10. D. Gusfield and J. Stoye. Linear time algorithms for finding and representing all the tandem repeats in a string. Technical Report CSE-98-4, Department of Computer Science, UC Davis, 1998.
11. S. Huddleston and K. Mehlhorn. A new data structure for representing sorted lists. *Acta Informatica*, 17:157–184, 1982.

12. F. K. Hwang and S. Lin. A simple algorithm for merging two disjoint linearly ordered sets. *SIAM Journal on Computing*, 1(1):31–39, 1972.

13. S. Karlin, M. Morris, G. Ghandour, and M.-Y. Leung. Efficient algorithms for molecular sequence analysis. *Proceedings of the National Academy of Science, USA*, 85:841–845, 1988.

14. R. Kolpakov and G. Kucherov. Maximal repetitions in words or how to find all squares in linear time. Technical Report 98-R-227, LORIA, 1998.

15. S. R. Kosaraju. Computation of squares in a string. In *Proceedings of the 5th Annual Symposium on Combinatorial Pattern Matching (CPM)*, volume 807 of *Lecture Notes in Computer Science*, pages 146–150, 1994.

16. G. M. Landau and J. P. Schmidt. An algorithm for approximate tandem repeats. In *Proceedings of the 4th Annual Symposium on Combinatorial Pattern Matching (CPM)*, volume 684 of *Lecture Notes in Computer Science*, pages 120–133, 1993.

17. M.-Y. Leung, B. E. Blaisdell, C. Burge, and S. Karlin. An efficient algorithm for identifying matches with errors in multiple long molecular sequences. *Journal of Molecular Biology*, 221:1367–1378, 1991.

18. M. G. Main and R. J. Lorentz. An $O(n \log n)$ algorithm for finding all repetitions in a string. *Journal of Algorithms*, 5:422–432, 1984.

19. M. G. Main and R. J. Lorentz. Linear time recognition of squarefree strings. In A. Apostolico and Z. Galil, editors, *Combinatorial Algorithms on Words*, volume F12 of *NATO ASI Series*, pages 271–278. Springer, Berlin, 1985.

20. E. M. McCreight. A space-economical suffix tree construction algorithm. *Journal of the ACM*, 23(2):262–272, 1976.

21. K. Mehlhorn. *Sorting and Searching*, volume 1 of *Data Structures and Algorithms*. Springer-Verlag, 1994.

22. K. Mehlhorn and S. Näher. *The LEDA Platform of Combinatorial and Geometric Computing*. Cambridge University Press, 1999. To appear. See http://www.mpi-sb.mpg.de/~mehlhorn/LEDAbook.html.

23. M.-F. Sagot and E. W. Myers. Identifying satellites in nucleic acid sequences. In *Proceedings of the 2nd Annual International Conference on Computational Molecular Biology (RECOMB)*, pages 234–242, 1998.

24. J. Stoye and D. Gusfield. Simple and flexible detection of contiguous repeats using a suffix tree. In *Proceedings of the 9th Annual Symposium on Combinatorial Pattern Matching (CPM)*, volume 1448 of *Lecture Notes in Computer Science*, pages 140–152, 1998.

25. E. Ukkonen. On-line construction of suffix trees. *Algorithmica*, 14:249–260, 1995.

26. P. Weiner. Linear pattern matching algorithms. In *Proceedings of the 14th Symposium on Switching and Automata Theory*, pages 1–11, 1973.

A Dynamic Data Structure for Reverse Lexicographically Sorted Prefixes

Hidetoshi Yokoo*

Department of Computer Science, Gunma University
Kiryu 376-8515, Japan
yokoo@cs.gunma-u.ac.jp

Abstract. This paper proposes a simple data structure, called a *prefix list*, which maintains all prefixes of a string in reverse lexicographic order. It can be on-line incrementally constructed in time and space linear in the string length. It is strongly related to suffix trees and suffix arrays, and may share applications with these existing structures. A suffix array can be built via the corresponding prefix list in linear time. Particular applications of the prefix list lie in source-coding problems that require on-line right-to-left string matching. We apply the prefix list to on-line estimation of source entropy and to context-based symbol-ranking text compression algorithms.

1 Introduction

We propose a simple data structure, called the *prefix list*, which can store all prefixes of a string in reverse lexicographic order. A prefix list can be on-line incrementally constructed in time and space linear in the string length. We can apply it to string matching problems and to data compression algorithms.

The proposed data structure is deeply related to such index structures as suffix trees [4], [10], [12] and suffix arrays [8], [6]. The suffix array for a text is an array of integers which represent lexicographic orders of all suffixes of the text. It was proposed as a space-efficient alternative to the more ubiquitous suffix tree. Whether we use suffix trees or suffix arrays, we usually suppose a text to be static and fixed in the sense that we preprocess it to accept multiple queries afterwards. In particular, a suffix array must be constructed from scratch even if a bit of modification is added to the text. In some actual situations, we must answer index-based string-matching problems while incrementally reading a text. A suffix tree, which can be constructed in an on-line manner, serves as a strong tool in such situations. However, since strings are represented in one direction from the root to leaves on a suffix tree, it is difficult to match strings from right to left. We actually have some on-line string problems, in which we should match strings in that direction.

* Partially supported by the Kayamori foundation of informational science advancement and by the Okawa foundation for information and telecommunications.

M. Crochemore, M. Paterson (Eds.): CPM'99, LNCS 1645, pp. 150–162, 1999.

Such situations sometimes arise in the context modeling stage in text compression. In most symbolwise (predictive) text compression algorithms, an upcoming symbol is predicted on the basis of its *context*. Such a "context-based" method gathers previous contexts according to their similarities to the current context. This requires an on-line right-to-left string matching process. Actually, the prefix list presented in this paper was initially suggested as an adaptive implementation of the *context table*, which was proposed as a common basis for representing text compression algorithms [15]. The most straightforward application of the prefix list is the context-sorting text compression algorithm [14]. It virtually prepares a ranked list of all possible symbols, ordered from most likely to least likely, but actually gives ranks to symbol candidates by searching a sorted list of previous contexts, sorted in reverse lexicographic order. In Matias et al. [9], in which the implementation of similar algorithms was referred to as the HYZ compression problem, the authors proposed to augment suffix trees to solve the problem.

A prefix list represents every prefix in a string as a node in a doubly-linked linear list. It is similar to the suffix tree in that on-line incremental construction is possible, and to the suffix array in that lexicographic linear order is incorporated. It seems that we need $O(n^2)$ time to construct a prefix list from a text of length n. However, if the text is an output from a finite-order Markov source, the expected complexity is reduced to $O(n)$. For a pattern generated from the same source, we can match it with the text in time linear in the pattern length.

In the next section, we define the prefix list and give an on-line procedure for its construction. We show that we can build it from a Markovian text in linear time. In Section 3, we slightly augment the prefix list to apply it to estimating the entropy of an actual text. Section 4 briefly reviews the context-sorting text compression algorithm, which motivates the development of prefix list. Section 5 is a survey of other possible applications.

2   Proposed Data Structure and Its Construction

Let
$$S[1..n] = s_1 s_2 \cdots s_n \quad (s_i \in \Sigma,\ 1 \le i \le n) \tag{1}$$
be an n-symbol string over an ordered alphabet Σ of size $|\Sigma|$. We represent a substring $s_i \cdots s_j$ as $S[i..j]$ and define $S[i..j] = \varepsilon$, the empty string, for $i > j$. The *prefix* of a string $S[1..n]$ that ends at position i is $S[1..i]$, and the *suffix* that begins at position i is $S[i..n]$. The ith symbol s_i is also denoted by $S[i]$.

Based on the ordering relation on Σ, we can define its associated lexicographic order on the set of all strings. *Reverse lexicographic ordering* is lexicographic ordering of reversed strings. For example, the word '**dog**' reverse-lexicographically (*re-lexically*, for short) precedes the word '**deer**' since '**god**' lexically precedes '**reed**'. Our new data structure maintains a re-lexically sorted set of all the prefixes of $S[1..n]$. As an example, consider the string:

$$S[1..9] = \texttt{yabrecabr}, \tag{2}$$

$$S[1..0] = \qquad \varepsilon$$
$$S[1..7] = \qquad \text{yabreca}$$
$$S[1..2] = \qquad \text{ya}$$
$$S[1..8] = \text{yabrecab}$$
$$S[1..3] = \qquad \text{yab}$$
$$S[1..6] = \qquad \text{yabrec}$$
$$S[1..5] = \qquad \text{yabre}$$
$$S[1..9] = \text{yabrecabr}$$
$$S[1..4] = \qquad \text{yabr}$$
$$S[1..1] = \qquad \text{y}$$

Fig. 1. Re-lexically sorted list of prefixes of $S[1..9] =$ 'yabrecabr'.

sorted

ε	y
yabreca	b
ya	b
yabrecab	r
yab	r
yabrec	a
yabre	c
yabrecabr	s_{10}
yabr	e
y	a

Fig. 2. Inserting the next prefix. The upcoming symbol is compared with the following symbols.

in which the ten prefixes including the empty string are re-lexically sorted in the order shown in Fig. 1.

For a pair of two adjacent prefixes in a re-lexically sorted list of prefixes, if a prefix $S[1..j]$ immediately follows $S[1..i]$, then the prefix $S[1..j]$ is said to be the *immediate successor* of $S[1..i]$ and, conversely, $S[1..i]$ is said to be the *immediate predecessor* of $S[1..j]$. We can insert an upcoming prefix into the re-lexically sorted list of previously occurred prefixes in a simple way. Consider again the example in (2), which may be followed by some symbol s_{10}. If the same symbol as s_{10} has not appeared so far, then the reverse lexicographic order of $S[1..10]$ is determined only by its last symbol s_{10}. Otherwise, that is, if we have already had s_{10} in $S[1..9]$, then we can find the position of $S[1..10]$ by searching the so-far occurred symbols for the match with s_{10}. As shown in Fig. 2, starting from the position corresponding to s_{10}, we search bidirectionally the following symbols of the sorted prefixes for the same symbol as s_{10}. If we hit a symbol $s_i = s_{10}$ in the ↑ direction, we should insert $S[1..10]$ as the immediate successor of $S[1..i]$. Conversely, if we find the same symbol as s_{10} in the ↓ direction then we should insert $S[1..10]$ as the immediate predecessor of $S[1..i]$. These can be validated by the recursive property of reverse lexicographic order. As a specific example, suppose that we have $s_{10} =$ 'e' in the example in Fig. 2. Then, we immediately reach $s_5 =$ 'e' in the ↓ direction. This implies that we should insert $S[1..10] =$ 'yabrecabre' as the immediate predecessor of $S[1..5]$. Another case may have $s_{10} =$ 'a'. In this case, we hit either $s_7 =$ 'a' in the ↑ direction or $s_2 =$ 'a' in the ↓ direction. In any of both cases, the position of $S[1..10] =$ 'yabrecabra' in the re-lexically sorted list of these prefixes is known to be between $S[1..7]$ and $S[1..2]$.

The prefix list is natural realization of the above idea. It is implemented as a doubly-linked linear list in which each element, or *node*, contains one integer and three pointers. The three pointers are *pred* and *succ* list pointers used to organize the doubly-linked list and *next* pointer used to designate the next position in the input string. Every prefix in a string is represented by a node of the list. If a node corresponds to the ith prefix $S[1..i]$, then its integer field contains the value

Fig. 3. Node representing $S[1..P{\uparrow}.indx]$.

Fig. 4. Prefix list for '**yabrecabr**'.

of position index i. Its *pred* and *succ* pointers point to nodes corresponding to the immediate predecessor and immediate successor of $S[1..i]$, respectively. The *next* pointer in the node for $S[1..i]$ points to the node for $S[1..i+1]$. If a node corresponds to the entire string $S[1..n]$, then its *next* pointer is set to **nil**. The initial state of a prefix list consists of a single special node H, which represents the empty string. We may add an extra node T into the end of a prefix list in order to simplify some list operations. If we schematically represent a node pointed to by a pointer P as is shown in Fig. 3, in which the left ($\leftarrow$) and right ($\rightarrow$) arrows represent the *pred* and *succ* pointers, respectively, and the vertical arrow ($\downarrow$) represents the *next* pointer, then our sample string in (2) is represented by the list shown in Fig. 4.

As mentioned above, a prefix list can be constructed incrementally in an on-line manner. Assume that the list representing all prefixes of an initial segment $S[1..i]$ has been already constructed and that the $i+1$st prefix $S[1..i+1]$ is about to be inserted. Let P be a pointer that points to the just-inserted node for $S[1..i]$. If the upcoming symbol s_{i+1} alphabetically precedes or succeeds any symbol seen so far, then the node for $S[1..i+1]$ should be inserted into the right of the list head (H) or the left of the list tail (T), respectively. Otherwise, if the symbol s_{i+1} is not included in $S[1..i]$, then the list has a unique position Q where the corresponding node $Q{\uparrow}$ satisfies

$$S[Q{\uparrow}.indx] \prec s_{i+1} \prec S[Q{\uparrow}.succ{\uparrow}.indx]. \qquad (3)$$

Here, '$\prec$' denotes the alphabetic order on Σ. We should insert a new node between the two nodes pointed to by Q and $Q{\uparrow}.succ$.

If the same symbol as s_{i+1} has already appeared in $S[1..i]$, the inequalities in (3) may hold with equality. In this case, in the re-lexically sorted list of prefixes of $S[1..i+1]$, the immediate predecessor or successor of $S[1..i+1]$ has the same last symbol as s_{i+1}. If the immediate *predecessor* $S[1..j+1]$ of $S[1..i+1]$ has the same last symbol s_{j+1} as s_{i+1} ($0 \le j < i$), then $S[1..j]$ re-lexically precedes $S[1..i]$. The node corresponding to $S[1..j]$ should be the first node with the same following symbol s_{j+1} as s_{i+1} when traversing the list from the current node to the head. We can see whether the following symbol matches s_{i+1} by traversing the next pointer. Conversely, if the last symbol s_{j+1} of the immediate *successor* $S[1..j+1]$ of $S[1..i+1]$ is equal to s_{i+1}, then the node for $S[1..j]$ should be the first node satisfying $s_{j+1} = s_{i+1}$ when traversing the list from the current node to the

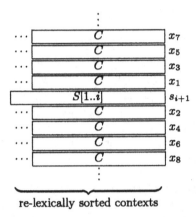

re-lexically sorted contexts

Fig. 5. Re-lexically sorted contexts and their following symbols.

tail. Thus, starting from the current node for $S[1..i]$, we search bidirectionally the list for the node for $S[1..j]$ while comparing the following symbols with s_{i+1}. Once we have found the node for $S[1..j]$, we can immediately reach the node for $S[1..j+1]$ via the next pointer. Then, the position where we should insert a new node representing $S[1..i+1]$ is adjacent to that node for $S[1..j+1]$.

Now, we show that a prefix list can be constructed in linear time if the string in question is drawn from a Markov source of finite order. In such a string, the kth symbol can be completely characterized by the conditional probabilities $\{\Pr(s_k|S[k-m..k-1]) \mid s_k \in \Sigma, S[k-m..k-1] \in \Sigma^m\}$, where $\Pr(s_k \mid S[k-m..k-1])$ is the conditional probability of s_k given $S[k-m..k-1]$. We say that the kth symbol s_k occurs in the *context* $S[k-m..k-1]$.

Suppose that, for sufficiently large i, we are about to insert the node corresponding to $S[1..i+1]$. To do it, we search the list bidirectionally for the match of s_{i+1}. We assume that the search is performed in both directions alternately. Figure 5 shows that the symbol-comparisons with s_{i+1} are done in x_1, x_2, x_3, $\ldots$ order. We evaluate the number of symbol-comparisons in this search. Let K_s be the total number of symbols compared until we reach the match $s_{i+1} = s$, and C denote the longest common context of those symbols. Then, the expected number of K_s is estimated as

$$E\{K_s\} = \frac{1}{\Pr(s_{i+1}=s \mid C)}. \tag{4}$$

Conversely, for any context C, the expected number of tested symbols over all possible upcoming symbols becomes

$$E\{K\} = \sum \Pr(s_{i+1}=s \mid C) \cdot E\{K_s\}$$
$$= |\{s \in \Sigma \mid \Pr(s \mid C) > 0\}|$$
$$\stackrel{\triangle}{=} \sigma_C \leq \sigma. \tag{5}$$

Table 1. Number of symbol-comparisons required to insert each prefix.

FILE	size (bytes)	number of distinct symbols	number of symbol-comparisons maximum	average
rand	500000	53	827	52.48
paper1	53161	95	36481	42.24
bib	111261	81	47442	29.69
alice29.txt	152089	74	34520	24.53
news	377109	98	200587	78.28
plrabn12.txt	481861	81	164435	23.11
book2	610856	96	459095	70.11
book1	768771	82	419926	33.10
obj2	246814	256	119013	124.55
pic	513216	159	287924	46.95
kennedy.xls	1029744	256	671660	124.36

Here,

$$\sigma \overset{\triangle}{=} |\{a \in \Sigma \mid P(a) > 0\}| \qquad (6)$$

denotes the number of symbols with non-zero probability. Therefore, the expected time complexity of the construction of a prefix list is linear in the string length with the coefficient of

$$\sum \Pr(C)\sigma_C \leq \sigma. \qquad (7)$$

Note that the equality in (5) holds when a data string is drawn from a memoryless source.

The above estimation is valid for Markovian data of finite order. In order to validate it on actual data, we performed simple measurement on the number of symbol-comparisons on artificial and natural data. In actual situations, σ, the number of symbols with non-zero probability, can be regarded as the number of distinct symbols occurred in the string. Thus, we make comparisons between the number of distinct symbols actually occurred and the number of symbol-comparisons required in the insertion of the prefixes. Table 1 shows some results of the measurement.

The first file, "rand," consists of lower- and upper-case letters and spaces, totally 53 distinct symbols. We assigned random but fixed probabilities to these symbols, and generated a sample sequence of 500000 symbols. Thus, the file can be thought of as a realization of a memoryless source. The other files come from the Calgary [2] and Canterbury [1] corpora, both of which are collected as standard data for the evaluation of text compression algorithms. The "size" column in Table 1 includes the length of each file. The number of distinct symbols in each file, which corresponds to σ, is shown in the third column. The fourth column represents the maximum number of symbol-comparisons required in searching the list for each symbol when we insert prefixes into the list. The last column is the average number of symbol-comparisons. Obviously, it never exceeds the corresponding number of distinct symbols in any file. On the file "rand," both numbers are almost the same, which is naturally expected on data from a memoryless source.

3 On-Line Computation of the Shortest Unique Substrings with an Application to Entropy Estimation

Figure 6 shows an extension of our data structure, where we add an auxiliary quantity to each node which represents the length of the longest common suffix of the strings corresponding to the node and to its immediate successor. This quantity serves as a measure for context similarity between two contexts which are adjacent to each other in a re-lexically sorted list of contexts. In this section, we describe on-line computation of the quantities and its application to the estimation of the entropy of a data source.

Let $S[1..j]$ denote the immediate successor of the ith prefix $S[1..i]$. Letting l_i be the maximum l such that $S[1..i+1-l] = S[1..j+1-l]$, we add it to the node corresponding to $S[1..i]$. Assume that the node has both immediate successor $S[1..j]$ and immediate predecessor $S[1..k]$ just after inserting that node ($1 \leq j < i$, $1 \leq k < i$). This implies that, immediately before the insertion of the node, the two nodes $S[1..k]$ and $S[1..j]$ are directly adjacent to each other. Suppose that these two nodes have had l_k and l_j, respectively, as shown in Fig. 7. The state in Fig. 7 may be changed into a new one by the insertion of the node for $S[1..i]$. As shown in Fig. 8, the value of l_j remains unchanged while the value of l_k may increase to l'_k. These satisfy both $l_i \geq l_k$ and $l'_k \geq l_k$. More specifically,

$$\text{If } l'_k > l_k \text{ then } l_i = l_k \text{ otherwise } l_i \geq l_k;$$
$$\text{If } l_i > l_k \text{ then } l'_k = l_k \text{ otherwise } l'_k \geq l_k.$$

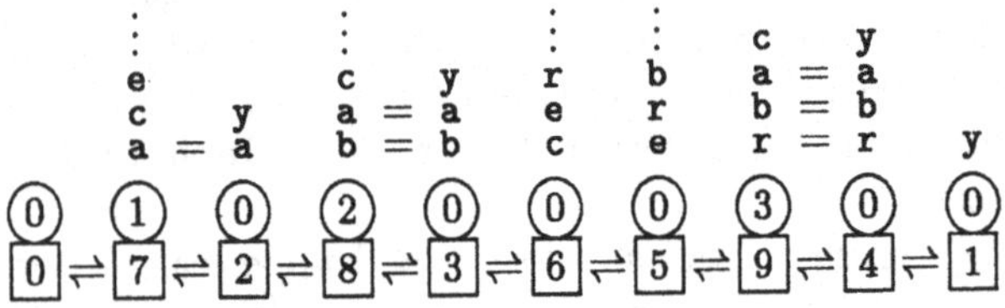

Fig. 6. Context similarities with the immediate successors (lengths of the common suffixes).

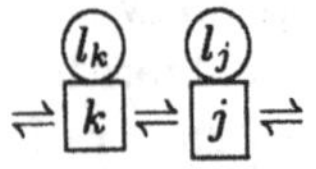

Fig. 7. A pair of adjacent nodes. 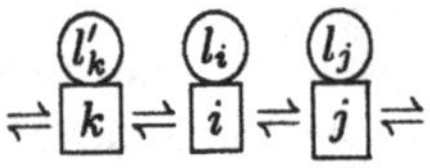   **Fig. 8.** After inserting the ith prefix.

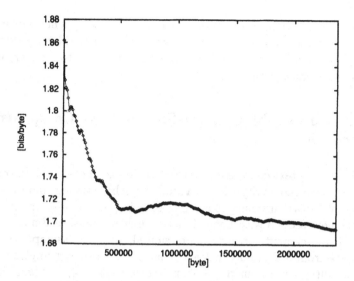

Fig. 9. On-line entropy estimation. The text is four Jane Austen's novels, consisting of 2,364,200 letters. The final result, 1.694 [bits/byte], estimated from the entire text, is 0.055 bits better than the estimate of Kontoyiannis et al.

We can use the above relations to minimize the number of actual comparisons required to compute l'_k and l_i.

Our first application of the prefix list is the estimation of entropy of actual data. As is well known in information theory, the data compression limit of a string is given by the entropy of its source. Of the methods for estimating entropy from sample data, the ones most related to our method are the SWE (Sliding-window Entropy) estimator and Grassberger's estimator [13]. Our estimate [15] from a string $S[1..n]$ is defined by

$$\widehat{H}_n = n \log n \Big(\sum_{i=1}^{n} L_i \Big)^{-1}, \tag{8}$$

where L_i is the minimum l such that a copy of the substring $S[i-l+1..i]$ does *not* appear anywhere else in the string. Thus, L_i represents the length of the shortest unique substring ending at position i. For the immediate predecessor $S[1..k]$ of $S[1..i]$, the value of L_i can be calculated as

$$L_i = \max\{l_i, l_k\} + 1.$$

Combining L_i with the equation (8), we can perform entropy estimation in an on-line manner.

Figure 9 shows an example of entropy estimation, where the estimate $\widehat{H}_n$ is plotted as a function of the input length n. The sample text is a concatenation of four Jane Austen's novels [7], which is the same as that used by Kontoyiannis et al. Compared with their results, we know that our method is not only efficient but also provides very good estimates. However, since the problem of estimating the entropy of English text itself is not the main focus of this paper, we will discuss our estimates elsewhere.

4 Implementing the Context-Sorting Text Compression Algorithm

The context-sorting text compression algorithm [14] is an on-line data compression method, which can be regarded as a kind of symbol-ranking compressors [5]. It is important in that it connects the block-sorting compression method mentioned in the next section with Lempel–Ziv-type dictionary-based methods [14], [9]. Although the context-sorting compression algorithm is asymptotically optimal for data from a finite-order Markov source, its existing implementation is naive and quite slow. In our previous implementation [14], we maintained re-lexically sorted contexts using a binary search tree. We had to limit the length of context and to consume time proportional to that bounded length. These have prevented us from introducing more sophisticated codes into the coding stage.

In the context-sorting method, we enumerate previous contexts in the order of their similarities to the current context. Then, we give ranks to distinct symbols in accordance with the orders of their contexts. The next symbol is encoded as its actual rank. Figure 10 shows an example of giving ranks to symbol candidates. In the figure, we assume that we have already encoded an initial segment ending with '··· to define' and are going to encode its following symbol. In this example, if the next symbol is 'm' then it is encoded as rank 0. The space '_' is encoded as rank 1, 'd' as rank 2, and so on. These ranks may be encoded by a fixed static code or an adaptive code. However, since the original implementation took much time in the ranking phase, it was difficult to use adaptive codes, which are generally slower than static ones. In our new implementation, we can combine an adaptive arithmetic code [2] with the prefix list. We no longer need

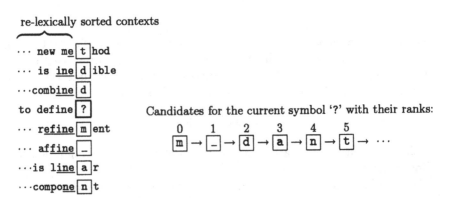

Fig. 10. Ranked candidates in the context-sorting compression algorithm.

to restrict the context length. The new implementation incorporating the prefix list runs more than ten times faster than the previous one with the bounded context length of 8 symbols.

Obviously, the context-sorting compression method stimulates the development of prefix list. We may compare the prefix list and the context-sorting compression method to two sides of the same coin. The essential component of the latter method is the calculation of symbol's rank. Therefore, when we design an improvement of the prefix list, we should consider not only its construction speed but also the possibility of efficient ranking of symbol candidates.

Although the context-sorting compression method is basically a symbolwise algorithm, it can be extended to Ziv–Lempel-type phrase-based compression methods [14]. The HYZ compression method [9] mentioned in the introductory section includes such an extension. Another example is the ACB algorithm of Buyanovsky [11], whose primal version is essentially the same as LZ77 [16]. What most distinguishes ACB from other LZ77 variants is its method of specifying the position of the longest match. If the longest match in the previous text begins with x_k in Fig. 5, where the current phrase begins with s_{i+1}, then the value of k is encoded as the match position. Therefore, we can apply the prefix list to the ACB algorithm to calculate the value of k.

5   Other Applications

String matching (Finding the longest common suffix): The (exact) string matching problem is to find a string called the *pattern* in a longer string called the *text*. We interpret it as a problem of finding a position to insert the pattern into the re-lexically sorted list of prefixes of the text. If we wish to find an occurrence of 'cabr' in $S[1..9]$ given in (2), it is enough to find the reverse lexicographic relationship:

$$S[1..5] = \text{'yabre'} \prec \text{'cabr'} \prec \text{'yabrecabr'} = S[1..9], \qquad (9)$$

where '$\prec$' denotes reverse lexicographic order. In this case, we can immediately know that the pattern in question appears as $S[6..9]$. If we have another pattern 'rabr', then a similar relation

$$S[1..9] = \text{'yabrecabr'} \prec \text{'rabr'} \prec \text{'yabr'} = S[1..4] \qquad (10)$$

holds. This time, there is no exact matching; instead the longest common suffix 'abr' can be found.

Thus, the problem of finding an occurrence of *pat* of length m in *text* of length n is conceptually the same as building a prefix list for

$$S[1..n + m + 1] = text\&pat, \qquad (11)$$

where the symbol $\&\notin \Sigma$ is a special delimiter that alphabetically precedes any symbol in Σ. Of course, there is no need for the actual insertion of prefixes ending in '&pat'. Actually, we first build a prefix list for the text. In order to expedite the matching process, we use an auxiliary array of pointers, which map symbols to nodes in the prefix list. The array element corresponding to a symbol s includes a pointer Q that points to the node such that

$$S[Q{\uparrow}.pred{\uparrow}.indx] \prec s = S[Q{\uparrow}.indx]. \tag{12}$$

Namely, the pointer Q points to the leftmost node among nodes representing prefixes with the same last symbol s. This array of pointers can also be used in the course of constructing a prefix list in order to check whether the next symbol has appeared in the initial segment seen so far. If the array element corresponding to the first symbol of the pattern is a nil pointer, then we know that the pattern is not contained in the text. Otherwise, we proceed to a procedure similar to the insertion of the rest of the pattern. If the pattern is generated from the same Markov source as for the text, this procedure runs linearly in the pattern length.

Suffix array construction: The suffix array for a string $S[1..n]$ is an array of the indexes from 1 to n, specifying the lexicographic orders of the suffixes of $S[1..n]$ [8], [6].

A prefix list maintains all prefixes in reverse lexicographic order. Thus, the construction of a suffix array is straightforward if we apply our prefix list construction procedure to a reversed string. It is sufficient to sequentially copy the indexes of a prefix list into a suffix array by traversing the list via *succ* links. Obviously, the conversion of a prefix list to a suffix array can be done in linear time.

In some applications of suffix arrays, information about the *longest common prefixes* (*lcps*) plays an important role [8]. The quantity l_i of the ith node can be used as the *lcp* of consecutive elements of a resulting suffix array.

Block-sorting data compression: The block-sorting data compression algorithm of Burrows and Wheeler [3] has received considerable attention in anticipation that it may outperform the Lempel–Ziv codes. Its operation begins with a special sort procedure, called the *BW transform*, which is followed by a sequential application of move-to-front heuristics and statistical encoding.

We can apply the prefix list to performing the BW transform. In our terms, the BW transform can be described in the following. Here, H and T denote the pointers to the list head and the list tail, respectively.

```
P ← H;
while (P ≠ T) {
    if (P↑.next = nil) output('&') else output(S[P↑.next↑.indx]);
    P ← P↑.succ;
}
```

This is not identical with the original definition of the transform but is essentially the same. The prefix list in Fig. 4 converts our sample string into

$$S'[1..10] = \text{ybbrrac\&ea}, \tag{13}$$

which in turn is encoded by a move-to-front coder. We are omitting the second half of the algorithm; see [3] for more details.

The BW transform is reversible; we can reconstruct $S[1..n]$ from its transformed string $S'[1..n+1]$. In order to explain the reverse transformation, we add subscripts to indicate the position of each symbol in $S'[1..n+1]$. In the above example, the input into the reverse transformation is $S'[1..10] = $ '$y_1 b_2 b_3 r_4 r_5 a_6 c_7 \&_8 e_9 a_{10}$'. First, we sort alphabetically the symbols in this string in a stable manner. Then, we have

$$S''[1..10] = \&_8 a_6 a_{10} b_2 b_3 c_7 e_9 r_4 r_5 y_1. \tag{14}$$

Second, we write

$$\pi(i) = j, \quad 1 \leq i, j \leq n+1 \tag{15}$$

for the symbol $S'[i]$ if it appears as the jth symbol $S''[j]$ in $S''[1..n+1]$. In the present example, we have $\pi(8) = 1$, $\pi(6) = 2$, $\pi(10) = 3$, and so on. Then, beginning with $S \leftarrow \varepsilon$ and $i \leftarrow 1$, we repeat

$$\begin{aligned} S &\leftarrow S \cdot S'[i]; \\ i &\leftarrow \pi(i) \end{aligned} \quad \left(\text{or equivalently,} \begin{aligned} i &\leftarrow \pi(i); \\ S &\leftarrow S \cdot S''[i] \end{aligned} \right)$$

n times to recover the original string in S.

The (forward) BW transform is more demanding than the corresponding inverse transform. Our prefix list performs the forward transform in linear time at least on a string that is drawn from a finite-order Markov source.

6   Conclusion

We have presented a conceptually simple data structure, called the *prefix list*. It is a linked-list representation of prefixes of a string sorted in reverse lexicographic order. It is quite similar to the suffix array in that lexicographic linear order is incorporated. While the suffix array has an off-line nature, a prefix list can be built in an on-line manner. This yields its characteristic applications. The prefix list provides a powerful tool to a class of context-based symbol-ranking data compression algorithms. We have also shown that the prefix list is applicable to other interesting problems.

Acknowledgments

I am grateful to Taro Yamagishi who implemented the proposed data structure and provided experimental results. Detailed comments by an anonymous referee should be acknowledged although they are not yet reflected on the paper.

References

1. Arnold, R. and Bell, T.: A corpus for the evaluation of lossless compression algorithms. DCC'97, Proc. Data Compression Conf., Snowbird, Utah (1997) 201–210
2. Bell, T. C., Cleary, J. G., and Witten, I. H.: Text Compression. Prentice Hall, Englewood Cliffs (1990)

3. Burrows, M. and Wheeler, D. J.: A block-sorting lossless data compression algorithm. SRC Research Report, 124 (1994)
4. Chen, M. T. and Seiferas, J.: Efficient and elegant subword-tree construction. In Apostolico, A. and Galil, Z. (eds.): Combinatorial Algorithms on Words, NATO ASI Series, Springer, Berlin (1984)
5. Fenwick, P. M.: Symbol ranking text compression with Shannon recodings. J. Universal Computer Science 3 (1997) 70–85. http://www.iicm.edu/jucs_3_2
6. Gonnet, G. H., Baeza-Yates, R. A., and Snider, T.: New indices for text: Pat trees and pat arrays. In Frakes, W. B. and Baeza-Yates, R. A. (eds.): Information Retrieval: Data Structures and Algorithms, Chap. 5. Prentice Hall, Englewood Cliffs (1992) 66–82
7. Kontoyiannis, I., Algoet, P. H., Suhov, Yu. M., and Wyner, A. J.: Nonparametric entropy estimation for stationary processes and random fields, with applications to English text. IEEE Trans. Inform. Theory 44 (1998) 1319–1327
8. Manber, U. and Myers, G.: Suffix arrays: A new method for on-line string searches. Proc. 1st Annual ACM–SIAM Symposium on Discrete Algorithms (1990) 319–327. Appeared also in SIAM J. Comput. 22 (1993) 935–948
9. Matias, Y., Muthukrishnan, S., Sahinalp, S. C., and Ziv, J.: Augmenting suffix trees with applications. Proc. ESA'98, European Symposium on Algorithms, Venice, Italy (1998)
10. McCreight, E. M.: A space-economical suffix tree construction algorithm. J. ACM 23 (1976) 262–272
11. Salomon, D.: Data Compression: The Complete Reference. Springer, New York (1998)
12. Ukkonen, E.: On-line construction of suffix trees. Algorithmica 14 (1995) 249–260
13. Wyner, A. D., Ziv, J., and Wyner, A. J.: On the role of pattern matching in information theory. IEEE Trans. Inform. Theory 44 (1998) 2045–2056
14. Yokoo, H.: Data compression using a sort-based context similarity measure. Computer Journal 40 (1997) 94–102
15. Yokoo, H.: Context tables: A tool for describing text compression algorithms. DCC'98, Proc. Data Compression Conf., Snowbird, Utah (1998) 299–308
16. Ziv, J. and Lempel, A.: A universal algorithm for sequential data compression. IEEE Trans. Inform. Theory IT-23 (1977) 337–343

A New Indexing Method
for Approximate String Matching*

Gonzalo Navarro and Ricardo Baeza-Yates

Dept. of Computer Science, University of Chile
Blanco Encalada 2120 - Santiago - Chile
{gnavarro,rbaeza}@dcc.uchile.cl

Abstract. We present a new indexing method for the approximate string matching problem. The method is based on a suffix tree combined with a partitioning of the pattern. We analyze the resulting algorithm and show that the retrieval time is $O(n^\lambda)$, for $0 < \lambda < 1$, whenever $\alpha < 1 - e/\sqrt{\sigma}$, where α is the error level tolerated and σ is the alphabet size. We experimentally show that this index outperforms by far all other algorithms for indexed approximate searching, also being the first experiments that compare the different existing schemes. We finally show how this index can be implemented using much less space.

1 Introduction

Approximate string matching is a recurrent problem in many branches of computer science, with applications to text searching, computational biology, pattern recognition, signal processing, etc.

The problem is: given a long text of length n, and a (comparatively short) pattern of length m, retrieve all the text segments (or "occurrences") whose *edit distance* to the pattern is at most k. The *edit distance* between two strings is defined as the minimum number of character insertions, deletions and replacements needed to make them equal. We define the "error level" as $\alpha = k/m$.

In the on-line version of the problem, the pattern can be preprocessed but the text cannot. The classical solution uses dynamic programming and is $O(mn)$ time [27, 28]. A number of algorithms improved later this result [34, 20, 16, 11, 35, 32, 12, 30, 36, 9, 8, 24]. The lower bound of the on-line problem (proved and reached in [12]) is $O(n(k + \log_\sigma m)/m)$, which is of course $\Omega(n)$ for constant m.

If the text is large even the fastest on-line algorithms are not practical, and preprocessing the text becomes necessary. However, just a few years ago, indexing text for approximate string matching was considered one of the main open problems in this area [35, 3]. Despite some progress in the last years, the indexing schemes for this problem are still rather immature.

There are two types of indexing mechanisms for approximate string matching, which we call "word-retrieving" and "sequence-retrieving". Word retrieving

* This work has been supported in part by Fondecyt grant 1-990627 and Fondef grant 96-1064.

M. Crochemore, M. Paterson (Eds.): CPM'99, LNCS 1645, pp. 163–185, 1999.

indices [22, 6, 2] are more oriented to natural language text and information retrieval. They can retrieve every *word* whose edit distance to the pattern is at most k. Hence, they are not able to recover from an error involving a separator, such as recovering the word "flowers" from the misspelled text "flo wers" or from "manyflowers", if we allow one error[1]. These indices are more mature, but their restriction can be unacceptable in some applications, especially where there are no words (as in DNA) or in agglutinating languages such as Finnish or German.

Our focus in this paper is sequence retrieving indices. Among these, we find two types of approaches.

A first type is based on simulating a sequential algorithm, but running it on the suffix tree [19, 1] or DAWG [14, 10] of the text instead of the text itself. Since every different substring in the text is represented by a single node in the tree or the DAWG, it is possible to avoid redoing the same work when the text has repetitions. Those indices take $O(n)$ space and construction time, but their construction is not optimized for secondary memory and is very inefficient in this case (see, however, [15]). Moreover, the structure is very inefficient in space requirements, since it takes 12 to 70 times the text size.

In [18, 33, 13], different algorithms that traverse the least possible nodes in the suffix tree (or in the DAWG) are presented. The idea is to traverse all the different tree nodes that represent "viable prefixes", which are text substrings that can be prefixes of an approximate occurrence of the pattern.

In [17], a simplified version of the above technique was independently proposed, consisting of a limited depth-first search (DFS) on the suffix tree. Since every substring of the text (i.e. every potential occurrence) can be found from the root of the suffix tree, it is sufficient to explore every path starting at the root, descending by every branch up to where it can be seen that that branch does not represent the beginning of an occurrence of the pattern. This algorithm inspects more nodes than the previous ones, but it is simpler. For instance, with an additional $O(\log n)$ time factor, the algorithm runs on suffix arrays, which take 4 times the text size instead of 12. This algorithm was analyzed in [4].

The second type of sequence-retrieving indices is based on adapting an on-line filtering algorithm. The filters are based in matching substrings of the patterns without errors, and checking for potential occurrences around those matches. The index is used to quickly find those substrings, and is based on storing some text q-grams (substrings of length q) and their positions in the text.

Different filtration indices [23, 31, 29, 7] differ mostly in how the text is sampled (distance between consecutive text samples, whether they overlap or not, etc.), in how the pattern is sampled, in how many matching samples are needed to verify their neighborhood in the text, etc. Depending on this and on q they achieve different space-time tradeoffs. In general, filtration indices are much smaller than suffix trees (1 to 10 times the text size), although they are less tolerant to the error level α. They can also be built in linear time.

---

[1] Although some, like Glimpse [22], can match the pattern inside a text word.

Somewhat special is [23], because it does not reduce the search to exact but to approximate search of pattern pieces. To search for a pattern of length $m \leq q-k$, all the maximal strings with edit distance $\leq k$ to the pattern are generated and searched in the set of q-grams. Later, all the occurrences are merged. Longer patterns are split in as many pieces as necessary to make them short enough.

In this paper we present a hybrid indexing scheme for this problem. It uses a suffix tree, where the pattern is partitioned in subpatterns which are searched with less errors in suffix tree. All the occurrences of the subpatterns are later verified for a complete match. The goal is to balance between the cost to search in the suffix tree (which grows with the size of the subpatterns) and the cost to verify the potential occurrences (which grows when shorter patterns are searched). This method shows experimentally to be by far superior to all other implemented proposals, and we show analytically that the average retrieval time can be made $O(n^{2(\alpha+H_\sigma(\alpha))/(1+\alpha)})$, where $H_\sigma(\alpha)$ is the base-σ entropy function. This is sub-linear for $\alpha < 1 - e/\sqrt{\sigma}$. This limit on α cannot probably be improved [8, 25]. We finally propose an alternative data structure to reduce the space requirements of the suffix tree, with little time penalty.

2   Combining Suffix Trees and Pattern Partitioning

We present now our alternative proposal. The general idea is to partition the pattern in pieces, search each piece in the suffix tree in the classical way, and check all the positions found for a complete match. We first consider how to search a piece in the suffix tree and later address the pattern partitioning issue.

2.1   DFS Using a Bit-Parallel Automaton

Let us consider the existing algorithms to traverse the suffix tree. While [33, 13] minimize the number of nodes traversed, [17] is simpler but inspects more nodes. We show that [17], thanks to its simplicity, can be adapted to use a node processing algorithm which is faster than dynamic programming, namely our on-line algorithm of [8][2]. The tradeoff is: we can explore less nodes at higher cost per node or more nodes at less cost per node. We show later experimentally that this last alternative is much faster when [8] is used to process the nodes.

We recall that the idea of [17] is a limited depth-first search on the suffix tree, starting at the root and stopping when it can be seen that the current text substring cannot start an approximate pattern occurrence. No text occurrence can be missed because every text substring can be found starting from the root.

More specifically, we compute the edit distance between the tree path and the pattern, and if at some node the distance is $\leq k$ we know that the text substring represented by the node matches the pattern. We report all the leaves of the suffix tree which descend from those nodes, since their text positions start with the matching substring. On the other hand, when we can determine that the

---

[2] Probably [24] would also fit well.

edit distance cannot be as low as k, we abandon the path. This surely happens at depth $m + k + 1$ but normally happens before.

We implement this traversal using our algorithm of [8] instead of dynamic programming. This algorithm uses bit parallelism to simulate a non-deterministic finite automaton (NFA) that recognizes the approximate pattern. We modify this automaton to compute edit distance (removing the initial self-loop it has in [8]).

Figure 1 shows the automaton to recognize **"patt"** with $k = 2$ errors. Every row denotes the number of errors seen. Every column represents matching a pattern prefix. Horizontal arrows represent matching a character (i.e. if the pattern and text characters match, we advance in the pattern and in the text). All the others increment the number of errors (move to the next row): vertical arrows insert a character in the pattern (we advance in the text but not in the pattern), solid diagonal arrows replace a character (we advance in the text and pattern), and dashed diagonal arrows delete a character of the pattern (they are empty transitions, since we advance in the pattern without advancing in the text). The automaton signals (the end of) a match whenever a rightmost state is active. If we do not care about the number of errors of the occurrences, we can consider final states those of the last full diagonal.

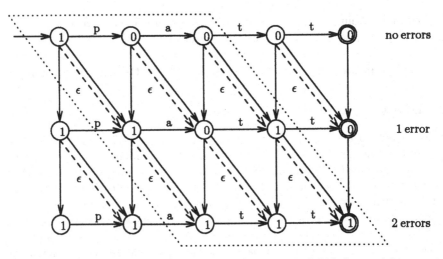

Fig. 1. An NFA for approximate string matching. Unlabeled transitions match any character. Dotted lines enclose the states actually represented in our algorithm.

Initially, the active states at row i are at the columns from 0 to i, to represent the deletion of the first i characters of the pattern. We do not need in fact to represent the initial lower-left triangle, since if a substring matches with initial insertions we will find (in other branch of the suffix tree) a suffix of it which does not need the insertions[3]. On the other hand, unlike [8], we need to represent the first full diagonal, since now it will not be always active. We start the automaton with only this first full diagonal active, and traverse the suffix tree path until the automaton runs out of active states or the lower right state is activated.

The simulation of this automaton needs $(m - k + 1)(k + 2)$ bits. If we call w the number of bits in the computer word, then when the previous number is $\leq w$ we can put all the states in a single computer word and work $O(1)$ per traversed node of the suffix tree. For longer patterns, the automaton is split in many computer words, at a cost of $O(k(m - k)/w)$. For moderate-size patterns this improves over dynamic programming, which costs $O(m)$ per suffix tree node.

This bit-parallel variation is only possible because of the simplicity of the traversal. For instance, the idea does not work on the more complex setup of [33, 13], since these need some adaptations of the dynamic programming algorithm that are not easy to parallelize. Note that this algorithm can be seen as a particular case of automaton searching over a trie [5].

2.2 Partitioning the Pattern

It is well known [33, 4] that the search cost using the suffix tree grows exponentially with m and k, no matter which of the two techniques we use (optimum traversal or DFS). Hence, we prefer that m and k are small numbers. We present in this section a new technique based in partitioning the pattern, so that the pattern is split in many sub-patterns which are searched in the suffix tree, and their occurrences are directly verified in the text for a complete match. We show in the experiments that this technique outperforms all the others.

This method is based on the pattern partitioning technique of [23, 8]. The core of the idea is that, if a pattern of length m occurs with k errors and we split the pattern in j parts, then at least one part will appear with $\lfloor k/j \rfloor$ errors inside the occurrence. In fact, the case $j = k + 1$ is the basis for the algorithm [9] and the q-gram index [7].

The new algorithm follows. We evenly divide the pattern in j pieces (j is unspecified by now). Then we search in the suffix tree the j pieces with $\lfloor k/j \rfloor$ errors using the algorithm of Section 2.1. For each match found ending at text position i we check the text area $[i - m - k..i + m + k]$.

The reason why this idea works better than a simple suffix tree traversal with the complete pattern is that, since the search cost on the suffix tree is exponential in m and k, it may be better to perform j searches of patterns of length m/j and k/j errors. However, the larger j, the more text positions have to be verified, and therefore the optimum is in between. In the next section we find analytically the optimum j and the complexity of the search

One of the closest approaches to this idea is Myers' index [23], which collects all the text q-grams (i.e. prunes the suffix tree at depth q), and given the pattern it generates all the strings at distance at most k from it, searches them in the

---

[3] If, after traversing a text substring s, a 1 finally exits from the lower-left triangle, then a suffix of s will do the same without entering into the triangle.

index and merges the results. This is the same work of a suffix tree provided that we do not enter too deep (i.e. $m + k \leq q$). If $m + k > q$, Myers' approach splits the pattern and searches the subpatterns in the index, checking all the potential occurrences. The main difference with our proposed approach is that Myers' index generates all the strings at a given distance and searches them, instead of traversing the structure to see which of them exist. This makes that approach degrade on biased texts, where most of the generated q-grams do not exist (in the experimental section we show that it works well on DNA but quite bad on English). Moreover, we split the pattern to optimize the search cost, while the splitting in Myers' index is forced by indexing constraints (i.e. q).

3   Analysis

3.1   Searching One Piece

An asymptotic analysis on the performance of a depth-first search over suffix trees is immediate if we consider that we cannot go deeper than level $m + k$ since past that point the edit distance between the path and our pattern is larger than k and we abandon the search. Therefore, we can spend at most $O(\sigma^{m+k})$ time, which is independent on n and hence $O(1)$. Another way to see this is to use the analysis of [5], where the problem of searching an arbitrary automaton over a suffix trie is considered. Their result for this case indicates constant time (i.e. depending on the size of the automaton only) because the automaton has no cycles.

However, we are interested in a more detailed average analysis, especially the case where n is not so large in comparison to σ^{m+k}. We start by analyzing which is the average number of nodes at level ℓ in the suffix tree of the text, for small ℓ. Since almost all suffixes of the text are longer than ℓ (i.e. all except the last ℓ), we have nearly n suffixes that reach that level. The total number of nodes at level ℓ is the number of different suffixes once they are pruned at ℓ characters. This is the same as the number of different ℓ-grams in the text. If the text is random, then we can use a model where n balls are thrown into σ^ℓ urns, to find out that the average number of filled urns (i.e. suffix tree nodes at level ℓ) is

$$\sigma^\ell \left(1 - \left(1 - 1/\sigma^\ell\right)^n\right) \;=\; \sigma^\ell \left(1 - e^{-\Theta(n/\sigma^\ell)}\right) \;=\; \Theta\left(\min\left(n, \sigma^\ell\right)\right)$$

which shows that the average case is close to the worst case: up to level $\log_\sigma n$ all the possible σ^ℓ nodes exist, while for deeper levels all the n nodes exist.

We also need the probability of processing a given node at depth ℓ in the suffix tree. In the Appendix we prove that the probability is very high for $\beta = k/\ell \geq 1 - c/\sqrt{\sigma}$ (Eq. (3)), and otherwise it is $O(\gamma(\beta)^\ell)$, where $\gamma(\beta) < 1$. The constant c can be proven to be smaller than $e = 2.718...$, and is empirically known to be close to 1. The $\gamma(x)$ function (Eq. (1)) is $1/(\sigma^{1-x} x^{2x} (1-x)^{2(1-x)})$, which goes from $1/\sigma$ to 1 as x goes from 0 to $1 - c/\sqrt{\sigma}$.

Therefore, we pessimistically consider that in levels

$$\ell \ \leq \ L(k) \ = \ \frac{k}{1 - c/\sqrt{\sigma}} \ = \ O(k)$$

all the nodes in the suffix tree are visited, while nodes at level $\ell > L(k)$ are visited with probability $O(\gamma(k/\ell)^{\ell})$, where $\gamma(k/\ell) < 1$. Finally, we never work past level $m + k$. We are left with three disjoint cases to analyze, illustrated in Figure 2.

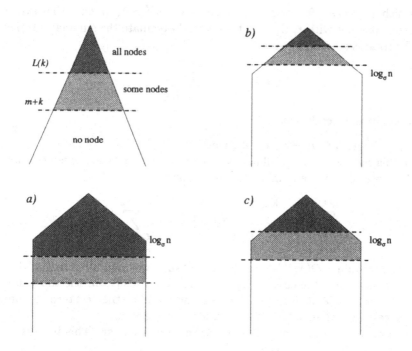

Fig. 2. The upper left figure shows the visited parts of the tree. The rest shows the three disjoint cases in which the analysis is split.

(a) $L(k) \geq \log_{\sigma} n$, i.e. $n \leq \sigma^{L(k)}$, or "small n"
   In this case, since on average we work on all the nodes up to level $\log_{\sigma} n$, the total work is n, i.e. the amount of work is proportional to the text size. This shows that the index simply does not work for very small texts, being an on-line search preferable as expected.

(b) $m + k < \log_{\sigma} n$, i.e. $n > \sigma^{m+k}$ or "large n"
   In this case we traverse all the nodes up to level $L(k)$, and from there on we work at level ℓ with probability $\gamma(k/\ell)^{\ell}$, until $\ell = m + k$. Under case (b), there are σ^{ℓ} nodes at level ℓ. Hence the total number of nodes traversed is

$$\sum_{\ell=0}^{L(k)} \sigma^\ell \; + \; \sum_{\ell=L(k)+1}^{m+k} \gamma(k/\ell)^\ell \sigma^\ell$$

where the first term is $O(\sigma^{L(k)})$. For the second term, we see that $\gamma(x) > 1/\sigma$, and hence $(\gamma(k/\ell)\sigma)^\ell > 1$. More precisely,

$$(\gamma(k/\ell)\sigma)^\ell \;=\; \frac{\sigma^k \ell^{2\ell}}{k^{2k}(\ell-k)^{2(\ell-k)}}$$

which grows as a function of ℓ. Since $(\gamma(k/\ell)\sigma)^\ell > 1$, we have that even if it were constant with ℓ, the last term would dominate the summation. Hence, the total cost in case (b) is

$$\sigma^{L(k)} \;+\; \frac{\sigma^k(1+\alpha)^{2(m+k)}}{\alpha^{2k}}$$

which is independent of n.

(c) $L(k) < \log_\sigma n \le m+k$, i.e. "intermediate n"
In this case, we work on all nodes up to $L(k)$ and on some nodes up to $m+k$. The formula for the number of visited nodes is

$$\sum_{\ell=0}^{L(k)} \sigma^\ell \; + \; \sum_{\ell=L(k)+1}^{\log_\sigma(n)-1} \gamma(k/\ell)^\ell \sigma^\ell \; + \; \sum_{\ell=\log_\sigma n}^{m+k} \gamma(k/\ell)^\ell n$$

The first sum is $O(\sigma^{L(k)})$. For the second sum, we know already that the last term dominates the complexity (see case (b)). Finally, for the third sum we have that $\gamma(k/\ell)$ decreases as ℓ grows, and therefore the first term dominates the rest (which would happen even for a constant γ).
Hence, the case $\ell = \log_\sigma n$ dominates the last two sums. This term is

$$n\gamma(k/\log_\sigma n)^{\log_\sigma n} = \frac{\sigma^k(\log_\sigma n)^{2\log_\sigma n}}{k^{2k}(\log_\sigma(n)-k)^{2(\log_\sigma(n)-k)}} = \frac{\sigma^k(\log_\sigma n)^{2k}}{k^{2k}}\,(1+o(1))$$

(this can be bounded by $(\sigma(1+1/\alpha)^2)^k$ by noticing that we are inside case (c), but we are interested in how n affects the growth of the cost).

The search time is then sublinear for $\log_\sigma n > \max(L(k), m+k)$, or which is the same, $\alpha < \max(\log_\sigma(n)/m\,(1-c/\sqrt{\sigma}), \log_\sigma(n)/m - 1)$. Figure 3 illustrates.

3.2   Pattern Partitioning

When pattern partitioning is applied, we perform j searches of the same kind of Section 2.1, this time with patterns of length m/j and k/j errors. We also need to verify all the possible matches.

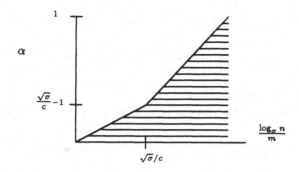

Fig. 3. Area of sublinearity for suffix tree traversal.

As shown in [8], the matching probability for a text position is $O(\gamma(\alpha)^m)$, where $\gamma(\alpha)$ is that of Eq. (1). From now on we use $\gamma = \gamma(\alpha)$. Using dynamic programming, a verification costs $O(m^2)$ [4]. Hence, our total search cost is

$$j \times suffix\_tree\_traversal(m/j, k/j) \;+\; j \times \gamma^{m/j} m^2 n$$

and we want the optimum j. First, notice that if $\gamma = 1$ (that is, $\alpha \geq 1 - c/\sqrt{\sigma}$), the verification cost is as high as an on-line search and therefore pattern partitioning is useless. In this case it may be better to use plain DFS. In the analysis that follows, we assume that $\gamma < 1$ and hence $\alpha < 1 - c/\sqrt{\sigma}$.

According to Section 3.1, we divide the analysis in three cases. Notice that now we can adjust j to select the best case for us.

(a) $\sigma^{L(k/j)} \geq n$, or $j \log_\sigma n \leq k/(1 - c/\sqrt{\sigma})$
In this case the search cost is $\Omega(n)$ and the index is of no use.

(b) $\sigma^{(m+k)/j} < n$, or $j \log_\sigma n > m + k$
In this case the total search cost is

$$j \left(\sigma^{L(k/j)} \;+\; \frac{\sigma^{k/j}(1+\alpha)^{2(m+k)/j}}{\alpha^{2k/j}} \;+\; \gamma^{m/j} m^2 n \right)$$

where the first two terms decrease and the last one increases with j. Since $a + b = \Theta(\max(a, b))$, the minimum order is achieved when increasing and decreasing terms meet. When equating the first and third terms we obtain that the optimum j is

$$j_1 \;=\; \frac{m}{\log_\sigma(m^2 n)} \left(\frac{\alpha}{1 - c/\sqrt{\sigma}} + \log_\sigma(1/\gamma) \right)$$

and the complexity (only considering n) is $O\left(n^{\alpha/(\alpha + (1 - c/\sqrt{\sigma}) \log_\sigma(1/\gamma))} \right)$.

---

[4] It can be done in $O((m/j)^2)$ time [23, 26], but this does not affect the result here.

On the other hand, if we equate the second and third term, the best j is

$$j_2 = \frac{m}{\log_\sigma(m^2 n)} (1 + 2((1+\alpha)\log_\sigma(1+\alpha) + (1-\alpha)\log_\sigma(1-\alpha)))$$

and the complexity is $O\left(n^{1-\log_\sigma(1/\gamma)/(1+2((1+\alpha)\log_\sigma(1+\alpha)+(1-\alpha)\log_\sigma(1-\alpha)))}\right)$.
In any case, we are able to achieve a sublinear complexity of $O(n^\lambda)$, where

$$\lambda = \max\left(\frac{\alpha}{\alpha+(1-c/\sqrt{\sigma})\log_\sigma(1/\gamma)}, \; 1 - \frac{\log_\sigma(1/\gamma)}{1+2((1+\alpha)\log_\sigma(1+\alpha)+(1-\alpha)\log_\sigma(1-\alpha))}\right)$$

Which of the two complexities dominates yields a rather complex condition that depends on the error level α, but in both cases $\lambda < 1$ if $\alpha < 1 - c/\sqrt{\sigma}$. If σ is large enough ($\sigma \geq 24$ for $c = e$), the complexity corresponding to j_2 always dominates. However, it is possible that j_1 or j_2 are outside the bounds of case (b) (i.e. they are too small). In this case we would use the minimum possible $j = (m+k)/\log_\sigma n$, and the third term would dominate the cost, for an overall complexity of $O(n^{1-\log_\sigma(1/\gamma)/(1+\alpha)})$. This complexity is also sublinear if $\alpha < 1 - c/\sqrt{\sigma}$.

(c) $\sigma^{L(k/j)} < n \leq \sigma^{(m+k)/j}$,   or $k/(1-c/\sqrt{\sigma}) < j\log_\sigma n \leq m+k$
The search cost in this intermediate case is

$$j\left(\sigma^{L(k/j)} + \frac{\sigma^{k/j}(\log_\sigma n)^{2k/j}}{(k/j)^{2k/j}} + \gamma^{m/j}m^2 n\right)$$

where the first two terms decrease with j and the last one increases. Repeating the same process as before, we find that the first and third term meet again at $j = j_1$ with the same complexity. We could not solve exactly where the second and third term meet. We found

$$j_3 = \frac{m(\alpha + 2\alpha\log_\sigma\log_\sigma n + \log_\sigma\frac{1}{\gamma} - 2\alpha\log_\sigma\frac{m}{j_3})}{\log_\sigma(m^2 n)} \approx \frac{m(\alpha + \log_\sigma\frac{1}{\gamma})}{\log_\sigma(m^2 n)}$$

and since the solution is approximate, the terms are not exactly equal at j_3. The second term is $O\left(n^{\alpha(1+2\log_\sigma(1/\gamma))/(\alpha+\log_\sigma(1/\gamma))}\right)$, slightly higher than the third. Again, it is possible that j_3 is out of the bounds of case (c) and we have to use the same limiting value as before.

The conclusion is that, despite that the exact formulation is complex, we have sublinear complexity for $\alpha < 1 - c/\sqrt{\sigma}$, as well as formulas for the optimum j to use, which is $\Theta(m/\log_\sigma n)$ with a complicated constant.

For larger α values the pattern partitioning method gives linear complexity and we need to resort to the traditional suffix tree traversal ($j = 1$). As shown in [8, 25], it is very unlikely that this limit of $1 - c/\sqrt{\sigma}$ can be improved, since there are too many real approximate occurrences in the text.

An interesting fact that is shown in the experiments is that in many cases the optima are out of bounds and hence the best is to put j in the limit of cases (b) and (c), just where the search of the subpieces become full searches. This shows that a technique that is simple and the best choice in most cases is to select $j = (m + k)/\log_\sigma n$, for a complexity of

$$O\left(n^{1-\frac{\log_\sigma(1/\gamma)}{1+\alpha}}\right) \;=\; O\left(n^{\frac{2(\alpha+H_\sigma(\alpha))}{1+\alpha}}\right)$$

where $H_\sigma(\alpha) = -\alpha\log_\sigma \alpha - (1-\alpha)\log_\sigma(1-\alpha)$ is the base-σ entropy function.

3.3 The Limits of the Method

Let us pay some attention to the limits of our hybrid method (Figure 4).

Since $j = \Theta(m/\log_\sigma n)$, the best j becomes 1 (i.e. no pattern partitioning) when $n > \sigma^{\Theta(m)}$ (this is because the cost of verifications dominates over suffix tree traversal). The best j is $\geq k+1$ for $n < \sigma^{\Theta(1/\alpha)}$. Since in this case we search the pieces with zero errors (i.e. $\lfloor k/(k+1)\rfloor = 0$, recall Section 2.2), the search in the suffix tree costs $O(m)$, and later we have to verify all their occurrences. This is basically what the q-gram index of [7] does, except it prunes the suffix tree at depth q.

Finally, the only case where the index is not useful is when n is very small. We can increase j to be more resistant to small texts, but the limit is $j = k+1$, and using that j the index ceases to be useful for $n < \sigma^{\frac{1}{1-c/\sqrt{\sigma}}} \leq \sigma^{1/\alpha}$. We have also to keep sublinear the cost of verifications, i.e. $n\gamma^{1/\alpha} = o(1)$, which happens for $\alpha < 1/\log_{1/\gamma} n$. This requires, in particular, that $m = \Omega(\log n)$.

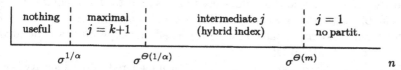

nothing useful	maximal $j = k+1$	intermediate j (hybrid index)	$j = 1$ no partit.
$\sigma^{1/\alpha}$	$\sigma^{\Theta(1/\alpha)}$	$\sigma^{\Theta(m)}$	n

Fig. 4. The j values to be used according to n.

This last consideration helps also to understand how is it possible to have a sublinear-time index based on filtering when there is a fixed matching probability per text position (γ^m), and therefore the verification cost must be $\Omega(n)$. The trick is that in fact we assume $m = \Omega(\log n)$, that is, we have to search longer patterns as the text grows. As we can tune j, we softly move to $j = 1$ (then eliminating verification costs) when n becomes large with respect to m. This "trick" is also present in the sublinearity result of Myers' index [23], and implicit in similar results on natural language texts [6, 25].

4 Experimental Results

We first validate some of the analytical results of the paper and later compare our indices against the other existing proposals. We used two different texts:

- DNA text ("h.influenzae"), a 1.34 Mb file. This file is called DNA in our tests, and H-DNA is the first half megabyte of it. In this case $\sigma = 4$.
- English literary text (from B. Franklin), filtered to lower-case and the separators converted into a single space. This text has 1.26 Mb, and is called FRA in the experiments. H-FRA is the first half megabyte of FRA. Given how the analysis uses the σ value, it is unrealistic to set it to the alphabet size, because the text is biased. It is much better to consider that $1/\sigma$ must be the probability that two random letters are equal. This sets $\sigma = 12.85$.

The texts are rather small, in some cases too small to appreciate the speedup obtained with some indices. This is because we had RAM problems to build suffix trees for larger texts. However, the experiments still serve to obtain basic performance numbers on the different indices.

We have tested short and medium-size patterns, searching with 1, 2 and 3 errors the short ones and with 2, 4 and 6 the medium ones. The short patterns were of length 10 for DNA and 8 for English, and the medium ones were of length 20 and 16, respectively[5]. We selected 1000 random patterns from each file and use the same set for all the k values of that length, and for all the indices.

4.1 Validating the Analysis

We first show that the suffix tree traversal has sublinear complexity. We built the suffix tree of incremental prefixes of FRA and DNA, from 100 Kb to 800 Kb (larger texts start to give I/O problems that disturb the CPU measures). According to our analysis, the m, k and σ values used correspond to intermediate text sizes (case (b) of Section 3.1) for $n = 4$Kb..4Mb on DNA and for $n = 40$Kb..8Gb on FRA. Hence, we are clearly in case (b) in all our experiments. The analysis predicts a complexity of $O((\log n)^{2k})$.

Figure 5 shows the user time as n grows, from where the sublinearity is clear. We have used least squares with the model $t = a \ln(n)^b$ to find out the empirical complexity and present it compared to the analytical complexity. The error of the approximation is always below 5%. We see that the analysis approximates reasonably the empirical results, despite the many simplifications done.

We consider now the optimal j value for pattern partitioning. Table 1 presents the query time using different j values in our index, for the FRA, H-FRA, DNA, and H-DNA texts. As it can be seen, there are big differences in time depending on j, and the optimum is a rather small j value (always 1 on short patterns). This matches reasonably our formulas. In fact, once properly rounded, our analysis recommends the correct j values. As mentioned before, the relevant value is always in the limit between cases (b) and (c).

Figure 6 shows the user time for long patterns, as n grows, using pattern partitioning with $j = 2$. This time we have used least squares with the model $t = an^b$. The error of the approximation is always below 2%. It can be seen

---

[5] This is because of the restrictions of Myers' index intersected with our interest in moderate-length patterns.

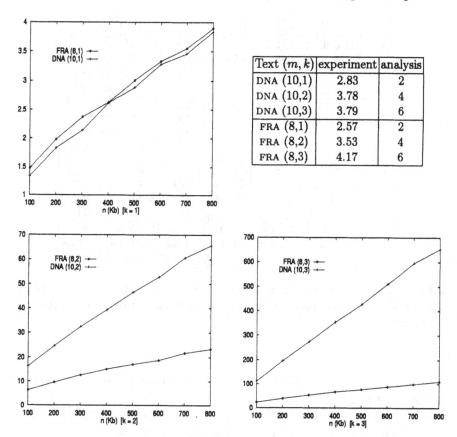

Text (m,k)	experiment	analysis
DNA $(10,1)$	2.83	2
DNA $(10,2)$	3.78	4
DNA $(10,3)$	3.79	6
FRA $(8,1)$	2.57	2
FRA $(8,2)$	3.53	4
FRA $(8,3)$	4.17	6

Fig. 5. User query time (in milliseconds) for short patterns as n grows for $k = 1$ to 3, using $j = 1$. On the top right, the empirical and analytical exponent of $\log n$.

that also in this case the analysis approximates reasonably the empirical results, slightly overestimating in most cases. The combination DNA $(20,6)$ is not included because it takes too long and already the case $(20,4)$ was clearly linear.

4.2 Comparison Against Others

We compare our index with the other existing proposals. However, as the task to program an index is rather heavy, we have only considered the other indices that are already implemented. The indices included in this comparison are

Myers': The index proposed by Myers [23]. We use the implementation of the author, which works for some m values only (that depend on σ and n).
Cobbs': The index proposed by Cobbs [13]. We use the implementation of the author, not optimized for space. The code is restricted to work on an alphabet of size 4 or less, so it is only built on DNA.

Text	$(10,1)$	$(10,2)$	$(10,3)$	$(20,2)$	$(20,4)$	$(20,6)$
DNA	**1: 6.81**	**1: 134.4**	**1: 1044**	1: 56.81	1: 1989	1: 10075
	2: 2391	2: 2585	2: 2756	**2: 15.80**	2: 1033	2: 9525
				3: 802.8	**3: 1010**	**3: 8841**
					4: 5862	4: 39077
H-DNA	**1: 2.71**	**1: 44.29**	**1: 394.6**	1: 23.72	1: 499.9	**1: 2308**
	2: 645.0	2: 715.3	2: 860.4	**2: 6.01**	**2: 305.2**	2: 2482
				3: 232.7	3: 305.9	3: 2464
					4: 1520	4: 10339

Text	$(8,1)$	$(8,2)$	$(8,3)$	$(16,2)$	$(16,4)$	$(16,6)$
FRA	**1: 6.11**	**1: 42.82**	**1: 215.2**	1: 35.98	1: 482.9	1: 2204
	2: 180.6	2: 1754	2: 19600	**2: 13.30**	**2: 88.22**	**2: 464.0**
				3: 90.71	3: 736.6	3: 4718
H-FRA	**1: 2.68**	**1: 14.28**	**1: 60.91**	1: 13.39	1: 126.4	1: 542.4
	2: 61.43	2: 601.1	2: 4920	**2: 5.30**	**2: 30.70**	**2: 146.4**
				3: 32.72	3: 255.5	3: 1538

Table 1. User query time (in milliseconds) for different (m, k) values (heading rows). Inside each cell we show the cost for different j values. The optimum is in boldface.

Samples(q): Our index based on q-grams presented in [7]. We show the results for $q = 4$ to 6.

Dfs(a/p): Our new index based on suffix trees. We show the results for the base technique (a) and pattern partitioning (p) with optimal j.

In particular, approximate searching on other q-gram indices [31] is not yet implemented and therefore is excluded from our tests. We know, however, that their space requirements are low (close to a word-retrieving index), but also that since the index simulates the on-line algorithm [30], its tolerance to errors is quite low (see [8, 25], for example).

All the indices were set to show the matches they found, in order to put them in a reasonably real scenario. We present the time to build the indices and the space they take in Table 2.

The first clear result of the experiment is that the space usage of the indices is very high. In particular, the indices based on suffix trees (Dfs and Cobbs') take 35 to 65 times the text size. This outrules them except for very small texts (for instance, building Cobbs' index on 1.34 Mb took 12 hours of real time in our machine of 64 Mb of RAM). From the other indices, Myers' took 7-9 times the text size, which is much better but still too much in practice. The best option in terms of space is the Samples index, which takes from 1.5 to 7 times the text size, depending on q and σ. The larger q or σ, the larger the index. Samples(5), which takes 2-5 times the text size, performs well at query time.

Compared to its size, Myers' index was built very quickly. The Dfs index, on the other hand, was built faster than Cobbs'. Notice that suffix trees are built

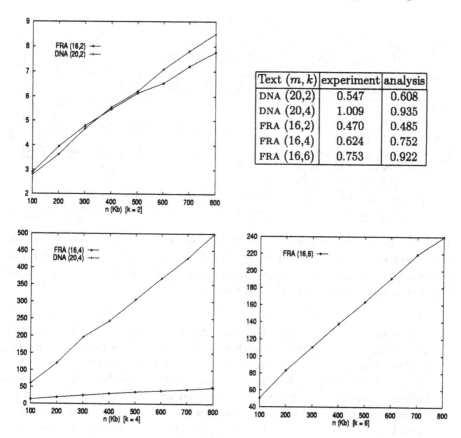

Text (m,k)	experiment	analysis
DNA $(20,2)$	0.547	0.608
DNA $(20,4)$	1.009	0.935
FRA $(16,2)$	0.470	0.485
FRA $(16,4)$	0.624	0.752
FRA $(16,6)$	0.753	0.922

Fig. 6. User query time (in milliseconds) for medium patterns as n grows for $k = 2$ to 6, using $j = 2$. On the top right, the empirical and analytical exponent of n.

quickly when they fit in RAM (as in the half-megabyte texts), but for larger texts the construction time is dominated by the I/O, and it increases sharply.

We consider now query time. Tables 3 and 4 present a comparison between the different indices, using for Dfs(p) the optimum j value of Table 1 (only for medium patterns, since for short ones Dfs(a) is always better). The system time is included because it is dominant in many cases. We include also the time of on-line searching for comparison purposes (we use the fastest on-line algorithm for each case). The results clearly show a number of facts.

- The indices work well only for moderate error levels. For larger texts the ratio indexed/on-line should improve. However, when I/O time is considered many indices seem useless, and it is not so clear that this improves for larger texts. This depends on the amount of main memory available, and is a consequence of most indices not being designed to work on secondary memory. This is a very important issue that has been rarely addressed.

Index	DNA	H-DNA	FRA	H-FRA
Myers'	5.84u+0.35s 10.68 Mb (7.97X)	2.08u+0.12s 4.50 Mb (9.00X)	5.22u+0.34s 9.39 Mb (7.46X)	2.01u+0.12s 4.18 Mb (8.35X)
Samples (4)	5.53u+0.19s 2.04 Mb (1.52X)	1.95u+0.10s 0.77 Mb (1.53X)	15.05u+0.41s 3.48 Mb (2.77X)	5.90u+0.24s 1.48 Mb (2.98X)
Samples (5)	7.37u+0.24s 2.48 Mb (1.85X)	2.62u+0.08s 0.94 Mb (1.87X)	20.82u+0.70s 5.18 Mb (4.11X)	8.70s+0.35s 2.32 Mb (4.65X)
Samples (6)	10.53u+0.32s 2.90 Mb (2.16X)	3.88u+0.13s 1.11 Mb (2.23X)	32.86u+1.34s 7.65 Mb (6.07X)	13.19u+0.97s 3.54 Mb (7.07X)
Cobbs'	108.70u+532.81s 87.99 Mb (65.67X)	30.50u+76.06s 32.93 Mb (65.85X)	n/a	
Dfs	30.89u+104.17s 52.25 Mb (38.99X)	6.48u+0.42s 19.55 Mb (39.10X)	28.46u+76.86s 44.66 Mb (35.45X)	6.43u+0.61s 17.66 Mb (35.32X)

Table 2. Times (in seconds) to build the indices and their space overhead. The time is separated in the CPU part ("u") and the I/O part ("s"). The space is expressed in megabytes, and also the ratio index/text is shown in the format rX, meaning that the index takes r times the text size.

- Our strategy Dfs(a) of using a simpler traversal algorithm on the suffix tree and in return using a faster search algorithm definitively pays off, since our implementation is 3 to 40 times faster than Cobbs', and it is the fastest choice for small m and k values. Independently of this fact, the suffix tree indices improve on larger alphabets, but they are much more sensitive to the growth of m or k. In fact, the differences between FRA and DNA are due to the different values of m used. The big problem with this type of index is of course the huge space requirements it poses.

- Myers' index behaves better on short patterns, when less splitting is necessary. It works well for DNA but it worsens on English text. We conjecture that the non-randomness may play a role here: the index takes internally $q \doteq \log_\sigma n$ to avoid searching a number of nonexistent samples that are at distance k or less from the pattern (in our case it took $q = 10$ for DNA and $q = 4$ for English). However, in biased texts like English, a lot of q-grams are not present anyway, and the index pays to search all of them. For DNA the index is a good alternative, since although it is up to 13 times slower than Dfs(a), it takes 4 times less space. It is also better than the Samples index when the pattern is short, but not when pattern partitioning is necessary.

- The Samples index reaches its optimum performance for q between 5 and 6, depending on the case. Unlike Myers', this index works better on English text than on DNA. In DNA it produces a small index (4 times smaller than Myers') but in general has worse search time. The index for $q = 5$ on English text is half the size of Myers' index, and it also obtains good results for medium patterns and low error levels.

- Dfs(p), which works on the same data structure of Dfs(a), improves over it when the patterns are not very short and the error level is not too high. When applicable, its query time is by far the lowest among all the indices.

Index	k	DNA $(m = 10)$	H-DNA $(m = 10)$	FRA $(m = 8)$	H-FRA $(m = 8)$
On-line	1	*131.0/21.35*	*55.01/15.24*	*59.74/17.31*	*29.99/9.00*
	2	*152.6/20.56*	*62.41/15.48*	*114.8/20.86*	*52.77/11.56*
	3	*188.7/20.96*	*84.20/15.33*	*142.2/20.56*	*60.90/13.76*
Myers'	1	0.29/1.74	0.64/2.15	7.04/8.04	6.17/7.29
	2	0.97/2.18	1.53/2.74	23.5/21.4	20.2/18.2
	3	6.29/6.79	8.17/8.10	22.4/20.8	20.9/18.5
Samples (4)	1	1.80/6.66	1.72/5.48	0.75/2.01	0.75/1.76
	2	9.33/26.7	9.10/23.4	3.30/9.54	2.69/2.68
	3	30.7/93.4	25.5/73.6	13.8/30.3	13.6/26.7
Samples (5)	1	0.91/2.81	0.93/2.38	0.75/1.91	0.77/1.74
	2	9.88/27.7	9.35/23.5	4.92/10.4	3.47/7.07
	3	36.4/97.2	30.9/77.3	23.9/38.9	21.5/33.5
Samples (6)	1	0.90/2.71	0.93/2.35	0.89/2.06	0.86/1.82
	2	11.3/29.4	10.9/24.6	6.81/12.5	4.81/8.99
	3	57.3/119	49.0/92.5	39.3/52.8	38.9/47.7
Cobbs'	1	0.83/1.98	1.85/3.67	n/a	n/a
	2	3.85/14.9	6.04/19.1		
	3	17.9/84.5	21.8/79.3		
Dfs(a)	1	**0.05/0.15**	**0.05/0.04**	**0.10/0.25**	**0.09/0.07**
	2	**0.88/2.72**	**0.71/0.57**	**0.37/0.96**	**0.27/0.22**
	3	5.53/16.9	4.69/3.96	1.51/4.39	1.01/0.82

Table 3. Query time for short patterns and for 1, 2 and 3 errors. The on-line algorithm shows time in milliseconds in the format "user/system", in italics. The indexed algorithms show the fraction they take of the time of the on-line algorithm. The format is "a/b", where a considers only user time and b considers both. The fastest indexed times are in boldface.

5   Conclusions and Future Work

We have presented a new indexing scheme for approximate string matching. The main idea is to split the pattern in pieces to be searched with less errors, and use a suffix tree to find their approximate matches in the text. Later, we verify all their matches for an occurrence of the complete pattern. The splitting technique balances between traversing too many nodes of the suffix tree and verifying too many text positions. We have also shown how to traverse the suffix tree efficiently in practice. We have proved analytically that the resulting index has sublinear retrieval time (of the form $O(n^\lambda)$, where $0 < \lambda < 1$ if the error level is moderate). Finally, we have presented the first (as far as we know) experimental results that compare the different implemented indexing schemes, which show that the proposed idea improves over all the previously implemented approaches.

A remaining problem is that the suffix tree data structure needs too much space. We plan to replace it by a suffix array [21]. The suffix array contains the leaves of the suffix tree in left-to-right order, or equivalently the pointers to all the text suffixes in lexicographical order. The space requirement is in practice 4 times the text size, which is reasonable. Suffix tree nodes (i.e. subtrees) correspond to suffix array intervals. Any movement in the suffix tree can be simulated in $O(\log n)$ time in the suffix array, and therefore the final complexity is multiplied by $O(\log n)$ and the condition for time sublinearity is not affected. Finally, we are still free to use the j value we like (unlike q-gram indices, which are limited by q). In particular, we can easily implement specialized pattern partitioning approaches for biased texts as in [7], where the partitioning minimizes the total number of text positions to verify.

Index	k	DNA $(m = 20)$	H-DNA $(m = 20)$	FRA $(m = 16)$	H-FRA $(m = 16)$
On-line	2	*184.6/22.18*	*75.16/16.61*	*60.59/17.56*	*29.91/9.48*
	4	*311.4/21.70*	*116.0/15.79*	*116.3/20.83*	*50.71/14.98*
	6	*779.2/21.42*	*297.4/15.77*	*205.6/20.58*	*92.36/13.37*
Myers	2	0.67/1.69	0.91/1.97	7.03/8.06	10.9/10.9
	4	5.13/5.50	5.61/5.74	32.7/29.2	31.9/26.3
	6	16.9/16.8	17.7/17.3	26.5/25.0	25.2/23.1
Samples (4)	2	1.55/5.10	1.60/4.55	0.44/0.95	0.63/1.03
	4	6.14/13.4	6.16/12.4	2.08/4.62	2.03/4.01
	6	9.10/25.4	9.48/27.9	9.59/18.6	8.85/16.1
Samples (5)	2	0.60/1.93	0.64/1.73	0.38/0.75	0.62/0.91
	4	5.26/11.3	5.77/11.9	2.21/4.87	2.15/4.19
	6	10.0/25.5	10.7/26.4	14.8/23.6	12.7/19.6
Samples (6)	2	0.31/0.83	0.41/0.84	0.39/0.70	0.60/0.91
	4	5.61/11.7	6.18/12.1	2.71/5.13	2.51/4.42
	6	15.2/31.5	15.3/30.8	22.9/31.1	19.3/25.3
Cobbs'	2	3.93/11.7	6.60/16.0	n/a	n/a
	4	***	69.5/171		
	6	***	***		
Dfs(a)	2	0.31/1.19	0.32/0.26	0.59/1.49	0.45/0.34
	4	6.39/30.8	4.31/3.79	4.15/14.4	2.49/1.93
	6	14.6/64.9	7.76/7.37	10.7/42.0	5.87/5.13
Dfs(p)	2	**0.09/0.23**	**0.08/0.07**	**0.22/0.43**	**0.18/0.13**
	4	3.24/6.42	2.63/2.32	**0.76/1.92**	**0.61/0.47**
	6	11.3/12.6	7.76/7.37	2.26/6.05	1.59/1.38

*** One single query took more than 2 hours of elapsed time.

Table 4. Query time for medium patterns and for $k = 2$, 4 and 6. The on-line algorithm shows time in milliseconds in the format "user/system", in italics. The indexed algorithms show the fraction they take of the time of the on-line algorithm. The format is "a/b", where a considers only user time and b considers both. The fastest indexed times are in boldface.

Acknowledgements

We thank the nice comments of two referees, which helped to improve this work. We also thank Erkki Sutinen for his code to build the suffix tree, and Gene Myers and Archie Cobbs for sending us their implemented indices.

References

[1] A. Apostolico and Z. Galil. *Combinatorial Algorithms on Words*. Springer-Verlag, New York, 1985.

[2] M. Araújo, G. Navarro, and N. Ziviani. Large text searching allowing errors. In *Proc. WSP'97*, pages 2–20. Carleton University Press, 1997.

[3] R. Baeza-Yates. Text retrieval: Theory and practice. In *12th IFIP World Computer Congress*, volume I, pages 465–476. Elsevier Science, September 1992.

[4] R. Baeza-Yates and G. Gonnet. All-against-all sequence matching. Dept. of Computer Science, University of Chile, 1990.

[5] R. Baeza-Yates and G. Gonnet. Fast text searching for regular expressions or automaton searching on a trie. *J. of the ACM*, 43, 1996.

[6] R. Baeza-Yates and G. Navarro. Block-addressing indices for approximate text retrieval. In *Proc. ACM CIKM'97*, pages 1–8, 1997.

[7] R. Baeza-Yates and G. Navarro. A practical q-gram index for text retrieval allowing errors. *CLEI Electronic Journal*, 1(2), 1998. http://www.clei.cl.

[8] R. Baeza-Yates and G. Navarro. Faster approximate string matching. *Algorithmica*, 23(2):127–158, 1999. Preliminary version in *Proc. CPM'96, LNCS 1075*.

[9] R. Baeza-Yates and C. Perleberg. Fast and practical approximate pattern matching. *Information Processing Letters*, 59:21–27, 1996.

[10] A. Blumer, J. Blumer, D. Haussler, A. Ehrenfeucht, M. Chen, and J. Seiferas. The samllest automaton recognizing the subwords of a text. *Theoretical Computer Science*, 40:31–55, 1985.

[11] W. Chang and J. Lampe. Theoretical and empirical comparisons of approximate string matching algorithms. In *Proc. CPM'92*, LNCS 644, pages 172–181, 1992.

[12] W. Chang and T. Marr. Approximate string matching and local similarity. In *Proc. CPM'94*, LNCS 807, pages 259–273, 1994.

[13] A. Cobbs. Fast approximate matching using suffix trees. In *Proc. CPM'95*, pages 41–54, 1995. LNCS 937.

[14] M. Crochemore. Transducers and repetitions. *Theoretical Computer Science*, 45:63–86, 1986.

[15] M. Farach, P. Ferragina, and S. Muthukrishnan. Overcoming the memory bottleneck in suffix tree construction. In *Proc. SODA'98*, pages 174–183, 1998.

[16] Z. Galil and K. Park. An improved algorithm for approximate string matching. *SIAM J. on Computing*, 19(6):989–999, 1990.

[17] G. Gonnet. A tutorial introduction to Computational Biochemistry using Darwin. Technical report, Informatik E.T.H., Zuerich, Switzerland, 1992.

[18] P. Jokinen and E. Ukkonen. Two algorithms for approximate string matching in static texts. In *Proc. MFCS'91*, volume 16, pages 240–248. Springer-Verlag, 1991.

[19] D. Knuth. *The Art of Computer Programming*, volume 3: Sorting and Searching. Addison-Wesley, 1973.

[20] G. Landau and U. Vishkin. Fast parallel and serial approximate string matching. *J. of Algorithms*, 10:157–169, 1989.

[21] U. Manber and G. Myers. Suffix arrays: a new method for on-line string searches. In *Proc. ACM-SIAM SODA'90*, pages 319–327, 1990.

[22] U. Manber and S. Wu. GLIMPSE: A tool to search through entire file systems. In *Proc. USENIX Technical Conference*, pages 23–32, Winter 1994.

[23] E. Myers. A sublinear algorithm for approximate keyword searching. *Algorithmica*, 12(4/5):345–374, Oct/Nov 1994.

[24] G. Myers. A fast bit-vector algorithm for approximate pattern matching based on dynamic progamming. In *Proc. CPM'98*, LNCS 1448, pages 1–13, 1998.

[25] G. Navarro. *Approximate Text Searching*. PhD thesis, Dept. of Computer Science, Univ. of Chile, December 1998. Technical Report TR/DCC-98-14. ftp://-ftp.dcc.uchile.cl/pub/users/gnavarro/thesis98.ps.gz.

[26] G. Navarro and R. Baeza-Yates. Improving an algorithm for approximate pattern matching. Technical Report TR/DCC-98-5, Dept. of Computer Science, Univ. of Chile, 1998. Submitted.

[27] S. Needleman and C. Wunsch. A general method applicable to the search for similarities in the amino acid sequences of two proteins. *J. of Molecular Biology*, 48:444–453, 1970.

[28] P. Sellers. The theory and computation of evolutionary distances: pattern recognition. *J. of Algorithms*, 1:359–373, 1980.

[29] F. Shi. Fast approximate string matching with q-blocks sequences. In *Proc. WSP'96*, pages 257–271. Carleton University Press, 1996.

[30] E. Sutinen and J. Tarhio. On using q-gram locations in approximate string matching. In *Proc. ESA'95*, LNCS 979, pages 327–340, 1995.

[31] E. Sutinen and J. Tarhio. Filtration with q-samples in approximate string matching. In *Proc. CPM'96*, LNCS 1075, pages 50–61, 1996.

[32] J. Tarhio and E. Ukkonen. Approximate Boyer-Moore string matching. *SIAM J. on Computing*, 22(2):243–260, 1993.

[33] E. Ukkonen. Approximate string matching over suffix trees. In *Proc. CPM'93*, pages 228–242, 1993.

[34] Esko Ukkonen. Finding approximate patterns in strings. *J. of Algorithms*, 6:132–137, 1985.

[35] S. Wu and U. Manber. Fast text searching allowing errors. *Comm. of the ACM*, 35(10):83–91, October 1992.

[36] S. Wu, U. Manber, and E. Myers. A sub-quadratic algorithm for approximate limited expression matching. *Algorithmica*, 15(1):50–67, 1996.

Appendix: Probability of Reaching a Suffix Tree Node

We need to determine which is the probability of the automaton being active at a given node of depth ℓ in the suffix tree. Notice that the automaton is active if and only if some state of the last row is active (recall Figure 1). This is equivalent to some *prefix* of the pattern matching with k errors or less the text substring represented by the suffix tree node under consideration.

We are therefore interested in the probability of a pattern prefix of length m' matching a text substring of length ℓ. This analysis is an extension of that of [8]. As Figure 7 illustrates, at least $\ell - k$ text characters text must match the pattern when $\ell \geq m'$, and at least $m' - k$ pattern characters must match the text whenever $m' \geq \ell$. Hence, the probability of matching is upper bounded by

$$\frac{1}{\sigma^{\ell-k}}\binom{\ell}{\ell-k}\binom{m'}{\ell-k} \qquad \text{or} \qquad \frac{1}{\sigma^{m'-k}}\binom{\ell}{m'-k}\binom{m'}{m'-k}$$

depending on whether $\ell \geq m'$ or $m' \geq \ell$, respectively (the combinatorials count all the possible locations for the matching characters in both strings). Notice that this imposes that $m' - k \leq \ell \leq m' + k$. We also assume $m' \geq k$, since otherwise the matching probability is 1. Since $k \leq m' \leq m$, we have that $\ell \leq m + k$, otherwise the matching probability is zero. Hence the matching probability is 1 for $\ell \leq k$ and 0 for $\ell > m+k$, and we are interested in what happens in between.

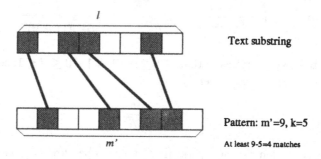

Fig. 7. Upper bound for the probability of matching. At least $\max(m' - k, \ell - k)$ characters must match, since otherwise it would not be possible to convert one string into the other.

Since we are interested in any pattern prefix matching the current text substring, we add up all the possible lengths from k to m:

$$\sum_{m'=k}^{\ell} \frac{1}{\sigma^{\ell-k}}\binom{\ell}{\ell-k}\binom{m'}{\ell-k} + \sum_{m'=\ell+1}^{m} \frac{1}{\sigma^{m'-k}}\binom{\ell}{m'-k}\binom{m'}{m'-k}$$

In the analysis that follows, we call $\beta = k/\ell$, where $\alpha/(1+\alpha) \leq \beta \leq 1$. We will prove that, after some depth ℓ in the suffix tree, the matching probability is $O(\gamma(\beta)^{\ell})$, for some $\gamma(\beta) < 1$. We begin with the first summation. We analyze its largest term (the last one), which is

$$\frac{1}{\sigma^{\ell-k}}\binom{\ell}{k}^2$$

and by using Stirling's approximation $x! = (x/e)^x \sqrt{2\pi x}(1 + O(1/x))$ we have

$$\frac{1}{\sigma^{\ell-k}}\left(\frac{\ell^{\ell}\sqrt{2\pi\ell}}{k^k(\ell-k)^{\ell-k}\sqrt{2\pi k}\sqrt{2\pi(\ell-k)}}\right)^2 \left(1 + O\left(\frac{1}{\ell}\right)\right)$$

which is

$$\left(\frac{1}{\sigma^{1-\beta}\beta^{2\beta}(1-\beta)^{2(1-\beta)}}\right)^{\ell} \ell^{-1}\left(\frac{1}{2\pi\beta(1-\beta)} + O\left(\frac{1}{\ell}\right)\right)$$

where the last step is done using Stirling's approximation to the factorial. This formula is of the form $\gamma(\beta)^{\ell} O(1/\ell)$, where we define

$$\gamma(x) = \frac{1}{\sigma^{1-x} x^{2x} (1-x)^{2(1-x)}} \tag{1}$$

The whole first summation is bounded by $\ell - k$ times the last term, which gives $(\ell - k)\gamma(\beta)^{\ell} O(1/\ell) = O(\gamma(\beta)^{\ell})$. Therefore the first summation is exponentially decreasing with ℓ if and only if $\gamma(\beta) < 1$, i.e.

$$\sigma > \left(\frac{1}{\beta^{2\beta}(1-\beta)^{2(1-\beta)}}\right)^{\frac{1}{1-\beta}} = \frac{1}{\beta^{\frac{2\beta}{1-\beta}}(1-\beta)^2} \tag{2}$$

It is easy to show analytically that $e^{-1} \le \beta^{\frac{\beta}{1-\beta}} \le 1$ if $0 \le \beta \le 1$, so it suffices that $\sigma > e^2/(1-\beta)^2$, or equivalently

$$\beta < 1 - \frac{e}{\sqrt{\sigma}} \tag{3}$$

is a sufficient condition for the largest (last) term to be $O(\gamma(\beta)^{\ell})$, as well as the whole first summation.

We address now the second summation, which is more complicated. In this case, it is not clear which is the largest term. We can see each term as

$$\frac{1}{\sigma^r}\binom{\ell}{r}\binom{k+r}{k}$$

where $\ell - k < r \le m - k$. By considering $r = x\ell$ ($x \in [1 - \beta, m/\ell - \beta]$) and applying again Stirling's approximation, we maximize the base of the resulting exponential, which is

$$h(x) = \frac{(x+\beta)^{x+\beta}}{\sigma^x x^{2x}(1-x)^{1-x}\beta^\beta}$$

Elementary calculus leads to solve a second-degree equation that has roots in the interval $[1-\beta, \infty)$ only if $\sigma \le \beta/(1-\beta)^2$. Since due to Eq. (3) we are only interested in $\sigma \ge 1/(1-\beta)^2$, $\delta h(x)/\delta x$ does not have roots, and the maximum of $h(x)$ is at $x = 1 - \beta$. That means $r = \ell - k$, i.e. the first term of the second summation, which is the same largest term of the first summation.

We conclude that the probability of being active at a node of level ℓ is upper bounded by

$$\frac{m-k}{\ell}\gamma(\beta)^{\ell}\left(1 + O\left(\frac{1}{\ell}\right)\right) = O\left(\gamma(\beta)^{\ell}\right)$$

and therefore Eq. (3) is valid for the whole summation. When $\gamma(\beta)$ is 1, the probability is very high: only considering the term $m' = \ell$ we have $\Omega(1/\ell)$.

Hence, the result is that the matching probability is very high for $\beta = k/\ell \ge 1 - e/\sqrt{\sigma}$, and otherwise it is $O(\gamma(\beta)^{\ell})$, where $\gamma(\beta) < 1$.

Although the e appeared via a bounding condition, we can see that this bound is tight: we take $\log_\sigma$ on both sides of the condition $\gamma(\beta) < 1$ and get

$$1 - \beta + 2(\beta \log_\sigma \beta + (1 - \beta) \log_\sigma (1 - \beta)) \;>\; 0$$

and by replacing $x = 1 - \beta$ and using $\ln(1 - x) = -x + O(x^2)$ we have

$$x \ln \sigma + 2(x \ln x - (1 - x)(x + O(x^2))) \;=\; x \ln \sigma + 2x \ln x - 2x + O(x^2) \;>\; 0$$

from where divide by x to obtain

$$x \;>\; \frac{e}{\sqrt{\sigma}} e^{O(x)} \;=\; \frac{e}{\sqrt{\sigma}} (1 + O(x)) \;=\; \frac{e}{\sqrt{\sigma}} (1 + O(1/\sqrt{\sigma}))$$

We conclude that the precise limit for $\beta = 1 - x$ is

$$\beta \;<\; 1 - \frac{e}{\sqrt{\sigma}} + O(1/\sigma)$$

As we show experimentally in [8], however, the real β limit is very close to the same formula if e is replaced by $c = 1.09$. The reason is that the bounding condition (Figure 7) we use is not strong enough: for instance, we could avoid replacements in the edit distance and the bound would be the same. In the paper we use a limit of the form $\beta = 1 - c/\sqrt{\sigma}$, knowing that we can prove $c \le e$ but in practice it holds $c \approx 1$.

The Compression of Subsegments of Images Described by Finite Automata

Juhani Karhumäki*[1], Wojciech Plandowski**[23], and Wojciech Rytter[34]

[1] Department of Mathematics, Turku University, Finland
karhumak@cs.utu.fi
[2] Turku Centre for Computer Science, DataCity 4th floor, Turku, Finland.
[3] Instytut Informatyki, Uniwersytet Warszawski
Banacha 2, 02–097 Warszawa, Poland
wojtekpl@mimuw.edu.pl, rytter@mimuw.edu.pl
[4] Department of Computer Science, University of Liverpool

Abstract. We investigate how the size of the compressed version of a 2-dimensional image changes when we cut off a part of it, e.g. extracting a photo of one person from a photo of a group of people. 2-dimensional compression is considered in terms of finite automata. Let n be the size of the smallest acyclic automaton which describes an image T. We show that the tight bound for the compression size of a subsegment (subimage) in the deterministic case is $\Theta(n^{2.5})$ and in the weighted case is $\Theta(n)$. We also show how to construct efficiently the compressed representation of subsegments given the compressed representation of the whole image. Two applications of subsegments compression are more efficient automata-compressed pattern-matching and the first polynomial time algorithm for the fully compressed pattern-checking problem for weighted automata.

1 Introduction

The compression size of images is of crucial importance in multimedia systems and in transferring large images in WWW. Deterministic and weighted finite automata are successful tools for compressing 2-dimensional images, see [3,4,5,7]. There are several software packages using this type of compression, see [9,4]. Finite automata can describe quite complicated images, for example deterministic automata can describe the Hilbert's curve with a given resolution, see [11], while weighted automata can describe even much more complicated curves, see also [3,4,5]. The objects considered are potentially exponentially compressed, so algorithms which apply decompression are theoretically not polynomial time algorithms. In practice exponential compression does not usually appear, nevertheless the compression ratio for two-dimensional images can be very high, especially compared with the one dimensional case (for example for images corresponding to fractals having short description). For one-dimensional words there

---

* Supported by Academy of Finland under grant 14047.
** Supported partially by the grant KBN 8T11C03915.

M. Crochemore, M. Paterson (Eds.): CPM'99, LNCS 1645, pp. 186–195, 1999.

exist polynomial-time deterministic algorithms for compressed and fully compressed pattern-matching [8,10,12], despite the fact that the uncompressed size of objects could be exponential. However these problems become much harder in the two-dimensional case. Our main result is a constructive proof of the fact that compressed size of subsegments of an automata-compressed images grows only polynomially. This contrasts with the exponential grow of compression size of subimages for compression in terms of recursive description, see [2]. Our alphabet is $\Sigma = \{0, 1, 2, 3\}$, the elements of which correspond to four quadrants of a square array, see Figure 1.

1	3
0	2

Fig. 1. Enumeration of the quadrants.

A word w of length k over Σ can be interpreted, in a natural way, as a unique address of a pixel x of a $2^k \times 2^k$ image (array), we write $address(x) = w$. The length k is called the *resolution* of the image. For a language $L \subseteq \Sigma^+$ denote by $Image_k(L)$ the $2^k \times 2^k$ black-and-white image such that the color of a given pixel x is black iff $address(x) \in L$. We consider also the weighted languages, formally they correspond to functions which associates with each word w a value $weight_L(w)$. A weighted language L over Σ and resolution k determine the gray-tone image $Image_k(L)$ such that the color of a given pixel x equals $weight_L(address(x))$. If all words in L are of the same length k then we can omit the subscript k and write $Image(L)$. Our description of the language is in terms of finite (unweighted or weighted) automaton A. We define $Image_k(A) = Image_k(L(A))$, where $L(A)$ is the language accepted by A.

Representation in terms of acyclic deterministic automata is equivalent to a representation by a 2-dimensional grammar, each production corresponds to the way of decomposing a square into 4 smaller subsquares of a same shape. For the automaton from Figure 2 we can define the subsquare corresponding to state $s1$ by ($\hat{\emptyset}$ denotes a blank subsquare of appropriate shape):

$$s_1 \rightarrow \begin{bmatrix} s_2 & \hat{\emptyset} \\ s_2 & s_2 \end{bmatrix}$$

We consider a subsegment image P and the host image T described by automata of sizes m and n. Denote by $Compress(P)$ and $Compress(T)$ the automata describing P and T, respectively.

Our main problem is the **Subsegment Compression Problem**:

Instance: $Compress(T)$ representation of a $2^k \times 2^k$ square image, a point x in T and an integer $k' < k$. Let R be a square $2^{k'} \times 2^{k'}$ subsegment of T whose left-upper corner is positioned at x.

Question: What is the size of $Compress(R)$? What is the complexity of computing $Compress(R)$.

Another problem is the pattern-checking, it consists in testing a fixed occurrence of a large compressed image.) It is co-NP complete for 2-dimensional compressions in terms of recursive generations, see [2]. We define the depth of the automaton as the longest path from the initial state to an accepting state. An acyclic automaton can be transformed to an equivalent automaton in which for each state q each path from the initial state to q has the same length. We say that a state q belongs to level t if all paths from the initial state to q are of length t. Here, the length of a path is the number of edges in the path. In our considerations we may restrict to acyclic automata since we consider only finite resolution images or more precisely finite resolution approximations of infinite resolution images.

Example. $Image(\{0, 1, 2\}^k) = S_k$ is the $2^k \times 2^k$ black-and-white square part of Sierpinski's triangle, see Figure 2 for the case $k = 4$. The corresponding smallest acyclic deterministic automaton accepting all paths describing black pixels has 5 states.

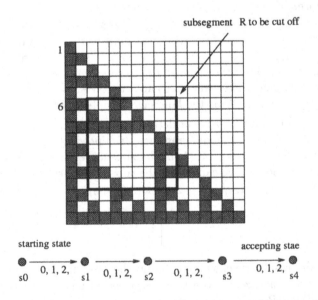

Fig. 2. The image S_4 and its smallest acyclic automaton. Edges which are not on accepting paths are disregarded.

2 The Subsegment Compression Problem for Deterministic Automata

Let $A = (\{0, 1, 2, 3\}, Q, q_0, \delta)$ be a deterministic acyclic automaton of depth n where Q is a set of states, $q_0 \in Q$ is the initial state, and $\delta : Q \times \Sigma^* \to Q$ is a partial transition function. The automaton A defines the language $L(A) = \{w : \delta(q_0, w)$ is defined $|w| = n$ and $\delta(q_0, w)$ is accepting$\}$. Note, that in the definition of an automaton we do not need to specify the single accepting state. For $q \in Q$, denote by $Image(q)$ the image which is generated by the automaton which is obtained from A by changing its initial state to q. Clearly, $Image(q_0) = Image(A)$.

A regular block of a $2^k \times 2^k$ image T is defined as follows. T is a regular block, and if B is a regular block then all its quadrants are also regular blocks. A square subsegment of the shape $2^t \times 2^t$ is said to be of *rank* t. Denote by $\hat{\emptyset}$ a square blank block consisting only of white pixels. We use the same notation for all possible sizes of $\hat{\emptyset}$, the size depends on the context. We can interpret the state q as a name of a regular block $X = Image(q)$, we write $name(X) = q$. If X is a blank block then we write $name(X) = \hat{\emptyset}$. We call states to be essential iff they are on a path to an accepting state.

Lemma 1. *Assume the whole image is not totally blank. The number of essential states of the smallest acyclic deterministic minimal automaton describing T equals the cardinality of different nonblank regular blocks of T.*

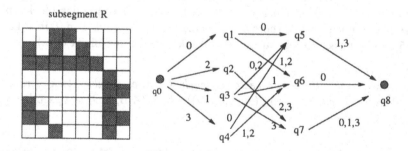

Fig. 3. The subsegment R of S_4 and the smallest acyclic automaton describing R.

We illustrate the lemma with the following example. The states at depth 2 of the automaton from Figure 3 corresponds to the blocks of the subsegment R in the way shown in Figure 4.

The crucial notion is that of a *pseudo-regular block*, defined as a square subsegment of a rank $t + 1$ of the corresponding image consisting of 4 adjacent *regular* blocks of rank T, the regular blocks themselves are also considered as pseudo-regular blocks. Define by $Pseudo\_Reg_t(T)$ the set of pseudo-regular blocks of

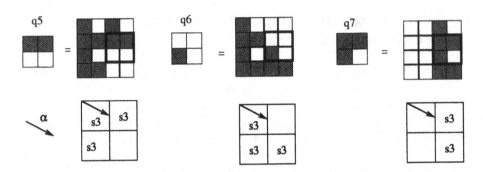

Fig. 4. Illustration of Lemma 1. Regular blocks of R (in bold) are parts of pseudo-blocks of T. There are 3 nonblank regular blocks of rank 1 in segment R from Figure 3, they correspond to the states of the corresponding automaton for R, each of these blocks is a part of a psuedo-regular block of rank 2 of T with the same vector α shown in the figure. We have $q5 = Sub(\alpha, s3, s3, \emptyset, s3)$, $q6 = Sub(\alpha, s3, s3, s3, \emptyset)$, $q7 = Sub(\alpha, \emptyset, \emptyset, s3, s3)$.

rank t of T. For a subblock X of a block X define the position of X in Y ($pos_X(Y)$) as the position of left-upper corner of Y inside X. For a block X of rank $t+1$ and vector α define $Sub_t(\alpha, X)$ to be the sublock Y of X such that $pos_X(Y) = \alpha$. Each pseudo-regular block X is identified by a composite name $(name(X_1), name(X_2), name(X_3), name(X_4))$ where $X_1, \ldots X_4$ are regular blocks which are quadrants of X, listed in the order corresponding to Figure 1. The compressed size of the subsegment can grow since the number of pseudo-regular blocks could be much larger than the number of regular blocks in the same image. For example there are 2 regular blocks of rank 1 in T in Figure 2, but 5 different pseudo-regular blocks of rank 1 (including blank ones).

Lemma 2.
For a given rank t there is a vector α_t such that each regular block of rank t in the subimage R equals $Sub(\alpha_t, (A, B, C, D))$ where (A, B, C, D) is a pseudo-regular block of T of rank $t+1$.

Theorem 1. *Assume the compression is in terms of deterministic automata. The compressed representation of a square subsegment R of T can be computed in $O(|Compress(T)|^{2.5})$ time.*

Proof.
The states of the automaton A' for the subsegment are tuples $(\alpha_k, (A, B, C, D))$ where (A, B, C, D) are pseudo-regular blocks of T of rank $k+1$, and α_k is a vector from Lemma 2. Due to Theorem 2 the number of pseudo-regular blocks is $O(n^{2.5})$, where $n = |Compress(T)|$.

We compute names of pseudo-regular blocks top down. However we consider only these pseudo-regular blocks which contain a nonblank block of the subsegment. If we know the names of pseudo-regular blocks of rank $t+1$ then each pseudo-regular block of rank t consists of 4 subblocks of rank t of a single pseudo-regular block of rank $t + 1$, the details will be given in the full version.

3   Tight Bounds for the Compression Size of Subimages

We need the following technical lemmas.

Lemma 3. *Let* $k_1, k_2 \ldots, k_r > 0$. *Then*

$$\sum_{i=2}^{r} \min(k_{i-1}^4, k_i^2 + k_{i+1}^2 + \ldots + k_r^2) \leq (k_1 + \ldots + k_r)^{2.5}.$$

Proof. We omit the technical proof (induction on r).

Lemma 4. *Each pseudo-regular block of rank i is a central block of a regular block of rank k, or a central block of two adjacent (horizontally or vertically) regular blocks of rank k, where $k \geq i$.*

Theorem 2. *For each subimage $\mathcal{R}$ of an image $\mathcal{T}$ described by a deterministic automaton of size n there is a deterministic automaton describing $\mathcal{R}$ of size $O(n^{2.5})$.*

Proof. By the construction of the proof of Theorem 1 it is enough to give an upper bound for the number of all pseudo-regular subsquares of $\mathcal{T}$.

Let ps_i be the number of pseudo-regular blocks of rank i for $1 \leq i \leq r$ and k_i be the number of regular blocks of rank i for $0 \leq i \leq r$. Then due to Lemma 4 the number of pseudo-regular blocks can be bounded by the number of pairs of regular blocks of rank at least i, hence we have

$$ps_i = O(k_i^2 + k_{i+1}^2 + \ldots + k_r^2)$$

In the same time each pseudo-regular block of rank i is composed of 4 regular blocks of rank $i - 1$, hence $ps_i \leq k_{i-1}^4$. Therefore

$$ps_i = O(\min(k_{i-1}^4, \ k_i^2 + k_{i+1}^2 + \ldots + k_r^2)$$

The conclusion of the theorem is now a consequence of Lemma 3.

In the proof of the lower bound we need two generations of size $O(n)$ which composed together give an object whose compression size is $\Omega(n^2)$. We use the following fact. Assume here the alphabet consists of all integeres and we have morphisms

$$h(i) = 2i\ 2i\ 2i + 1\ 2i + 1; \quad g(i) = 2i\ 2i + 1\ 2i\ 2i + 1;$$

We have

$$h^2(0) = 0011001122332233;$$
$$g^2(0) = 0101232301012323.$$

Lemma 5.
Let $u = h^k(0)$ and $w = g^k(0)$. Then all n^2 pairs $(u[i], w[i])$ are different for $1 \le i \le n^2$, where $n = 2^k$.

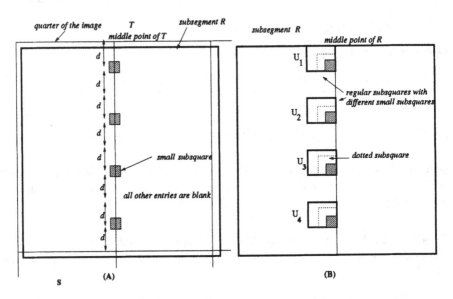

Fig. 5. The structure of the image $T = \mathcal{I}_k$ and the subsegment R (indicated in bold).

The structure of an image whose compression size is n and compression size of its subsegemnt is $\Omega(n^{2.5})$ is illustrated in Figure 5. Lemma 5 is used to generate n different subsquares on one side of the middle line and n different subsquartes on its other side in such a way that we have $\Omega(n^2)$ diferent pairs of these subsquares. We can "pump" these subsquares in such a way that we receive many different subsquares which are of size $O(2^{\sqrt{n}})$ and which are blank except small shaded subsquares. There are $\Omega(n^2)$ different (shaded) small subsquares of T, each one consiting of 4 regular subsquares touching the middle line. The distance between consecutive small subsquares is $2d$, where $d = 2^{\sqrt{(n)}}$. There are $\Omega(n^2)$ $d \times d$ regular subsquares $U_1, U_2, \ldots$ in the subsegment (touching the middle line). Each small shaded corner subsquare of U_i is different (there are $\Omega(n^2)$ of them), so there are together $\Omega(n^2 \cdot \log(|U_i|)) = \Omega(n^{2.5})$ different regular subsquares (dotted subsquares in the figure). Hence the compressed size of the subsegment should be $\Omega(n^{2.5})$.

Theorem 3 (lower-bound).
There is an infinite sequence of deterministic automata of square images T described by deterministic automata such that $|Compress(T)| = n$ and there is a square subsegment R of T satisfying $|Compress(R)| = \Omega(n^{2.5})$.

4   The Subsegment Compression Problem for Weighted Automata

For weighted automata the compression size of the subsegment grows only linearly, this surprising phenomenon is due to the fact that for weighted automata we can have many edges from the same state labelled with the same symbol, but having possibly different weights. This enables to do operations similar to matrix addition, such trick is not possible in the deterministic case. A weighted finite automaton describing an image is specified by (see [9] for details): set of states Q, the alphabet $\{0,1,2,3\}$, weight of edges given by the function $W_a : Q \times Q \to (-\infty, \infty)$ for edges labeled by the symbol a, for $a \in \{0,1,2,3\}$, a function $I : Q \to (-\infty, \infty)$ called *initial distribution function* and a function $F : Q \to (-\infty, \infty)$ called *final distribution function*.

The weight of a word $w = a_1 a_2 \ldots a_k$ is interpreted to give a color $W(w)$ for the pixel $entry(w)$. It is defined as $W(w) = I W_{a_1} W_{a_2} \ldots W_{a_k} F$.

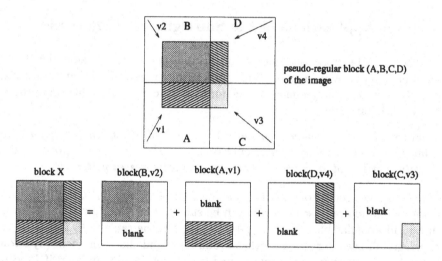

Fig. 6. Regular block X cut off from a pseudo-regular block of T (as a matrix) can be treated as the sum: $X = block(A, v1) + block(B, v2) + block(C, v3) + block(D, v4)$.

For a regular block Y of T and a vector v denote by $block(Y, v)$ the square which is identical to Y on the overlap of Y with the square of the same shape shifted from the corner of Y inside Y by the vector v, all other entries are blank (contain zeros), see Figure 6.

Theorem 4 (weighted automata).

Assume the compression is in terms of weighted automata. If $|Compress(T)| = n$ then for a square subsegment R of T we have $|Compress(R)| = O(n)$. The Sub-

segment Compression Problem for weighted automata for a square subsegment R can be computed in linear time.

Proof. (sketch)
Let A be the weighted automaton defining T, we identify its states with regular blocks. For each rank t a regular block of rank t of subsegment is located in a pseudo-regular block of T, where vectors $v1, v2, v3, v4$ are as in Figure 6.

We create the automaton A' for R. Its states are identified with $block(Y, v)$, where Y is a regular block of T and v is one of the vectors $v1, \ldots, v4$ which depend on the rank. Correctness is based on the fact that the subblock X can be treated as a matrix and it is the sum of 4 matrices, as shown in the figure. For each rank there are 4 different vectors, so the number of states of A' is $O(n)$. The construction goes top-down similarly as in Theorem 1 using the ideas from the proof of Theorem 4.2 in [11], where *atomic dependencies* are decomposed into smaller ones. In our case atomic dependencies correspond to subblocks of the type $block(Y, v)$. We omit the details.

5   Two Applications of the Subsegment Compression

We sketch two simple consequences of the subegment compression. The first is an improvement (and simplification) upon a similar result in [11] and the second one gives the first polynomial time algorithm for compressed checking problem for weighted automata.

Theorem 5. *There is an algorithm for compressed pattern-matching for deterministic automata working in time $O(n^{2.5}m)$ where n is the compressed representation of T and m is the total size of an uncompressed pattern P.*

Proof. In the process of construction the representation for a subsegment we compute pseudo-regular blocks. It can be shown that P occurs in T if it occurs in a psudo-regular block of rank $t + 1$, where t is the rank of P. Hence we can construct all pseudo-regular blocks of rank $t + 1$ and check for them (by known linear time algorithms for uncompressed two-dimensional matching) if P occurs in one of them.

Theorem 6. *There is a polynomial time algorithm for the fully compressed checking problem for weighted automata.*

Proof.
We use the following result due to [6].
Claim. The equivalence of two weighted automata can be checked in polynomial time.

We can construct the compressed representation of the subsegment R of T which is of the same shape as the patern P and starts at the same location. Then we check equality of the images R, P in polynomial time due to the claim.

References

1. A. Amir and G. Benson, Efficient two dimensional compressed matching, *Proc. of the 2nd IEEE Data Compression Conference* 279-288 (1992).
2. P. Berman, M. Karpinski, L. Larmore, W. Plandowski, W. Rytter, The compexity of pattern matching of highly compressed two-dimensional texts, Combinatorial Pattern Matching 1997, in Springer Verlag
3. K. Culik and J. Karhumaki, Finite automata computing real functions, *SIAM J. Comp* (1994).
4. K. Culik and J. Kari, Image compression using weighted finite automata, *Computer and Graphics* 17, 305-313 (1993).
5. D. Derencourt, J. Karhumäki, M. Letteux and A. Terlutte, On continuous functions computed by real functions, *RAIRO Theor. Inform. Appl.* 28, 387-404 (1994).
6. S. Eilenberg, *Automata, Languages and Machines*, Vol.A, Academic Press, New York (1974).
7. K. Culik and J. Kari, *Fractal image compression: theory and applications*, (ed. Y. Fisher), Springer Verlag 243-258 (1995).
8. M. Farach and M. Thorup, String matching in Lempel-Ziv compressed strings, in *STOC'95*, pp. 703-712.
9. J. Kari, P. Franti, Arithmetic coding of weighted finite automata, *RAIRO Theor. Inform. Appl.* 28 343-360 (1994).
10. L. Gąsieniec, M. Karpiński, W. Plandowski and W. Rytter, Efficient Algorithms for Compressed Strings, in *SWAT'96* (1996).
11. J.Karhumaki, W.Plandowski, W. Rytter, Pattern matching for images generated by finite automata, FCT'97, in LNCS Springer Verlag 1997
12. M. Karpinski, W. Rytter and A. Shinohara, Pattern-matching for strings with short description, in *CPM'95* (1995).

Ziv Lempel Compression of Huge Natural Language Data Tries Using Suffix Arrays

Strahil Ristov [1]   Eric Laporte [2]

[1] Ruđer Bošković Institute
Laboratory for stohastic signals and processes research
Zagreb, Croatia
ristov@rudjer.irb.hr
[2] Institut Gaspard Monge
Centre d'études et de recherches en
informatique linguistique
Université de Marne-la-Vallée, France
laporte@bastille.univ-mlv.fr

Abstract. We present a very efficient, in terms of space and access speed, data structure for storing huge natural language data sets. The structure is described as LZ (Ziv Lempel) compressed linked list trie and is a step further beyond directed acyclic word graph in automata compression. We are using the structure to store DELAF, a huge French lexicon with syntactical, grammatical and lexical information associated with each word. The compressed structure can be produced in O(N) time using suffix trees for finding repetitions in trie, but for large data sets space requirements are more prohibitive than time so suffix arrays are used instead, with compression time complexity O(N log N) for all but for the largest data sets.

1   Introduction

Natural language processing has been existing as a field since the origin of computer science. However, the interest for natural language processing increased recently due to the present extension of Internet communication, and to the fact that nearly all texts produced today are stored on, or transmitted through a computer medium at least once during their lifetime. In this context, the processing of large, unrestricted texts written in various languages usually requires basic knowledge about words of these languages. These basic data are stored into large data sets called lexicons or electronic dictionaries, in such a form that they can be exploited by computer applications like spelling checkers, spelling advisers, typesetters, indexers, compressors, speech synthesizers and others. The use of large-coverage lexicons for natural language processing has decisive advantages: Precision and accuracy: the lexicon contains all the words that were explicitly included and only them, which is not the case with recognizers like spell [5]. Predictability: the behavior of a lexicon-based application can be deduced from the explicit list of words in the lexicon. In this context, the storage and lookup of large-coverage dictionaries can be costly. Therefore, time and space efficiency is crucial issue.

M. Crochemore, M. Paterson (Eds.): CPM'99, LNCS 1645, pp. 196-211, 1999.

Trie data structure is a natural choice when it comes to storing and searching over sets of strings or words. In the contemporary usage of the term, a trie for a set of words is a tree in which each transition represents one symbol (or a letter in a word), and nodes represent a word or a part of a word that is spelled by traversal from the root to the given node. The identical prefixes of different words are therefore represented with the same node and space is saved where identical prefixes abound in a set of words - a situation likely to occur with natural language data. The access speed is high, successful look up is performed in time proportional to the length of word since it takes only as many comparisons as there are symbols in the word. The unsuccessful search is stopped as soon as there is no letter in the trie that continues the word at a given point, so it is even faster.

When sets of strings are huge a simple trie can grow to such proportions that its size becomes a restrictive factor in applications. A huge data structure that can't fit into main memory means slower searching on disk, furthermore if the structure is small enough to fit into cache memory the search speed is increased. Numerous researchers did a lot of work on compacting tries, reducing the size and increasing the search speed. As there are many possible uses of a trie, most of the compaction methods are optimized according to specific application requirements. When data must be handled dynamically (databases, compilers) trie has to support insertion and deletion operations as well as a simple lookup; the best results in trie compaction, however, are achieved with static data. Few examples of work on dynamic trie compaction are [3], [7], [8], [15]. Static tries are used successfully in a number of important applications (natural language processing, network routing, data mining) and the efforts in static trie compression are both numerous and justified. Although researchers usually try to establish as good trade-off between speed and size as possible, in most of the work emphasis is on one of the two. Two examples of work where the speed is of main concern are [2] where search speed is increased by reducing the number of levels in a binary trie and [1] where trie data structures are constructed in such manner that they accord well with computer memory architecture. When the size of the structure is of primary concern the work is usually focused on automata compression. With natural language data significant savings in memory space can be obtained if the dictionary is stored in a directed acyclic word graph (DAWG), a form of a minimal deterministic automaton, where common suffixes are shared [4], [12], [13], [17].

Majority of European languages belong to a family of languages where (i) most of the words belong to a set of several morphologically close words (inflectional languages), and (ii) the differences between two such morphologically close words is usually a suffix substitution (suffixal inflection). That accounts for good results with automata minimization, on the average a substantial portion of a word is overlapped with other words' prefixes and suffixes. However, this works well only for simple word lists used mainly in spelling checkers, for most other applications (dictionaries, lexicons, translators) some additional data (lexical tags, index pointers) has to be attached to the word sharply reducing the overlapping of the suffixes. The additional data can be efficiently incorporated in the trie by more complex implementation [16] or by using the hashing transducers. The hashing transducer of a finite set of words was discovered and described independently in [13] and [17]. This scheme implements a one-to-one correspondence between the set of N words and the set of integers from 1 to N, the words being taken in alphabetical order. The user can obtain the number from the word and the word from the number in linear time in the length of the word, independently of the size of the lexicon therefore producing a perfect hashing. The transducer has the same states and the same transitions as the minimal automaton, but an integer is associated to each transition. The number of a word is the sum of the integers on the path that recognizes the word. Once the number of a word is known, a table is looked up in order to obtain the data associated with the word.

In this paper we investigate a new method of static trie compaction that reduces the size beyond that of minimal finite automaton and allows incorporating the additional data in the trie itself. This involves coding the automaton so that not only common prefixes or suffixes are shared, but also the internal patterns. The procedure is best described as a generic Ziv Lempel compression of a linked list trie. Final compressed structure is formally more complex and has less states than minimal finite automata used in [4] and [13]. Particularly attractive feature is a high repetition rate of structural units in compressed structure that enables space efficient coding of the nodes. The idea has been informally introduced in [18] and [19]. Here we shall describe the method in more detail and demonstrate how it performs when used for storing DELAF, a huge lexicon of French words. We also present some compaction results for various natural language data sets. For the sets on which previous work has been reported in the literature our results are significantly better.

In section 2 we present our method and introduce notation we use throughout the article. Two essentially similar algorithms for compression are described in section 3, the first one is simpler and slower, the second one much faster but requires more space. We also explain some heuristic for simplification of the algorithms and propose a related problem as an open problem in theory of NP completeness. In section 4 we describe experimental data sets, among them a huge French lexicon, and present compression results. Conclusion is in section 5.

2   Overview of the Linked List Trie LZ Compression

A trie T is a finite automaton and is as such defined with the quintuple $T = \{Q, A, q_0, \delta, F\}$, where Q is a finite set of states, A is an alphabet of input symbols, $q_0 \in Q$ is the initial state, δ is a transition function from Q x A to Q and $F \subseteq Q$ is the set of accepting or final states. When trie T is produced from a set of words W, then W is the language recognized by T.

Natural language data usually produce very sparse tries that lend themselves to various possibilities for space reduction with retained high access speed. Sparseness of a tree is a strong indication for employing the linked list data structure in representation of the nodes. When linked list is used it is convenient to associate symbols of alphabet with the levels rather than with the transitions in the trie. In this case levels are represented with lists of structural units where four pieces of information (Fig. 1a) are assigned to each unit:

1. a symbol (letter) $a \in A$;
2. a binary flag f indicating whether a word ends at this point (corresponding to a final state);
3. a binary flag c indicating whether there is a continuation of valid sequence of symbols past the current unit to the next level below;

4. a pointer l to the next unit at the same level (if null, there are no more elements on the current level); if we use addressing in number of units, the size bound for l is the number of units in T.

A linked list trie is then represented with a sequence or a string of units. Now, the units themselves can be regarded as symbols that make up a new alphabet U and the implemented trie structure can be defined as a string.

DEFINITION: Linked list trie LLT is a string of symbols u from alphabet U. If we denote by N the number of structural units in LLT then:

$$LLT = u_0 u_1 u_2 \dots u_N \mid u_i \in U, N = |LLT| \qquad \text{where}$$

$$u_i = a_i f_i c_i l_i \mid a_i \in A, f_i \in \{0, 1\}, c_i \in \{0, 1\}, \ 0 \le l_i \le N$$

To illustrate this, in Fig. 1b units of the trie from Fig. 1a are replaced with a new set of symbols yielding a string representation of LLT. Of course, when each of their parts are identical, two units are identical too and consequently represented with the same symbol.

As on any string, some compression procedure can be attempted now on LLT. Particularly natural approach is to use LZ paradigm of replacing repeated substrings with pointers to their first occurrences in the string [23]. The general condition for compression is that the size of pointer must be less than the size of the replaced substring. We used the constant and equal size units for representation of the elements of U and the pointers so that compression is achieved whenever repeated substring is of size 2 or more elements. In Fig. 1c repeated substrings are replaced with information in parenthesis about the position of the first occurrence of repeated substring and it's size. The first number designates the position in (compressed) string and second the length of replaced substring. Note that the first occurrence of a substring can include a pointer to the previous first occurrence of a shorter substring.

DEFINITION: Let ls_i be the length of i-th substituted substring in LLT and K be the number of substitutions. Then, reduction in space $R = \Sigma \ (ls_i - 1)$, for $i = 1 - K$. Let LLTC denote the compressed linked list trie such as that of Fig. 1c. The size N_c of compressed structure is then $N_c = |LLTC| = N - R$, and the compression ratio $C = 1 - N_c/N$.

All size values are given in number of structural units. For the example of Fig. 1c, $R = 9$ and $C = 1 - 11/20 = 45\%$.

The sequence in Fig. 1c is a simplified representation of a compressed trie structure; look up for the input is not performed sequentially as it may seem suggested by the Figs. 1b and 1c, but still by following trie links. Only now when, in reading the structure, at the position P_1 a pointer unit (P_0, ls_1) is encountered, reading procedure jumps to the position P_0, and after ls_1 units read, jumps back to the position $P_1 + 1$.

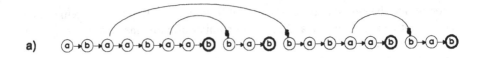

a)

b)   **a b c a b d a e b a e b a b d a e b a e**

c)   **a b c (1, 2) d a e b (6, 2) b (4, 8)**

d)

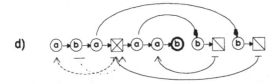

Fig. 1. a) A trie of four words {abaabaab, abaabbab, abbabaab, abbabbab} is presented in a graphical arrangement that points out its sequential features. Final states are indicated by thick circles; horizontal arrows represent c flags; inflected arrows represent l pointers. Structure is traversed by following the arrows and comparing the current input symbol with one in the trie, if symbols don't match and there is no l pointer from the current unit then input is rejected. The input sequence is accepted if it leads to a final state. b) LLT represented with new set of symbols; identical units are replaced with the same symbol. c) Compressed representation of LLT string. The first number in parenthesis is the position of the first occurrence of repeated/replaced substring, the second number is the substring's length. d) Implementation of compressed structure includes two types of pointers: $\square$ signs indicate pointers that replace whole branches and $\boxtimes$ sign stands for pointer that replaces only a portion of a branch and carries the information about its length (2 in this case). Inflected arrows below indicate the paths the reading procedure must follow in the structure. Full lines indicate one-way directions, dashed lines indicate directions implied by $\boxtimes$ pointer.

The actual implementation of LLT compression is more complex than in straightforward application of a LZ procedure on a string in Fig. 1c where there's no difference in treatment of repeated substrings. The underlying structure of LLT is that of a tree and this divides repeated substrings of LLT into two categories depending on whether the repeated substring represents a complete branch of a tree or just a portion of a branch. Only for this latter case should the pointers carry the information about the number of replaced units; when the whole branch is replaced, every possible continuation of the current input is contained in the first occurrence of the substring and there is no need for coming back to the original position of a pointer. Second and third pointers of Fig. 1c replace whole branches of the trie and the first one substitutes only a part of a branch. This LLTC sequence with two types of pointers then might look like this: abc(1,2)daeb(6,_)b(4,_) where "_" indicates that there is no possible need for coming back. The Fig. 1d shows how actually the structure of Fig 1a is compressed with two different types of pointers.

DEFINITION: Let's call *one-way* pointers pointers that replace whole branches and *two-way* pointers those that replace only parts of branches. Let's say that a substring s $= u_1 u_2 ... u_{ls}$ of LLT is *closed* if no unit $u_i \in$ s contains a $l_i \in u_i$ pointer that points outside s, and there is no continuation to the next level from the last unit of s. That is:

   s is closed if $\forall\ l_i \in u_i \in$ s | value $(l_i) \leq$ position (u_{ls}) and $c_{ls} = 0$.

Otherwise let's call s *open*. One-way pointers replace closed repeated substrings, two-way pointers replace open repeated substrings.

THEOREM: Replacing every closed repeated substring of LLT with one-way pointers produces DAWG for a given set of words W.

Proof: DAWG for a set of words W is the minimal finite automaton recognizing all the words in W. Minimization is obtained by merging all the equivalent states of the automaton.

   If two states are reached by sequences s_1 and s_2 they are equivalent if for every sequence z holds that if $s_1 z$ is in W then $s_2 z$ is also, and if $s_1 z$ is not in W neither is $s_2 z$. Since substrings of LLT replaced with one-way pointers are identical it is obvious that they carry identical partial transition function, and since they are closed there exist no other unknown suffixes so the repeated states are indeed equivalent.

The additional compression, above that of automata minimization, is achieved with introduction of two-way pointers capable of replacing open substrings of LLT. It is worth noting that the formal complexity of compressed structure is then higher than that of finite automaton. States replaced by two-way pointers are not equivalent in the finite automata sense and some conditional branching is introduced in the procedure of reading the structure. For example, after reading b in the second position on Fig. 1d further direction depends on whether this is the first time read or the read directed by the pointer at the fourth position. This type of decision is beyond the power of finite automata.

3   Algorithms

We first present a simple quadratic algorithm for producing LLT from W and then replacing repetitions with pointers. Denote with $s_i \in$ LLT a substring of units starting at the position i in LLT. Let E be a relation of substring prefix equality on LLT such that $s_i E s_j$ means that there are at least two first units of s_i and s_j that are equal. That is: $s_i E s_j => u_i ... u_k = u_j ... u_k$ and $k \geq 2$. Let R be the relation of substring substitutability where $s_i R s_j$ means that s_j can be replaced with the pointer to s_i. For the algorithmic complexity reasons R covers smaller class of LLT substrings than E; this will be explained a bit latter. The algorithm is then as follows:

ALGORITHM A:

```
sort W
build LLT(W)
for i = 1 to N – 3
```

for j = i+2 to N − 1
    if s_iEs_j
        if s_iRs_j
            check whether substitutable substrings are open or closed
            replace s_j with the appropriate pointer
end

Building of LLT(W) is a straightforward procedure of building a trie that can be done in O(N) time and will not be explained here. Initial sorting of W is the simplest way of preventing following situations to occur: Let M be the number of words in W, w_m | m < M denote m-th word in W, and LLT_m a linked list trie with m words built into it. If $w_{(m+1)}$ has a prefix w_k that is also a word of W such that k < m and w_m is not a prefix of $w_{(m+1)}$ then there is no place for the suffix of $w_{(m+1)}$ that is the difference between $w_{(m+1)}$ and w_k. This suffix should find its place right after w_k, but since there is already at least one word not prefix of $w_{(m+1)}$ in the structure, this place is occupied. Situation like this would require usage of additional pointers in the construction of the trie and it is more economical instead to arrange the input order in a way to avoid this. The simplest way to do this is to sort W before building LLT(W), then if words exist in W that are prefixes of other words they are all grouped together and any existing prefix of $w_{(m+1)}$ is at the end of LLT_m.

The central part of presented algorithm has clear quadratic time complexity. Double loop of comparing each position in LLT with every other to check whether they are the starting positions of equal substrings takes $N^2/2$ iterations. (The inner loop is only shifted to the right by two − the minimum size of substitutable substrings.) The procedures of checking whether repeated substrings are open or closed and replacing them with pointers are done only once for each replaced substring so they add to the overall complexity only a linear factor proportional to R. The average input sorting procedure is done in O(M log M) time and the total time complexity for producing LLTC from W is then O(M log M + N + N^2 + R) with O(N^2) being by far the most important bound. In practice this simple procedure is fast enough for smaller data sets such are smaller simple word lists with high prefix repetition rate that produce smaller tries. Unfortunately, for bigger sets of entries that do not share too many common prefixes, and therefore produce huge tries, the exhaustive quadratic procedure is not feasible.

3.1   Speed Up via Suffix Matching

Speed up is possible and in fact a linear time bound can be achieved using suffix tree for finding repetitions in LLT. The idea of assisting LZ compression with suffix tree search has firstly been presented in [21]. A suffix tree of all suffixes in LLT can be built in O(N) time, all the repetitions in LLT are then associated with the nodes in the suffix tree and easily found in linear time [11]. The problem with building suffix trees is that they require to much space when alphabet is large as is the alphabet of all different units of LLT, and for this case a better approach is to use suffix arrays [14]. A suffix array for LLT is an array of starting positions in LLT of sorted suffixes of LLT. Sorting is on the average done in O(N log N) time and then all repeated substrings are grouped together in suffix array.

Now the problem rests of finding the best candidates for replacement with pointers among the substrings grouped together. The simplest way to do this is to delimit groups of suffixes in suffix table that have at least two first elements identical and then to perform quadratic search only on elements in the group. These groups should be sorted according to the suffix starting position in LTT so that search and replace procedure can be done in consecutive order from the beginning of the structure. This is important because it avoids considerable expense of keeping track of all the changes in the structure that can interfere with incoming replacements. Overall, this is much faster way to find possible candidates for the substitution with pointers than the exhaustive quadratic search of Algorithm A. The procedure is then:

ALGORITHM B:

1:   sort W
2:   build LLT(W)
3:   build suffix_array(LLT(W))
4:   define partitions of suffix_array(LLT(W)) that comprise two or more entries
     with identical first two units
5:   sort the suffixes in partitions according to their position in LLT(W)
6:   from the first to the last element in partitions compare each element to every
     other from the same group | check whether substitutable substrings are open
     or closed | replace substitutable substrings with the appropriate pointers

Time complexity of comparing substrings at suffixes' starting positions to possible candidates for replacement within the groups is still quadratic but with much smaller base. If there are G different groups of suffixes with identical beginnings in suffix_array(LLT(W)) and SG_i, $i = 1 - G$ is the number of elements in i-th group, the time complexity of step 6 is $O(\Sigma SG_i^2)$. For real data the size of any group is much smaller than N so this improves strongly on time requirements of Algorithm A. The price is paid in space used for suffix array and tables needed for storing and searching groups. There is also the sorting of the groups procedure of step 5 that requires $O(\Sigma SG_i \log SG_i)$ time so the total time complexity of Algorithm B is $O(M \log M + N + N \log N + \Sigma SG_i \log SG_i + \Sigma SG_i^2 + R)$. When running the experiments it is apparent that steps 3 to 5 consume most of the running time of Algorithm B for values of N up to a million. Only for tries with more units quadratic time complexity of step 6 becomes increasingly important. However, these are the values of N where the difference in complexity of Algorithms A and B matters the most. For the biggest tries that we experimented with (N = 16 million) estimated run time of Algorithm A is 250 times longer.

If some additional structures are used to mark already replaced substrings then ΣSG_i^2 factor can be improved to ΣRG_i^2 where RG_i is the number of substitutions actually performed in i-th group. This has not been justified experimentally since Algorithm B already uses considerably more space than Algorithm A and for large N values the size of additional structures may become a restricting factor.

3.2   Bounds for Compression of LLTC

LLTC produced by Algorithms A or B is not necessarily the smallest possible structure of this sort recognizing W. There exist one obvious structural limitation for compression – a constant size of unit, and some algorithmic limitations that are imposed for the sake of the algorithmic simplicity.

Size of Structural Units. If the size of structural unit is kept constant, which immensely simplifies and speeds up the look up procedure, then the bound for the size of each unit is the size of units holding the largest numerical information. There are two types of structural units in LLTC: the *symbol* units, same as those of LLT that carry the symbol code a, f and c flags and the l pointer, and the *pointer* units that are either one- or two–way pointers replacing repeated substrings in LLTC. The size limit for symbol unit in bits is given by $\lceil \log A \rceil + 1 + 1 + \lceil \log N_c \rceil$ and this limit is forced onto pointer units too. Pointer units carry information about the address of the first occurrence of substituted substring, about its length (if two-way) and some information that distinguishes them from symbol units. In symbol units either f or c flag or both must be 1 (true) because the word can only end with the current symbol or be continued to the next one. Therefore combination of two zeros for f and c flags is impossible in symbol units and this is used as an indication that the current unit is a pointer. The bound for the size of the address of the first occurrence of replaced substring is $\lceil \log N_c \rceil$ again, so this leaves $\lceil \log A \rceil$ bits in pointer units for storing the length of replaced substring for two-way pointers. This was enough for every data set we have experimented with so far. LLTC normally supports embedded pointers, i.e. a pointer can point to a sequence of units that contains another pointer, and this can have many levels. For reasons of space economy we are storing in two-way pointers only the number of units that has to be followed on the first level which is usually considerably smaller than the full length of the replaced substring. Apart from this little trick there is another reason why $\lceil \log A \rceil$ bits are enough for two-way pointer information - the longest substituted substrings are usually closed and are therefore replaced with one-way pointers. The problem with constant size units is in that when N_c is big, most of the l pointers are much smaller in value and a considerable amount of space is wasted. If this becomes critical it is always possible to use variable size coding of units or, which should be the best solution for the overall reduction of redundancy in LLTC, to use additional table for minimal size coding of units described latter in section 3.3.

Algorithmic Complexity Constraints on Possible Substring Substitution. There are three algorithmic limitations to compression of LLTC arising from its underlying tree structure and they are defined with the following rules:

Rule 1. If the repeated substrings overlap, then shorten them so that they don't.

Rule 2. If $s_i = u_i...u_k...u_{i+ls}$ is a repeated substring and $l_k \in u_k$ has value(l_k) > i+ls+1 then shorten s_i to $s_{iR} = u_i...u_{k-1}$.

Rule 3. If $s_i = u_i...u_k...u_{i+ls}$ is a repeated substring and there exists $l_h \in u_h \mid h < i$, such that value(l_h) = k $\mid$ i+1 $\leq$ k $\leq$ i+ls, then shorten s_i to $s_{iR} = u_i...u_{k-1}$.

The above three rules account for the aforementioned difference between classes of equal and substitutable substrings of LLT. If these rules are not observed situations would be occurring that would require complicated procedures to solve while at the same time not improving much on the compression. If overlapping of replaced substrings is allowed it would take great pains to avoid never-ending loops and the savings in space would be only one unit per occurrence. (If overlapping is allowed *pattern.pattern.pattern* can be replaced with *pattern.pointer*, and if not, with *pattern.pointer.pointer* with only the cost of one pointer increase in space.) Hence the Rule 1.

Rule 2 prevents the substitution of a substring s_i that contains a l pointer pointing out of s_i by more than one. This is necessary because it is possible that substring of LLT between the end of s_i and the position the l pointer points to can latter be replaced with another pointer unit and then the value of l won't be correct anymore. To account for that a complicated and time costly checking procedure should be employed and the savings would be at most two units per occurrence. (If $k = i + ls$ then only unit u_{i+ls} is not included in the substituted substring, if $k = i + ls - 1$ then the loss is two units $u_{i+ls-1}u_{i+ls}$, and if $k < i + ls - 1$ then the part of s_i behind u_k is a new repeated substring and can be replaced with a new pointer so the loss is again only two units.)

Rule 3 for the similar reasons shortens s_i up to the position pointed to by some l pointer positioned before s_i. If s_i is replaced then this l value wouldn't be correct anymore and the necessary checking would be unjustifiably costly. Analogously to Rule 2 the loss in compression is at most two units per occurrence.

It should be noted that situations where Rules 1, 2 and 3 come to effect occur seldom enough in natural language data that we have been experimenting with so far. Apparently, application of these rules worsens the compression by not more than 3%.

Input Ordering Problem. Apart from that, there exists a serious algorithmic impediment in optimization of LLTC compression introduced by the order of input words when building LLT. Fortunately, this has only a theoretical importance and carries a little weight in practice. Let us consider a special case where W can be divided into a set of distinct partitions W_i, $W_i \in W$, such that every word in W_i has the same length L_i and differs from other words in W_i only in the last letter. Let P_i denote a sequence of units in LLT that represents a common prefix of words in W_i, then length(P_i) = L_i - 1. Let u_{Lik} denote the unit representing the last letter in word $w_{ik} \in W_i$ where $k = 1 - Ki$, $Ki = |W_i|$. Suppose that no word in W is a prefix of another word in W, then when a linked list trie is built each subset W_i produces a LLT branch of type $P_iu_{Li1}u_{Li2}u_{Li3}...u_{LiKi}$. Units corresponding to the last letters in words are connected with l pointers of value one and are identical in every aspect but for the symbol content throughout all the subsets W_i. Ordering of the sequence of u_{Li} units has no bearing on the content of LLT, it is determined by the ordering of input words which can be arbitrary since no word of W is a prefix of another word in W. Now, the problem is how to order sequences of u_{Li} units in such a way as to obtain the highest possible compression achieved by replacing substitutable substrings in LLT with pointers. We haven't been able to find an efficient solution for this problem and we suspect it is NP-hard. We haven't been able to prove that neither so we propose this as an open problem in theory of NP completeness. Reduced for the simplicity it can be stated as:

INSTANCE: Finite set of variables V and a collection T of triples (v_j, v_k, v_l) from V. For each triple holds a statement

$$v_j \angle v_k \quad \text{and} \quad v_j \angle v_l$$

where $\angle$ stands for any transitive, asymmetrical and irreflexive relation such as 'smaller than', 'bigger than', 'has lower/higher rank' etc.

QUESTION: Is there an assignment of values to variables in V such that the number of statements (or triples) that are satisfied is not less than a given integer $I \leq |T|$? $\square$

The order of input words may therefore have influence on how well the linked list trie is compressed. With actual natural language data this is not an important factor, the lexicographical sort of input results in highly repetitious LLT structure and this normally solves the problem well enough. When we investigated possible variations between worst and best case orderings on actual data the difference in size of compressed structures could never be above 2%.

3.3   Minimal Size Unit Coding with Table Lookup

An interesting and exploitable feature of LLTC is a high repetition rate of identical units throughout the structure. Apparently, lexicographic sort of input records combined with employed linked list representation produces a high level of structural unit repetitions in both LLT and LLTC. This effect gets more pronounced with larger data sets. For example, in a compressed trie of over 2 million elements only about 200,000 units are different. A simple and very effective coding of the units is therefore possible for reducing redundancy in the structure. If all the different units are stored eparately in a table of size ND × (unit size), where ND is the number of different units, then LLTC can be represented with an array of N pointers of size $\lceil \log ND \rceil$ bits. On top of this, up to two bits per table unit can be saved by using their position in table instead of flags. In most cases table coding leads to important savings in space and the time needed for table lookup only about halves the search speed, as indicated by our experiments.

The compressed structures produced with Algorithms A or B are very compact and fast to search. Typical access speed for LLTC is measured in tens of thousands of found words per second. This is fast enough for any real time application, even for those that rely on an exhaustive search in space of similar words. In the following section we describe some actual data sets and present results of compaction experiments.

4   Data Sets and Experimental Results

4.1   Natural Language Lexicon

A simple spell-checker needs only to recognize whether a word belongs in the vocabulary of the language or not. In that case, the states of the automaton recognizing a word set are classified as final or non-final. For most other applications, correct words need to be assigned a lexical tag with a grammatical content: part of speech (noun, verb...), inflectional features (plural, 3rd person...), lemma (e.g. the infinitive for a verb). For instance, *woods* should be assigned a tag like *wood*.N:p (i.e. the noun wood in the plural). A minimal automaton can still represent a dictionary that assigns tags to words. Two methods are used to allow for tags in the dictionary. In the first [17], [20], tags are associated to states; the automaton has multiple finalities, i.e. the number of finalities is not necessarily 2 (final/non-final) but the number of tags. In the second method [12], tags are considered as parts of dictionary items. In both cases, minimization is still possible and time efficiency is preserved, but the minimization is less efficient in space, since common suffixes are no longer shared when the words have different tags (e.g. *ds* in the noun *woods* and in the verb *adds*).

When the linguistic information in the tags is limited to basic grammatical information, the number of possible different tags can remain small and these solutions are still optimal. The limit is reached when more elaborate information is included into the tags, namely syntactic information (number of essential complements of verbs, prepositions used with them, distribution of subjects and complements). When this information is provided systematically, the number of different tags comes close to the number of words, and beyond because this level of description requires more sense distinctions [9]. Consequently, the minimal automaton grows nearly as large as the trie. However, the variety of labels used in tags is more limited and there exists a substantial amount of substring repetition in lexical entries. For this reason LLTC structure seems like a natural choice for storing lexicons.

We used LLTC for compressing a comprehensive dictionary of French, the DELAF [6]. This dictionary lists 600,000 inflected forms of simple words. It is used by the INTEX system of lexical analysis of natural-language texts [22]. Linguistic information are attached to each form: parts of speech (noun, verb...); inflectional features (gender, tense...); lemma (e.g. the infinitive in the case of a verbal form); syntactic information about verbs. In case of ambiguities, appropriate sense distinctions are made. The syntactic information attached to verbal forms is derived from the lexicon-grammar of French, a systematic inventory of formal syntactic constraints: number of essential complements, prepositions used with them, distribution of subjects and complements etc. [10]. The size of DELAF in text format is 21 Mbytes and a typical example of three entries in DELAF is presented in Fig. 2.

abandon,.N:ms

abandonna,abandonner.V+t+32CL+32H+36R+38LR+38L1+6+9:IPA3s/abandonner.V+
{s'~}+i+7:IPA3s/abandonner.V+i+31H+35R:IPA3s

abandonnai,abandonner.V+t+32CL+32H+36R+38LR+38L1+6+9:IPA1s/abandonner.V
+{s'~}+i+7:IPA1s/abandonner.V+i+31H+35R:IPA1s

Fig. 2. Three entries in DELAF lexicon of French words with attached grammatical, syntactical and lexical data.

Three things are obvious from this example: first, the amount of repeated substrings is high; second, a simple DAWG would be of little use since the endings of entries are highly diversified (i.e. there are not too many equivalent states in finite automaton produced from DELAF); and third, a trie produced from entries such as those on Fig 2 will be huge. The first two facts speak in favor of trying to store DELAF in LLTC, but the third presents a problem. A huge LLT means a huge N and the quadratic part of compression algorithm becomes important. In fact, with Algorithm B the compression time for LLT(DELAF) was 5.5 hours on a 333 MHz PC running Linux. In Table 1 we present all the relevant numbers for experiments with DELAF and other data sets.

The compressed size with table unit coding is 5.5 Mbytes. This is a considerable improvement over currently used format with tags stored separately that is over twice that size. Reduction in size can be important in integrated applications where lexicon is only a part of the system (computer-aided translation, natural language access to databases, information retrieval). The five and half hour compression time is acceptable for this instance because it is unlikely that data sets of this type will be updated on the run. The search speed is high enough for every possible application.

4.2   Other Data Sets

In order to demonstrate the potential of our method for compressing static dictionaries we present in Table 1 experimental results for seven additional natural language data sets. Six are publicly available and some compression results have already been published for two of them. Here are the brief descriptions:

- DELAF word forms: all the simple French word forms without any additional information, extracted from DELAF
- Calgary book1 7-tuples: a list of all successive seven-tuples from book1 of Calgary corpus; the compressed size of this set as reported in [7] is about 2.5 M
- words: a list of English words found in /usr/dict/words on Unix systems (older release); the compressed size of this set as reported in [13] is 112 K
- linux.words: a list of English words found in /usr/dict/linux.words on Linux systems
- Moby words simple: a list of simple English words from http://www.dcs.shef.ac.uk/research/ilash/Moby/mwords.html
- Moby words compound: a list of compound English words from http://www.dcs.shef.ac.uk/research/ilash/Moby/mwords.html
- Moby words all: combined simple and compound word lists of above.

The compression times and search speeds were measured on 333 MHz P II PC under Linux OS. The compression times given are for Algorithm B steps 3 to 6, i.e. without initial sorting of input entries and building the trie. Search speed is calculated by measuring the time needed for reading all the input words from disk and looking them up in the compressed structure loaded in the main memory. The first, most densely populated, level of the compressed trie is accessed through the array of starting positions for each letter instead of searching the list. This speeds up the search for up to 20% with the space overhead of only 512 bytes for the array (if long integers are used as pointers to starting positions of different letters in LLTC).

In standard coding of LLTC units node sizes are rounded to a whole byte for optimum speed and simplicity. In some cases this is a considerable waste; for instance, Moby data largest pointer units require 26 bits, leaving 6 bits per 4 byte unit unused. In structures with minimal coding all elements are coded with minimum number of bits. Only a small overhead of few bytes is necessary for denoting table and array element sizes, as well as the distribution of various pointers in the table.

Table 1. Experimental results for various natural language data sets

data set	No. of entries	ASCII size (K)	No. of nodes in trie	No. of nodes in compressed trie	LLTC size with rounded nodes (node size)	table coded minimal LLTC size	compression time (seconds)	search speed (words per second)	
								standard	table coded
DELAF	609,454	21,430	16 M	2.14 M	8.4 M (4)	5.5 M	5.5 hours	30 K	15 K
DELAF word forms	609,454	6,942	1.2 M	130,100	385 K (3)	292 K	164	70 K	35 K
Calgary book1 7-tuples	768,764	6,005	759,200	328,500	1.3 M (4)	785 K	44	45 K	15 K
words	25,486	205	81,700	41,300	120 K (3)	83 K	7	70 K	35 K
linux.words	45,402	399	114,800	47,950	140 K (3)	95 K	12	70 K	35 K
Moby words simple	354,984	3,626	986,000	338,500	1.33 M (4)	780 K	90	80 K	30 K
Moby words compound	256,772	3,134	1.35 M	474,300	1.85 M (4)	1.11 M	189	80 K	30 K
Moby words all	611,756	6,759	2.27 M	780,200	3 M (4)	1.9 M	380	60 K	25 K

5   Conclusion

Experimental results presented in Table 1 show that our method exhibits considerable potential for storing natural language data, for inflected languages more than for non-inflected - the French word forms set compresses considerably better than the sets of English words. Still, it performs well for every set tested. The only data sets we could find with previously published results (words and 7-tuples) compress better than previously reported. One would expect that increased number of words would always lead to a better overlapping of substrings. It is therefore somewhat surprising that combined sets of Moby simple and compound words do not compress better than when separated. Also, although we are satisfied with the final result, the huge number of different tags in DELAF did not compress as well as we expected. When partitions of DELAF (even as small as 10,000 entries) are compressed separately the compression ratio is roughly the same as for the whole set. Obviously, with LLTC compression, as with any compression method, the degree of success depends on the actual data. Overall, we believe that presented method of LZ linked list trie compression can be successfully used for storing and accessing data in various natural language related applications.

Acknowledgements

We thank the LADL (Laboratoire d'automatique documentaire et linguistique, University Paris 7) for providing DELAF and anonymous referees for helpful comments.

References

1. A. Acharya, H. Zhu and K. Shen, Adaptive Algorithms for Cache-efficient Trie Search, ACM and SIAM Workshop on Algorithm Engineering and Experimentation ALENEX 99, Baltimore, Jan. 1999.
2. A. Andersson and S. Nilsson, Improved Behaviour of Tries by Adaptive Branching. *Information Processing Letters*, Vol. 46, No. 6, 295-300, 1993.
3. J. Aoe, K. Morimoto and T. Sato, An efficient implementation of trie structures, *Software-Practice and Experience*, Vol. 22, No. 9, 695-721, 1992.
4. A.W. Appel and G.J. Jacobson, The world's fastest scrabble program, *Communications of the ACM*, Vol. 31, No. 5, 1988.
5. J. Bentley, A spelling checker, *Communications of the ACM*, Vol. 5, No. 28, 456-462, 1985.
6. B. Courtois, Un système de dictionnaires électroniques pour les mots simples du français, in Langue Française 87, Dictionnaires électroniques du français, eds. Blandine Courtois and Max Silberztein, Larousse, Paris, 11-22, 1990.
7. J.J. Darragh, J. G. Cleary and I. H. Witten, Bonsai: A Compact Representation of Trees, *Software-Practice and Experience*, Vol. 23, No. 3, 277-291, 1993.

8. J. A. Dundas, Implementing dynamic minimal-prefix tries, *Software-Practice and Experience*, Vol. **21**, No. 10, 1027-1040, 1991.

9. M. Gross, La construction de dictionnaires électroniques, *Ann. Télécommun.* Vol. 44, No. 1-2, 4-19, 1989.

10. M. Gross, Constructing Lexicon-Grammars, in Computational Approaches to the Lexicon, eds. B.T.S. Atkins and A. Zampolli, Oxford University Press, 1994.

11. D. Gusfield, Algorithms on Strings, Trees, and Sequences, Cambridge University Press, 1997.

12. T. Kowaltowski, C. Lucchesi and J. Stolfi, Finite automata and efficient lexicon implementation, Technical report IC-98-2, University of Campinas, Brazil, 1998.

13. C. Lucchesi and T. Kowaltowski, Applications of finite automata representing large vocabularies, *Software-Practice and Experience*, Vol. 23, No. 1, 15-30, 1993.

14. U. Manber and G. Myers, Suffix arrays: a new method for on-line search, *SIAM Journal on Computing*, Vol. 22, No. 5, 935-948, 1993.

15. K. Morimoto, H. Iriguchi and J. Aoe, A method of compressing trie structures, *Software-Practice and Experience*, Vol. **24**, No. 3, 265-288, 1994.

16. T. D. M. Purdin, Compressing tries for storing dictionaries, *Proceedings of the 1990 Symposium on Applied Computing*, Fayetteville, Apr. 1990.

17. D. Revuz, Dictionnaires et lexiques: Méthodes et algorithmes, Ph.D. thesis, CERIL, Université Paris 7, 1991.

18. S. Ristov, Space saving with compressed trie format, *Proceedings of the 17th International Conference on Information Technology Interfaces*, eds. D. Kalpić and V. Hljuz Dobrić, 269-274, Pula, Jun 1995.

19. S. Ristov, D. Boras and T. Lauc, LZ compression of static linked list tries, *Journal of Computing and Information Technology*, Vol. 5, No. 3, 199-204, Zagreb, 1997.

20. E. Roche, Analyse syntaxique transformationnelle du français par transducteurs et lexique-grammaire, Ph.D. thesis, CERIL, Université Paris 7, 1993

21. M. Rodeh, V.R. Pratt and S. Even, A linear algorithm for data compression via string matching, *Journal of the ACM*, Vol. 28, No. 1, 16-24, 1981.

22. M. Silberztein, INTEX: a corpus processing system, *Proceedings of COLING-94*, Kyoto, 1994.

23. J. Ziv and A. Lempel, A universal algorithm for sequential data compression, *IEEE Transactions on Information Theory*, Vol. IT-23, No. 3, 337-343, 1977.

Matching of Spots in 2D Electrophoresis Images. Point Matching Under Non-uniform Distortions

Tatsuya Akutsu[1], Kyotetsu Kanaya[2], Akira Ohyama[2], and Asao Fujiyama[3]

[1] Human Genome Center, Institute of Medical Science, University of Tokyo
4-6-1 Shirokanedai, Minato-ku, Tokyo 108-8639, Japan
takutsu@ims.u-tokyo.ac.jp
[2] Department of Bioscience Systems, Mitsui Knowledge Industry Co., Ltd.
2-7-14 Higashinakano, Nakano-ku, Tokyo 164-8555, Japan
{kanaya,akr}@hydra.mki.co.jp
[3] National Institute of Genetics
1111 Yata, Mishima-city, Shizuoka 411-8540, Japan
afujiyam@lab.nig.ac.jp

Abstract. In this paper, we study pattern matching of points under non-uniform distortions. First we give a natural definition for the problem. Next we present a simple polynomial time algorithm for the one-dimensional case of the problem, whereas we prove that it is NP-hard in two (or more) dimensions. Then we present a practical heuristic algorithm for finding a matching between two sets of spots obtained by the two-dimensional gel electrophoresis technique, which is a special but important case of the problem.

1 Introduction

Matching of spatial point sets (i.e., comparing two sets of points) is an important pattern matching problem, and thus many studies have been done in computational geometry [1,3,7] and pattern recognition [4,11].

In most studies in computational geometry [1,3,7], only uniform transformations (e.g., translations, rigid motions and/or scalings) were considered. However, in some applications, non-uniform distortions may occur and thus pattern matching based on local similarity is important. Pattern matching of spots obtained by the two-dimensional gel electrophoresis technique is an important example of such applications [2,6,12], where we are also developing a system named DDGEL [8] for analysis of two-dimensional gel electrophoresis image obtained from genomic DNA by means of the RLGS (Restriction Landmark Genomic Scanning) method [5]. In this application, positions of spots are distorted non-uniformly and thus the methods developed in computational geometry are not directly applicable.

On the other hand, in pattern recognition (and in image analysis of electrophoresis data), many studies have been done for pattern matching under non-uniform distortions [2,4,6,11,12] although most of them are heuristic.

M. Crochemore, M. Paterson (Eds.): CPM'99, LNCS 1645, pp. 212–222, 1999.

Appel *et al.* considered transformation based on second-order and third-order polynomials in order to cope with non-uniform distortions appearing in electrophoresis image data [2]. However, their method (and many of other methods for electrophoresis image analysis) uses so-called *landmarks* in order to find polynomials, where landmarks are spot pairs intensively marked in both images by the user and selected as putative matching pairs.

Several groups applied Delaunay graphs (Delaunay triangulations) and/or relative neighborhood graphs for point matching under non-uniform distortions [4,6,11,12]. In most of such studies, a Delaunay graph (or a relative neighborhood graph) is first computed from each set of points, and then a maximum common subgraph (or a similar structure) between two graphs is computed. However, finding a maximum common subgraph is time consuming (it is NP-hard in general) and thus various heuristics are employed in Delaunay based approaches. It is natural question whether or not such a time consuming procedure is essential for point matching under non-uniform distorsions. This question is a theoretical motivation of this study.

This paper consists of two parts: theoretical part and practical part.

In the theoretical part, we give a natural definition for point matching under non-uniform distortions, where similar formalizations are given in [6,11]. We present a simple polynomial time algorithm for the one-dimensional case of the problem, which is similar to well-known DP (dynamic programming) algorithms for approximate string matching and sequence alignment. On the other hand, we prove that the problem is NP-hard in two or more dimensions. This result answers the above question: time consuming search procedures such as finding a maximum common subgraph are essential for point matching under non-uniform distortions unless $P = NP$.

In the practical part, we show a heuristic method for spot matching for two-dimensional electrophoresis gel image data. Although this method is heuristic, it uses an algorithm which is an extension of the DP algorithm for 1-D (one-dimensional) case. The method is implemented in the DDGEL system and is being tested using real gel image data.

2   Point Matching Under Non-uniform Distortion

2.1   Definition of the Problem

Let $P = \{p_1, \ldots, p_m\}$ and $Q = \{q_1, \ldots, q_n\}$ be point sets in d-dimensions, respectively. We call a set of pairs $M = \{(p_{i_1}, q_{j_1}), \ldots, (p_{i_l}, q_{j_l})\}$ a *matching* if $(\forall h \neq k)(p_{i_h} \neq p_{i_k} \text{ and } q_{j_h} \neq q_{j_k})$.

Definition 1. (Point Matching Under Non-uniform Distortion)
Point matching under non-uniform distortion is, given a positive real ϵ, two point sets $P = \{p_1, \ldots, p_m\}$ and $Q = \{q_1, \ldots, q_n\}$ in d-dimensional Euclidean space, find a maximum matching $M = \{(p_{i_1}, q_{j_1}), \ldots, (p_{i_l}, q_{j_l})\}$ (i.e., a matching M with the maximum cardinality) satisfying

$$(\forall k)(\forall h \neq k)(\; \frac{1}{1+\epsilon} \; < \; \frac{|q_{j_h} - q_{j_k}|}{|p_{i_h} - p_{i_k}|} \; < \; 1+\epsilon\;),$$

where $|p - q|$ denotes the Euclidean distance between p and q.

Note that P and Q can be exchanged in the above definition because $\frac{1}{1+\epsilon} < \frac{x}{y} < 1 + \epsilon$ if and only if $\frac{1}{1+\epsilon} < \frac{y}{x} < 1 + \epsilon$. Note also that, in this definition, local similarity must be preserved because error for two point pairs must be small if distances between points are small.

2.2   A Simple DP Algorithm for 1-D Case

For 1-D case, point matching under non-uniform distortion can be solved by the following simple dynamic programming algorithm, where only the procedure for computing the scores of point pairs is shown. In the following, we assume that points are already sorted in the ascending order (i.e., $p_1 < p_2 < \cdots < p_m$, $q_1 < q_2 < \cdots < q_n$).

> **for** $i = 1$ **to** m **do** $D[i, 1] \leftarrow 1$;
> **for** $j = 1$ **to** n **do** $D[1, j] \leftarrow 1$;
> **for** $i = 2$ **to** m **do**
>   **for** $j = 2$ **to** n **do**
>   **begin**
> $maxD \leftarrow 0$;
>   **for** $k = 1$ **to** $i - 1$ **do**
>   **for** $h = 1$ **to** $j - 1$ **do**
>   **if** $\frac{1}{1+\epsilon} < \frac{|q_j - q_h|}{|p_i - p_k|} < 1 + \epsilon$ **and** $D[k, h] > maxD$
>   **then** $maxD \leftarrow D[k, h]$;
> $D[i, j] \leftarrow maxD + 1$;
>   **end**;

It is obvious that the algorithm works in $O(m^2 n^2)$ time. The correctness of the algorithm follows from the proposition below.

Proposition 1. *If both* $\frac{1}{1+\epsilon} < \frac{|q_{j_2} - q_{j_1}|}{|p_{i_2} - p_{i_1}|} < 1 + \epsilon$ *and* $\frac{1}{1+\epsilon} < \frac{|q_{j_3} - q_{j_2}|}{|p_{i_3} - p_{i_2}|} < 1 + \epsilon$
hold where $i_1 < i_2 < i_3$ *and* $j_1 < j_2 < j_3$, *then* $\frac{1}{1+\epsilon} < \frac{|q_{j_3} - q_{j_1}|}{|p_{i_3} - p_{i_1}|} < 1 + \epsilon$ *holds.*

2.3   NP-Hardness Result for 2-D Case

In this subsection, we prove the following theorem.

Theorem 1. *Point matching under non-uniform distortion is NP-hard in d-dimensions where $d \geq 2$.*

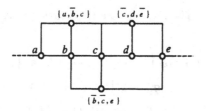

Fig. 1. Example of a grid embedding of a planar graph for a 3SAT instance $\{\{a, \bar{b}, c\}, \{\bar{c}, d, \bar{e}\}, \{\bar{b}, \bar{c}, e\}\}$

Proof. Since details are complicated, we only show a sketch of the proof. We use a polynomial time reduction from PLANAR 3SAT [10]. Let $C = \{c_1, c_2, \ldots, c_N\}$ be an instance of PLANAR 3SAT over the set of variables $V = \{v_1, v_2, \ldots, v_K\}$, where we assume that each clause c_i consists of 3 literals. Note that, in PLANAR 3SAT, graph $G(V \cup C, E)$ must be planar (see Fig. 1), where $E = \{\{v_i, c_j\} | v_i \in c_j \text{ or } \overline{v_i} \in c_j\} \cup \{\{v_i, v_{i+1}\}\} \cup \{\{v_1, v_K\}\}$.

From this instance, we construct an instance (P, Q, ϵ) of the point matching problem. The construction will be made up of several components, which can be partitioned into three parts, grouped according to their intended function: "truth-setting" components, "satisfaction-testing" components, and "routing" components.

First we describe "satisfaction-testing" components (see also Fig. 2) since these are the core parts of the construction. For each clause c_i, we construct a set of points $T_i = \{p_i^0, p_i^1, p_i^2, p_i^3, q_i^1, q_i^2, q_i^3, q_i^{1t}, q_i^{2t}, q_i^{3t}, q_i^{1f}, q_i^{2f}, q_i^{3f}\}$, where

$$p_i^0 = (0, 0),$$
$$p_i^1 = (0, L), \quad p_i^2 = (-\tfrac{\sqrt{3}L}{2}, -\tfrac{L}{2}), \quad p_i^3 = (\tfrac{\sqrt{3}L}{2}, -\tfrac{L}{2}),$$
$$q_i^1 = (0, -\alpha L), \quad q_i^2 = (\tfrac{\sqrt{3}\alpha L}{2}, \tfrac{\alpha L}{2}), \quad q_i^3 = (-\tfrac{\sqrt{3}\alpha L}{2}, \tfrac{\alpha L}{2}),$$
$$q_i^{1t} = (0, (1-\alpha)L), \quad q_i^{2t} = (-\tfrac{\sqrt{3}(1-\alpha)L}{2}, -\tfrac{(1-\alpha)L}{2}), \quad q_i^{3t} = (\tfrac{\sqrt{3}(1-\alpha)L}{2}, -\tfrac{(1-\alpha)L}{2}),$$
$$q_i^{1f} = (0, (1+\alpha)L), \quad q_i^{2f} = (-\tfrac{\sqrt{3}(1+\alpha)L}{2}, -\tfrac{(1+\alpha)L}{2}), \quad q_i^{3f} = (\tfrac{\sqrt{3}(1+\alpha)L}{2}, -\tfrac{(1+\alpha)L}{2}),$$

and each T_i is to be translated to an appropriate position.

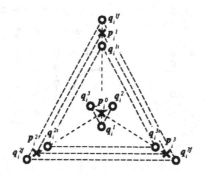

Fig. 2. "Satisfaction-testing" component

Here we let $\epsilon = 2\alpha$. Then we can see the following relations hold for small α ($\alpha < 0.22$):

$$(\forall j)(\ |q_i^{jf} - q_i^j| = (1+\epsilon)L\ ), \quad (\forall j)(\ \left(\tfrac{1}{1+\epsilon}\right)L < |q_i^{jt} - q_i^j| < (1+\epsilon)L\ ),$$
$$(\forall j)(\forall k \neq j)(\ \left(\tfrac{1}{1+\epsilon}\right)L < |q_i^{jt} - q_i^k| < (1+\epsilon)L\ ),$$
$$(\forall j)(\forall k \neq j)(\ \left(\tfrac{1}{1+\epsilon}\right)L < |q_i^{jf} - q_i^k| < (1+\epsilon)L\ ).$$

From these, the followings must be satisfied (in order to make $|M| = |P|$):

(a) if $(p_i^0, q_i^1) \in M$, then $(p_i^1, q_i^{1t}) \in M$, $(p_i^2, q_i^{2t}) \in M$ or $(p_i^2, q_i^{2f}) \in M$, and, $(p_i^3, q_i^{3t}) \in M$ or $(p_i^3, q_i^{3f}) \in M$,

(b) if $(p_i^0, q_i^2) \in M$, then $(p_i^2, q_i^{2t}) \in M$, $(p_i^1, q_i^{1t}) \in M$ or $(p_i^1, q_i^{1f}) \in M$, and, $(p_i^3, q_i^{3t}) \in M$ or $(p_i^3, q_i^{3f}) \in M$,

(c) if $(p_i^0, q_i^3) \in M$, then $(p_i^3, q_i^{3t}) \in M$, $(p_i^1, q_i^{1t}) \in M$ or $(p_i^1, q_i^{1f}) \in M$, and, $(p_i^2, q_i^{2t}) \in M$ or $(p_i^2, q_i^{2f}) \in M$.

Here we assume that $c_i = \{v_1, v_2, v_3\}$. Then, case (a) corresponds to a case where v_1 is satisfied, case (b) corresponds to a case where v_2 is satisfied, and case (c) corresponds to a case where v_3 is satisfied.

Next we describe "truth-setting" components. For each variable v_i, points as shown in Fig. 3(a) are constructed, where the points are partitioned into three subsets: P_i, Q_i^t, Q_i^f. It is easy to define coordinates so that if at least one point in P_i corresponds to some point in Q_i^t (resp. Q_i^f), then P_i must correspond to Q_i^t (resp. Q_i^f).

Next we describe "routing" components. Here we assume that a grid embedding of $G(V \cup C, E)$ is already obtained as shown in Fig. 1. Note that a grid embedding of size $O(N) \times O(N)$ can be computed in linear time from $G(V \cup C, E)$ [9]. According to this embedding, we connect "truth-setting" components to "satisfaction-testing" components. In oreder to connect "truth-setting" components to "satisfaction-testing" components, the following constructions are important: (i) copying a truth assignment; (ii) inverting a truth assignment on v_i

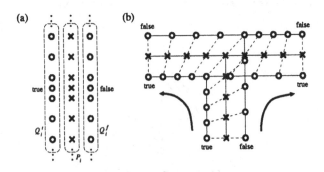

Fig. 3. (a) "Truth-setting" component and (b) "Routing-component" for copying a truth assignment

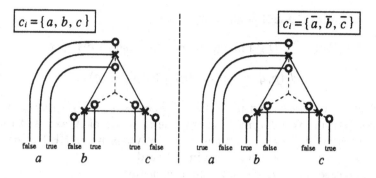

Fig. 4. Connection from "truth-setting" components to a "satisfaction-testing" component

(i.e., creating $\overline{v_i}$); (iii) connecting a truth assignment to a "satisfaction-testing" component. (i) can be done as in Fig. 3(b) and (ii) can be done in a similar way. (iii) can be done as in Fig. 4, where each bold line denotes a sequence of points.

Then, it can be proved for an appropriate value of ϵ that there exists a maximum matching M satisfying $|M| = |P|$ if and only if there exists a truth assignment satisfying all clauses in C.

Although we omit details, the total number of created points is polynomially bounded and thus the construction can be done in polynomial time.   □

3   A Practical Algorithm for 2-D Gel Image Data

Although the DP algorithm in Section 2.2 is valid only for 1-D case, the idea can still be used as a heuristic for matching of 2-D points. In this section, we describe a pattern matching method based on such a heuristic algorithm. The method is implemented in the α-version of the DDGEL system (http://bonsai.ims.u-tokyo.ac.jp/cgi-bin/ddtop/cgi-bin/index.cgi) [8], where DDGEL is an image analysis system for 2-D gel electrophoresis obtained from DNA by means of the RLGS (Restriction Landmark Genomic Scanning) method [5].

The matching method consists of two major steps: *finding an initial matching* and *finding the final matching*. In the first step, we find a rough matching between two point sets. In the second step, we first transform P according to the result of the initial matching and then we compute and refine matchings. The heuristic algorithm mentioned above is used in the first step. The first step corresponds to the matching by *landmarks* used in several practical systems [2]. But, in our case, we do not need landmarks.

Note that, in analysis of 2-D gel images, the *spot detection* step is also required in order to extract spots from the original gel images, where we use standard image processing techniques for this step. Since spot detection is out of the scope of this paper, we do not describe the details of spot detection.

In the followings, for a point $\boldsymbol{p}$, $(\boldsymbol{p})_x$ (resp. $(\boldsymbol{p})_y$) denotes x-coordinate (resp. y-coordinate) of $\boldsymbol{p}$. If $(\boldsymbol{p})_x > (\boldsymbol{q})_x$ and $(\boldsymbol{p})_y > (\boldsymbol{q})_y$ hold, we write $\boldsymbol{p} \succ \boldsymbol{q}$.

3.1   Finding an Initial Matching

In point matching of 2-D gel image, we consider the L_1 distance

$$d_1(p, q) = |(p)_x - (q)_x| + |(p)_y - (q)_y|,$$

instead of L_2 distance used in Section 2, because a 2-D gel image is usually obtained by using two enzymes: one for the direction of X-axis and the other for the direction of Y-axis. Although we do not yet prove an NP-hardness result for this case, this case seems to remain NP-hard.

For this case, we have the following proposition:

Proposition 2. *If both* $\frac{1}{1+\epsilon} < \frac{d_1(q_{j_2}, q_{j_1})}{d_1(p_{i_2}, p_{i_1})} < 1+\epsilon$ *and* $\frac{1}{1+\epsilon} < \frac{d_1(q_{j_3}, q_{j_2})}{d_1(p_{i_3}, p_{i_2})} < 1+\epsilon$ *hold where* $p_{i_3} \succ p_{i_2} \succ p_{i_1}$ *and* $q_{j_3} \succ q_{j_2} \succ q_{j_1}$, *then* $\frac{1}{1+\epsilon} < \frac{d_1(q_{j_3}, q_{j_1})}{d_1(p_{i_3}, p_{i_1})} < 1+\epsilon$ *holds.*

Based on this, we can obtain a longest sequence $((p_{i_1}, q_{j_1}), \ldots, (p_{i_l}, q_{j_l}))$ satisfying $(\forall h \neq k)(\frac{1}{1+\epsilon} < \frac{d_1(q_{j_h}, q_{j_k})}{d_1(p_{i_h}, p_{i_k})} < 1+\epsilon)$, $(\forall h)(p_{i_{h+1}} \succ p_{i_h})$ and $(\forall h)(q_{j_{h+1}} \succ q_{j_h})$, by means of a simple DP algorithm similar to that in Section 2.2.

However, in this case, we obtain a matching only for points in a *diagonal*-like region. Such a matching is not sufficient for an initial matching. Therefore, we use the following DP algorithm for finding an initial matching. In this case, there is no longer any theoretical guarantee for the obtained matchings. However, it worked well when the number of insertions and deletions of points was not large and the distortion was not very large.

```
for i = 1 to m do
    for j = 1 to n do
        begin
            score ← 0;  maxscore ← 0;   count ← 0;
            for each p_k ∈ neighbors(p_i) do
                for each q_h ∈ neighbors(q_j) do
                    if d_1(p_k - p_i, q_h - q_j) < α_1 · |p_k - p_i| then
                        begin
                            if S[k][h] > maxscore then maxscore ← S[k][h];
                            count ← count + 1;   skip to next p_k;
                        end;
            S[i][j]  ←  maxscore + 2 · count + 1;
        end
```

In the above, $neighbors(p)$ denotes the set of K_1-nearest points p' of p such that $p \succ p'$, where K_1 is a constant (we use $K_1 = 6$ in the current implementation). α_1 is also a constant depending on the size and the density of input point sets, where we use $\alpha_1 = 0.4$ in the current implementation. In order to find matching pairs in non-diagonal regions, we use $maxscore + 2 \cdot count + 1$ as a score, where there is no concrete reason for using this value. The procedure works in $O(mn)$ time for a constant K_1. Note also that only the procedure for computing scores of point pairs is described in the above. A set of point pairs (i.e., an initial matching) is obtained from the scores by a *traceback*-like method, where we omit details here.

3.2   Finding the Final Matching

In the second step, we compute the final matching based on the initial matching found in the first step. As in the first step, there is no theoretical guarantee for the obtained matchings. Since several heuristics are used in the second step, we only describe an outline of the procedure. We call that p is *locally similar* to q, if the following condition is satisfied:

$$|\{\, p_i|\; p_i \in neighbors(p) \text{ and } (\exists q_j \in neighbors(q))$$
$$(d_1(p_i - p, q_j - q) < \alpha_2 \cdot |p_i - p|) \,\}| \; > K_2$$

where α_2 and K_2 are constants depending on the size and the density of input point sets. We use $\alpha_2 = 0.25$ and $K_2 = 4$ in the current implementation. In this case, $neighbors(p)$ is the set of 10-nearest neighbors of p.

(1) From the set of pairs $\{(p_{i_1}, q_{j_1}), \ldots, (p_{i_l}, q_{j_l})\}$ found in the first step, compute the affine transformation $\quad TR : (x, y) \longrightarrow (ax + b, cy + d) \quad$ such that $\sum_k |TR(p_{i_k}) - q_{j_k}|^2$ is minimized, by means of the least squares fitting method. Then, we apply TR to P.

(2) Apply the DP algorithm to P and Q again and execute (1) again.

(3) For each point $p \in P$, find a corresponding point $q \in Q$ (if there exists) such that $|p - q| < D$ and p is locally similar to q (currently we use $D = 20$).

(4) Apply the *local transformation* to each point $p \in P$, where the local transformation is computed from $neighbors(p)$ and $neighbors(q)$ by means of the least squares fitting method.

(5) Repeat (3) and (4) for several times.

3.3   Examples

Here, we show examples where the above method was applied to point sets obtained from real RLGS image data. Although we applied the method to several data, we only show two typical examples here. The method is implemented on a SUN Ultra-2 workstation with 300MHz CPU and 640MByte main memory using C-language.

The first example (see Fig. 5) is an easy case: the number of insertions and deletions of points is small and the distortion is small. In this case, 1052 pairs of matching points were found where P consisted of 1128 points and Q consisted of 1148 points. It took 32.7 sec. in total.

The second example (see Fig. 6) is a rather difficult case: either the number of insertions and deletions or the distortion is not small.

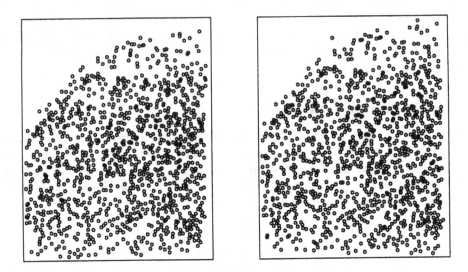

Fig. 5. Example 1. The point set on the left hand side (P) consists of 1128 points and the point set on the right hand side (Q) consists of 1148 points. In this case, 1052 matching pairs were found.

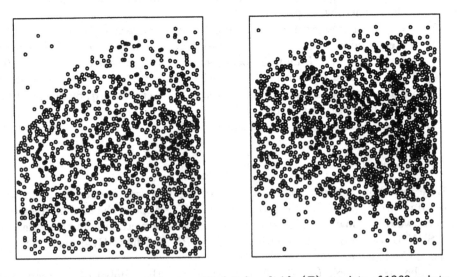

Fig. 6. Example 2. The point set on the left hand side (P) consists of 1363 points and the point set on the right hand side (Q) consists of 1682 points. In this case, 824 matching pairs were found (see also Fig. 7).

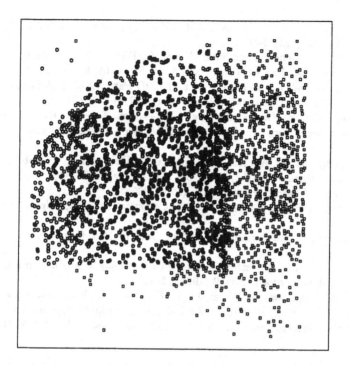

Fig. 7. Result of point matching in Example 2, where one point set (P) is transformed by a non-uniform transformation.

A good matching was still found in this case: 824 pairs of matching points were found where P consisted of 1363 points and Q consisted of 1682 points. It took 57.7 sec. in total. The result of the matching is shown in Fig. 7, where P is transformed by a non-uniform transformation generated by the matching program.

Acknowledgment

We thank Prof. Hisao Tamaki in Meiji University for valuable suggestions. He suggested us the use of PLANAR 3SAT for proving the NP-hardness result in 2-dimensions.

References

1. Alt, H., Melhorn, K., Wagner, H., Welzl, E.: Congruence, Similarity and Symmetries of Geometric Objects. Discrete and Computational Geometry **3** (1988) 237–256
2. Appel, R.D., Vargas, J.R., Palagi, P.M., Walther, D., Hochstrasser, D.F.: Melanie II - A Third Generation Software Package for Analysis of Two-dimensional Electrophoresis Images: II. Algorithms. J. Electrophoresis **18** (1997) 2735–2748
3. Cardoze, D.E., Schulman, L.J.: Pattern Matching for Spatial Point Sets. In: Proc. 39th IEEE Symp. Found. Comput. Sci. (1998) 156–165

4. Finch, A.M., Wilson, R.C., Hancock, E.R.: Matching Delaunay Graphs. Pattern Recognition **30** (1997) 123–140

5. Hatada, I. *et al.*: A Genomic Scanning Method for Higher Order Organisms Using Restriction Sites as Landmarks. Proc. Natl. Acad. Sci. USA. **88** (1991) 9523–9527

6. Hoffmann, F., Kriegel, K., Wenk, C.: Matching 2D Patterns of Protein Spots. In: Proc. 14th ACM Symp. Computational Geometry (1998) 231–239

7. Irani, S., Raghavan, P.: Combinatorial and Experimental Results for Randomized Point Matching Algorithms. In: Proc. 12th ACM Symp. Computational Geometry (1996) 68–77

8. Kanaya, K., Ohyama, A., Akutsu, T., Fujiyama, A.: Development of Web Interface of Image Analysis System DDGEL for 2d Gel Electrophoresis. In: Genome Informatics 1998, Universal Academy Press, Tokyo (1988) 336–337

9. Kant, G.: Drawing Planar Graphs Using the Canonical Ordering. Algorithmica **16** (1996) 4–32

10. Lichtenstein, D.: Planar Formulae and Their Use. SIAM J. Computing **11** (1982) 329–343

11. Ogawa, H.: Labeled Point Pattern Matching by Delaunay Triangulation and Maximal Cliques. Pattern Recognition **19** (1986) 35–40

12. Takahashi, K., Nakazawa, M., Watanabe, Y., Konagaya, A.: Fully-Automated Spot Recognition and Matching Algorithms for 2-D Gel Electrophoretogram of Genomic DNA. In: Genome Informatics 1998, Universal Academy Press, Tokyo (1988) 161–172

Applying an Edit Distance to the Matching of Tree Ring Sequences in Dendrochronology*

Carola Wenk

Institut für Informatik
Freie Universität Berlin
Takustr. 9, D-14195 Berlin
wenk@inf.fu-berlin.de

Abstract. In dendrochronology wood samples are dated according to the tree rings they contain. The dating process consists of comparing the sequence of tree ring widths in the sample to a dated master sequence. Assuming that a tree forms exactly one ring per year a simple sliding algorithm solves this matching task.
But sometimes a tree produces no ring or even two rings in a year. If a sample sequence contains this kind of inconsistencies it cannot be dated correctly by the simple sliding algorithm. We therefore introduce a $\mathcal{O}(\alpha^2 mn + \alpha^4(m + n))$ algorithm for dating such a sample sequence against an error-free master sequence, where n and m are the lengths of the sequences. Our algorithm takes into account that the sample might contain up to α missing or double rings and suggests possible positions for these kind of inconsistencies. This is done by employing an *edit distance* as the distance measure.

1 Introduction

1.1 Dendrochronology

The tree ring structure in wood samples is important in many research areas, for instance in archaeology, climatology, geomorphology and glaciology. The reason for that is that the growth of a tree and therefore its rings depend on the environmental conditions that the tree has been exposed to, so that the tree rings build an archive of these environmental conditions. The science that deals with the dating of tree rings in order to answer questions related to natural history is called dendrochronology. The name is derived from the greek words *dendron* (wood), *chronos* (time) and *logos* (the science of).

A tree ring is a growth layer that the tree forms under its bark during the vegetation period. It consists of big, thinwalled cells that are built at the beginning of the growth period and of thin, thickwalled cells built at the end. The first type of cell ensures the food supply to the shoots, whereas the other type accounts for the stability of the stem. Since the second type of cell looks much

---

* Part of a research project supported by Deutsche Forschungsgemeinschaft, grant AL 253/4-2

M. Crochemore, M. Paterson (Eds.): CPM'99, LNCS 1645, pp. 223–242, 1999.

darker than the first type, it is possible to visually detect the border between two successive tree rings. In areas with an annual vegetation and winter period a tree usually adds exactly one tree ring per year.

In dendrochronology a wood sample is characterized by the sequence of its tree ring widths [1]. Since trees growing under similar conditions (especially climatic conditions like rainfall) build similar tree rings, it is possible to successfully compare certain tree ring sequences. In fact, the usual way of dating tree ring sequences in dendrochronology is to compare the undated sequence to a dated sequence. This procedure, called *crossdating*, is a fundamental task in dendrochronology.

1.2   Crossdating

Assuming that the trees being considered have built exactly one ring each year, a crossdating can be performed by sliding the sample along the master sequence starting and ending with a certain constant minimum overlap of e.g. 50 rings. At each position the distance (according to a predefined distance measure) between the overlapping parts of the sequences is computed and the position yielding the best distance is proposed as the correct dating position. The most common distance measures are the *t-value*, and the so-called *Gleichläufigkeitskoeffizient* (percentage of slope equivalence).

Let $x = x_0, \ldots, x_{N-1}$ and $y = y_0, \ldots, y_{N-1}$ be the two sequences being compared in one step of the algorithm. Then the *t*-value (*Student's t*) is defined by

$$t = r\sqrt{\left(\frac{N-2}{1-r^2}\right)} \tag{1}$$

where r is the correlation coefficient

$$r = \frac{\sum\limits_{i=0}^{N-1} (x_i - \bar{x})(y_i - \bar{y})}{\sqrt{\sum\limits_{i=0}^{N-1} (x_i - \bar{x})^2 \sum\limits_{i=0}^{N-1} (y_i - \bar{y})^2}} \tag{2}$$

$$= \frac{N \sum\limits_{i=0}^{N-1} x_i y_i - \sum\limits_{i=0}^{N-1} x_i - \sum\limits_{i=0}^{N-1} y_i}{\sqrt{(N \sum\limits_{i=0}^{N-1} x_i^2 - (\sum\limits_{i=0}^{N-1} x_i)^2)(N \sum\limits_{i=0}^{N-1} y_i^2 - (\sum\limits_{i=0}^{N-1} y_i)^2)}} \tag{3}$$

---

[1] Depending on the application there are also other tree ring characteristica than the width used (see e.g. [11]), but in this article we will regard tree ring widths only.

with the arithmetic means $\bar{x}$ and $\bar{y}$. The *Gleichläufigkeitskoeffizient Glk* is the percentage of slope equivalence of the two sequences,

$$\text{Glk} \quad = \frac{1}{N-1} \sum_{i=0}^{N-2} \chi(x_{i+1} - x_i = y_{i+1} - y_i) \tag{4}$$

$$= \frac{1}{N-1} \sum_{k=-1}^{1} \sum_{i=0}^{N-2} \chi(x_{i+1} - x_i = k) \cdot \chi(y_{i+1} - y_i = k) \tag{5}$$

with the characteristic function $\chi(a = b) = \begin{cases} 1, & \text{if} \quad a = b \\ 0, & \text{if} \quad a \neq b \end{cases}$.

A sequential computation of all distances takes $\theta(nm)$ time, where n and m are the lengths of the master and the sample sequences, respectively. Considering (3) a non-sequential computation of all correlation coefficients depends on the efficient calculation of the *correlation terms* $\sum x_i y_i$, since all other terms can be computed in linear time. Note that the inner sum in (5) is also a correlation term. Employing the Fast Fourier Transform (FFT) all such correlation terms can be computed in time $\theta((n+m)log(n+m))$ instead of the brute force $\theta(nm)$, see e.g. [6],[7],[2]. Due to the discretization of the slope of the sequences the Gleichläufigkeitskoeffizient usually gives less information than the t-value.

Since the data is very noisy it usually does not suffice to simply date the sample sequence according to the crossing position which yields the best distance. The results of the matching algorithm are always visually checked by a dendrochronologist.

Before the above described comparison can be made, the tree ring sequences have to be filtered. This so-called *standardization* process cleans the data from individual trends, which usually are long term trends. Thus only general trends which occur in several tree ring sequences remain. Typically high-pass filters like the percentage of a five-year running mean or the logarithmic difference are used.

1.3   Missing and Double Rings

However, the assumption made above that a tree ring sequence contains exactly one value per year is not always true. First of all mistakes during the measurement of the ring widths happen, especially if the rings are very thin. Moreover, due to bad growing conditions a tree might also not build a ring around the whole stem or even not at all, which can result in a *missing ring* in the tree ring sequence. Climatic changes can also cause a tree to build two rings a year, a *double ring*.

If the sequences to be compared contain missing or double rings most matching algorithms do not produce satisfying results since they do not take into account the transposition in time which is caused by a missing or a double ring. The usual approach to date a sample sequence which may contain inconsistencies against a clean master sequence is to split up the sample into shorter parts and to date each part on its own (either manually or using Cofecha [4]). Finally a possible position for a ring insertion or merging is manually concluded. Cofecha [4] is a quality control tool which checks a set of dated samples for mutual dating consistency by splitting up each sequence into small pieces and comparing these to the other sequences. This leads to a lot of information to be evaluated. The information needed to deduce a possible missing ring (i.e. when the pieces to the right of the missing ring position date all to one year later) is then available, but a missing ring is not explicitly proposed.

2 Edit Distances in an α-Box

2.1 A Simple Edit Distance

Let $A = a_0, ..., a_{n-1}$ and $B = b_0, ..., b_{m-1}$ be two standardized tree ring width sequences, where A may contain missing or double rings (representing the sample sequence) whereas B is known to be a clean reference sequence (representing a part of the master sequence). In order to get a notion of how much A differs from B we look for a transformation transforming A by inserting rings (which compensates a missing ring) or merging two rings into one (which compensates a double ring) into a sequence close to B. Closeness is defined by taking the sum of the squared differences. The transformations allowed are described by *transformation sequences* over the alphabet $\{I, M, N\}$ where I stands for *insert* M for *merge* and N for *identity operation*.

Let for example be $A = 3, 2, 1, 2, 5, 3$ and consider a transformation sequence $\tau = MMNIIN$, see Fig. 1. The transformation is performed sequentially from left to right by merging 3 and 2 to 5 (M), merging 1 and 2 to 3 (M), not changing 5 (N), inserting a ring which is done by taking the average $\mu = \frac{5+3}{2} = 4$ of the two surrounding rings (I), inserting another ring in the same manner (I) and finally not changing the last ring 3 (N).

τ		M		M	N	I	I	N
A		3	2	1	2	5		3
$\tau(A)$		5		3	5	4	4	3

Fig. 1. An example of a sequence A, a transformation sequence τ and the transformed sequence $\tau(A)$.

The *simple edit distance* $D_{simp}(A, B)$ is defined by

$$D_{simp}(A, B) = \begin{cases} \min_{\tau \in T_{n,m}} \sum_{v=0}^{m-1} (\tau(A)_v - B_v)^2 & \text{, if } T_{n,m} \neq \emptyset \\ \text{not defined} & \text{, otherwise} \end{cases} \quad (6)$$

where $T_{n,m}$ contains all transformation sequences (which we identify with a transformation each) that transform a sequence of length n into a sequence of length m. We call a transformation sequence which minimizes the sum *optimal*. For a transformation $\tau \in T_{n,m}$ let γ_τ be the number of merge operations, ι_τ the number of insert operations and ν_τ the number of identity operations in τ. Then from the definition of τ follows $n = 2 * \gamma_\tau + \nu_\tau$ as well as $m = \gamma_\tau + \iota_\tau + \nu_\tau = n + \iota_\tau - \gamma_\tau$. Making use of these properties it is easy to show that $T_{n,m}$ is non-empty if and only if $n \leq 2m$. This notion of an edit distance is based on the edit distance for strings (see [3], [12]) or on the dynamic time warping in speech recognition (see [10], [9]) respectively.

The profit of taking the sum of the squared distances as a minimization criterion is the existence of a recurrence which leads to an efficient computation of the simple edit distance. Define $D_{simp}(i,j) := D_{simp}(A[0..i-1], B[0..j-1])$ to be the simple edit distance of the prefixes of A and B for $i \leq 2j$. Then the definition of the transformations by transformation sequences implies the existence of the following recurrence:

$$D_{simp}(0,0) = 0$$

$$D_{simp}(i,j) = \min \left\{ \begin{array}{l} D_{simp}(i-2, j-1) + (a_{i-2} + a_{i-1} - b_{j-1})^2, \\ D_{simp}(i-1, j-1) + (a_{i-1} - b_{j-1})^2, \\ D_{simp}(i, j-1) + (\mu(i-1) - b_{j-1})^2 \end{array} \right\} \qquad (7)$$

for all $0 \leq i \leq n$, $0 \leq j \leq m$ with $i \leq 2j$.

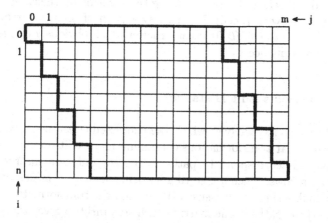

Fig. 2. Matrix for the dynamic programming computation of the simple edit distance $D_{simp}(A, B)$. The left cut-off is caused by the condition $i \leq 2j$, the right by $(n - i) \leq 2(m - j)$.

Although the transformation space is exponentially big a dynamic programming approach allows to compute $D_{simp}(A, B)$ in $\theta(nm)$ time and space. This is accomplished by sequentially filling an $(n + 1) \times (m + 1)$ matrix (see Fig. 2) in which cell (i,j) contains the value $D_{simp}(i,j)$. The condition $i \leq 2j$ and the symmetrical condition $(n-i) \leq 2(m-j)$ cut two corners off the matrix which represent undefined or for the computation of $D_{simp}(n,m)$ unnecessary values, respectively. According to (7) a value is computed out of at most three values (see Fig. 3). The value $D_{simp}(A,B) = D_{simp}(n,m)$ is placed in cell (N,M).

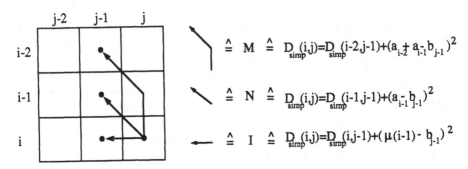

Fig. 3. Dynamic programming computation of the simple edit distance according to (7).

An optimal transformation can be retrieved from the filled matrix by backtracking the performed computation. This is done by starting in cell (n,m) and recursively checking which of the three possible cells contributed its value to the examined cell (either by recalculating the sums or by checking a previously saved pointer/arrow to a cell). In this way a path of cells (or arrows, see Fig. 3) from cell (n,m) to cell $(0,0)$ is constructed which obviously corresponds to a transformation sequence.

2.2   Van Deusen's Edit Distance

The transformation space over which the simple edit distance is minimized includes in particular transformations containing many edit operations. Transformations like this correspond to paths in the computation matrix with many non-diagonal arrows. Since a tree ring sequence usually contains only very few missing or double rings, Van Deusen [1] reduced the transformation space by allowing only those paths in the matrix which stay inside a given strip of constant width around the diagonal starting at $(0,0)$; see Fig. 4.

The width of a strip is given by a parameter α which denotes the width on each side of the diagonal. The in this way reduced transformation space contains only transformations in which the edit operations are locally balanced. Yet there are still transformations with many edit operations possible. For instance if the transformation sequence alternates between a merge and an insert operation the conforming transformation path still stays inside a strip of width 1 around the diagonal.

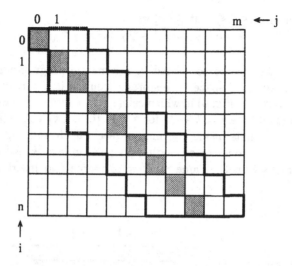

Fig. 4. Van Deusen's computation matrix with a strip of width $\alpha = 2$ to each side of the diagonal. The diagonal has been shaded.

2.3 α-Box Edit Distance

A straight forward improvement of Van Deusen's edit distance is the following notion which we call α-*box edit distance* or k-*edit distance*. This type of edit distance has been proposed for strings by Sankoff and Kruskal [5]. Since the number of edit operations contained in an optimal transformation should be small, the idea is to regard transformations and edit distances depending on the number of edit operations. We therefore define the k-*edit distance* as follows:

$$D(A, B, k) = \begin{cases} \min_{\tau \in \mathcal{T}_{n,m,k}} \sum_{v=0}^{m-1} (\tau(A)_v - B_v)^2 & \text{, if } \mathcal{T}_{n,m,k} \neq \emptyset \\ \text{not defined} & \text{, otherwise} \end{cases} \tag{8}$$

where $\mathcal{T}_{n,m,k}$ is the set of all transformation sequences transforming a sequence of length n into a sequence of length m using exactly k edit operations. Let again γ_τ be the number of merge operations, ι_τ the number of insert operations and ν_τ the number of identity operations in τ. Then from the definition of τ follow $m = \gamma_\tau + \iota_\tau + \nu_\tau = n + \iota_\tau - \gamma_\tau$ and $\iota_\tau + \gamma_\tau = k$. It is then easy to show that $\mathcal{T}_{n,m,k}$ is non-empty if and only if $m \geq k$ and $m - n = k - 2\gamma$ for a $\gamma \in \{0, \ldots, k\}$. We define $D(i, j, k)$ to be the k-edit distance between the prefixes $A[0..i-1]$ and $B[0..j-1]$. Just as in the case of the simple edit distance the k-edit distance satisfies the following recurrence:

$$D(0, 0, 0) = 0$$

$$D(i, j, k) = \min \left\{ \begin{array}{l} D(i-2, j-1, k-1) + (a_{i-2} + a_{i-1} - b_{j-1})^2, \\ D(i-1, j-1, k) + (a_{i-1} - b_{j-1})^2, \\ D(i, j-1, k-1) + (\mu(i-1) - b_{j-1})^2 \end{array} \right\} \tag{9}$$

for all $0 \leq i \leq n,\ 0 \leq j \leq m$
with $j \geq k$ and $j - i = k - 2\gamma$ for a $\gamma \in \{0, \ldots, k\}$.

The α-edit distance can be computed in a dynamic programming manner in $\theta(\alpha^2 \min(n, m))$ time and space. The storage required is a part of an $(\alpha + 1) \times (n+1) \times (m+1)$ box (see Fig. 6) in which cell (i, j, k) contains the value $D(i, j, k)$. Due to the condition $j - i = k - 2\gamma$ for a $\gamma \in \{0, \ldots, k\}$ the defined values of $D(i, j, k)$ form diagonals inside the matrix (see Fig. 5 and Fig. 6). For each $\gamma \in \{0, \ldots, k\}$ there is one corresponding diagonal in level k. All edit distances in one diagonal contain the same number of merge and insert operations as shown in Fig. 6.

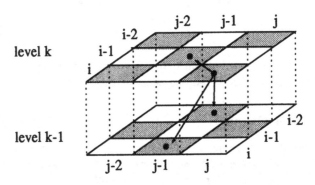

Fig. 5. Dynamic programming computation of the k-edit distance according to (9). A projection of the two levels onto one level results in the matrix shown in Fig. 3.

According to (9) the value $D(i, j, k)$ is computed out of at most three values (see Fig. 5) whereby a change of the k-level is performed only in the case of an edit operation (merge or insert). The computation is carried out by filling the diagonals level by level, thereby touching each cell only a constant number of times, until finally filling cell (n, m, α). Note that this computation box (α-box) contains especially all k-edit distances between A and B with $0 \leq k \leq \alpha$. In each k-level there are at most $k + 1$ diagonals and each diagonal contains at most $\min(n, m) + 1$ cells. Therefore there are $(\min(n, m) + 1) \sum_{k=0}^{\alpha} (k + 1) = \mathcal{O}(\alpha^2 \min(n, m))$ cells to be filled.

Often an optimal transformation contains two opposite edit operations (insert/merge or merge/insert) almost successively. A pair of edit operations like this has no global effect on the edited sequence, but only a local effect during the short time interval between the two edit operations. Given a threshold (e.g. 10) for the minimum number of years between two opposite edit operations, 10 cells on the diagonal after an edit operation are marked so that the opposite edit operation is not allowed when calculating the edit distances for these cells.

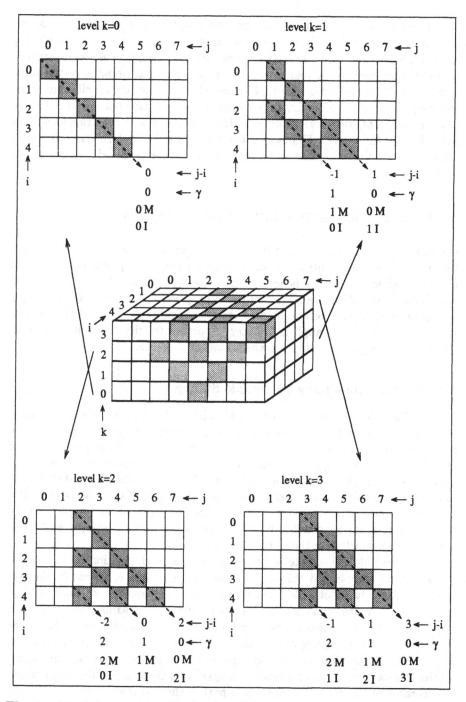

Fig. 6. Dynamic programming box needed for the computation of the 3-edit distance between two sequences of length 4 and 7. The defined cells have been shaded. The number of merge (*M*) and insert (*I*) operations corresponding to the diagonals have been written next to each diagonal.

Although theoretically there can be several optimal transformations associated with one edit distance (this corresponds to more than one arrow leaving one cell), we always choose exactly one optimal transformation per edit distance (or cell), since due to the real valued input data an exact equality of the sums is unlikely. Then instead of storing in each cell a pointer to the cell which contributed its value to the sum, we collapse a path of diagonal pointers to one pointer (*shortcut*) directly pointing to the position where the transformation path changes the k-level. That way a traceback of the transformation path of a cell in level k needs $\theta(k)$ time and a transformation can be saved for later use in $\theta(k)$ space.

3   Crossdating Employing k-Edit Distances

Crossdating is usually performed by sliding the sample sequence across the master sequence and computing distances between the overlapping parts at each crossing position. The same is done in the algorithm presented here using the k-edit distance and therefore taking into account possible missing or double rings. The parameter α must be specified by the user in advance. As a postprocessing step a new heuristic is applied in order to further restrict the number of edit operations being contained in a transformation sequence.

3.1   Simple Crossdating by Sample Sliding

Let us take a closer look at the α-box which has to be filled in order to compute the α-edit distance between the sample and a specific (coherent) piece of the master sequence. Assume the master piece starts at year y. Then the last row of each k-level, $0 \le k \le \alpha$, contains all k-edit distances between the sample and all prefixes of the master piece, while each last column contains all k-edit distances between the master piece and all prefixes of the sample. If the master piece is a suffix of the master sequence (or if its length is the length of the sample plus α), the α-box contains all possible k-edit distances between the sample and those master pieces which date the sample to year y.

When we slide the sample across the master, at each position computing an α-box for the sample and the suffix of the master sequence, especially the last row and last column of each level, we obtain all information we need to date the sample: The last rows contain results comparing the sample to the master sequence at different offsets, where each offset is represented in one α-box. The last columns are interesting only in the case that the sample partly overlaps the end of the master sequence, so that a prefix of the sample must be considered. In the case where the master piece is longer than the sample, the last column degenerates to one cell which is already part of the last row.

After having computed all edit distances in the last rows and last columns of each level in each α-box, we need to compare them in order to find an acceptible dating. A simple comparison is not meaningful, since the number of terms adding up to a k-edit distance in (8) varies according to the number and kind of edit operations performed and also according to a partial overlap of the sequences. So we need to normalize the edit distance by dividing by the number of added terms which is the length of the transformed sample sequence. However, a simple comparison of all normalized edit distances (which means sorting them and taking the smallest as the best) proved not to be useful. The reason for that is that the normalization removes the information about the length of the sequence (i.e. the number of summands), so that shorter sequences cause a better edit distance more easily than longer sequences do.

Since the t-value is length-dependant and a commonly used distance measure in dendrochronology (which dendrochronologists are familiar with) we take the t-value between the transformed sequence and the master piece as the judging criterion. The correlation coefficient can be implicitly calculated during the box filling process at no extra cost asymptotically. So we sort all t-values of the last row and last column of each box and output the biggest t-values, each together with an optimal transformation (i.e. the positions of possible missing and double rings) and the corresponding dating proposal (offset) to the user.

3.2   Heuristic Postprocessing of the Results

The algorithm described so far simply sorts all results in the end without taking the number of edit operations into account. Therefore the best results will often contain too many edit operations. A standard approach to that problem is to penalize edit operations either by a multiplicative or an additive term. Unfortunately this also affects edit operations at correct positions. Indeed, it seems that in respect to penalties incorrect edit operations are somehow more robust. We therefore decided not to penalize the edit operations, but we compare the obtained results in a heuristic postprocessing step. We store all edit distances of all last rows and last columns of each level of each α-box in an overall result structure. Then usually a good dating appears several times among the best results. Those similar results then differ only in some edit operations, whereby they usually share some edit operations (the correct ones) and include some more edit operations which improve the edit distance a little but which are incorrect.

The heuristic we developed then tries in two phases to identify some possible redundant results and deletes those from the overall result structure. For each edit distance result (that is an edit distance in the last row or column of a box-level) we do a redundancy check within the box plus another check concerning some neighboring boxes.

In the first check phase it is tested, if the normalized distance is significantly smaller than each normalized edit distance associated with each last cell on the diagonals on the transformation path. This captures the idea that one good match often appears several times, where the different occurences share some correct edit operations and include some more incorrect edit operations. An additional edit operation should therefore be admitted only if it improves (hence decreases) the edit distance significantly. A normalized edit distance e is said to be significantly smaller than the normalized edit distance e_{comp}, if $e/e_{comp} < 0.9$.

If an edit distance did not pass this check it is deleted from the overall result structure and not compared to other edit distances anymore.

The second check phase is established to eliminate those inter-box redundant results, which date the sequence incorrectly by a few years according to superfluous edit operations at the beginning of the transformation sequence. Call the subsequence of the transformation sequence which includes only the insert and merge operations an *edit sequence*. For every prefix of the edit sequence the time transposition it induces is calculated (a merge operation corresponds to a transposition one year to the left, an insert operation one year to the right). The α-box at the transposed position is checked if it contains an edit distance e_{comp} with an edit sequence equal to the remaining suffix of the edit sequence being checked. Now the normalized edit distance e whose edit sequence probably contains an unnecessary prefix is deleted if $e/e_{comp} \geq 0.9$.

3.3   Crossdating Algorithm

Figure 7 shows the crossdating algorithm. The standardization can be done in $\theta(m+n)$ time and space. The number of α-boxes to be filled is $\mathcal{O}(n+m)$, and since we need $\theta(\alpha^2 \min(n,m))$ time to fill one α-box (see Par. 2.3), we can fill them all in $\mathcal{O}(\alpha^2 mn)$ time. We do not need to store all α-boxes since we need only the last row and the last column of each level of every α-box. There is one

---

Crossdating algorithm:

Standardization of the master and the sample sequence
For all overlap positions of the sample in the master
    Fill α-box
    For all cells in the last row and last column of each level
        Normalize edit distance
        Compute optimal transformation
        Redundancy check 1: Check with normalized edit distances on
                                 transformation path.
        If edit distance is not redundant:
            Store the distance, the t-value, the transformation and the offset
            number in an overall result structure.
Redundancy check 2: Remove inter-box-redundant results.
Sort all results in the result structure by decreasing t-value.
Display the best results (those with the highest t-values).

Fig. 7. Crossdating algorithm

last row or last column entry for each diagonal, so that we only have to count the diagonals of which there are at most $(k+1)$ in every k-level. So we have $\mathcal{O}((m+n)\sum_{k=0}^{\alpha}(k+1)) = \mathcal{O}(\alpha^2(m+n))$ edit distances to be stored. But for each edit distance we also store its corresponding optimal transformation which needs $\theta(k)$ space, so that the space which is altogether needed to store all results sums up to $\mathcal{O}((m+n)(1+\sum_{k=1}^{\alpha}(k+1)k)) = \mathcal{O}(\alpha^3(m+n))$. Additionally there is $\mathcal{O}(\alpha^2\min(n,m))$ space for one α-box needed to fill a box.

The first redundancy check needs $\theta(k)$ time for each edit distance which sums up to $\mathcal{O}((m+n)(1+\sum_{k=1}^{\alpha}(k+1)k)) = \mathcal{O}(\alpha^3(m+n))$ altogether. The second redundancy check needs $\mathcal{O}(k^2)$ time each, hence together $\mathcal{O}((m+n)(1+\sum_{k=1}^{\alpha}(k+1)k^2)) = \mathcal{O}(\alpha^4(m+n))$. For the redundancy checks there is asymptotically no more space needed. The sorting of all results takes $\mathcal{O}(\alpha^2(m+n)\log(\alpha^2(m+n)))$ time. (Since we are interested in some of the best results only we actually do not have to sort all results, but sorting does not affect the asymptotical running time.) So altogether the algorithm needs $\mathcal{O}(\alpha^2mn + \alpha^4(m+n))$ time and $\mathcal{O}(\alpha^3(m+n))$ space.

4 Test Results

4.1 Implementation

The crossdating algorithm has been implemented in C++ in a command-line-oriented Unix environment. A program executable can be obtained from the author.

In practice a crossdating program outputs several good matchings (e.g. the best 5 or 10), and the dendrochronologist visually checks if one of them represents the correct dating. Likewise the program we have implemented allows the user to subsequently evaluate the results according to different criteria. That is, once the results have been computed, the best results in a certain time interval, those having a bigger minimum overlap or those for a lower value for α can be queried. The computation time of such modified result queries by scanning the list of the sorted results is linear in the number of results, thus $\mathcal{O}(\alpha^2(m+n))$. Note that our program especially computes all those t-values that are computed during the simple sliding algorithm. They can be accessed by querying the results for $\alpha = 0$. Our program therefore generalizes the simple sliding algorithm.

Several tests have been performed which we will present in the following three paragraphs: Tests with randomly generated missing or double rings, tests on data containing real missing rings and finally runtime tests. However, the automatized tests do not cover the interactive program properties.

4.2 Randomly Generated Disturbances

The program was tested on collections of already dated samples. In each collection one sample was randomly disturbed by deleting or splitting up some values, and this sample was then tried to date against the mean sequence of the remaining sequences in the collection.

Tests were performed mainly for the parameter values $\alpha = 2$, a minimum overlap of 50 and a redundancy threshold of 10 (see end of Par. 2.3). For each sample a random disturbance has been carried out 5 times. Some test results are shown in Tables 1, 2, 3, 4 and 5. Column *date* shows the percentage of those data sets in the collection for which the correct dating has been found. *Date & edit* shows the percentage of those data sets for which the correct dating and the correct type of editation in an interval of radius 10 around the correct position have been found. The column *k* shows the average number of proposed edit operations for those results for which the correct date and editation has been found. For standardization the percentage of a five-year running mean (also called *floating average*; in the following tables abbreviated with *float-ave*) or the difference of logarithms (abbreviated with *log*) were used.

Table 1. Test results without manipulation of the data.

	date	date	k	date	date	k
	$\alpha = 0$	$\alpha = 2$		$\alpha = 0$	$\alpha = 2$	
kieftest	96 %	80 %	0.3	96 %	86 %	0.5
germ001	98 %	80 %	0.2	98 %	83 %	0.2
germ003	100 %	94 %	0.1	100 %	88 %	0.1
germ004	96 %	79 %	0.7	96 %	88 %	0.9
germ006	100 %	88 %	0.4	100 %	94 %	0.0
cana030	94 %	72 %	0.3	94 %	89 %	0.0
az052	100 %	80 %	0.1	100 %	80 %	0.4
az526	100 %	100 %	0.1	100 %	95 %	0.8
SET01	94 %	69 %	0.7	94 %	63 %	0.5
SET02	96 %	71 %	0.8	96 %	63 %	0.5
oh004	100 %	72 %	0.1	100 %	79 %	0.1
swed302	98 %	82 %	0.0	98 %	89 %	0.1

Float-ave preprocessing   Log preprocessing

The data used was supplied by the following sources: The *kieftest* data is tree ring width data from a German pine, which was supplied by Deutsches Archäologisches Institut[2]. The *SET01* and *SET02* data are files which come with the crossdating program TSAP [8] (which does not search for missing or double rings during the crossdating). The other data was taken from the ITRDB[3], where the *germ* data sets are from German oaks, the *cana* data from Canadian white spruce, the *az* data from Arizona where *az526* is ponderosa pine, the *oh* data from white oak from Ohio and the *swed* data from scotch pine from Sweden.

Table 1 shows how many samples of each tested collection the algorithm dates correctly when no random disturbances have been performed. With $\alpha = 0$ this equals a simple sliding algorithm concerning *t*-values which dates 97% samples correctly on the average[4]. With $\alpha = 2$ the percentage of correct datings decreases

---

[2] We like to thank Dr. K.-U. Heußner from Deutsches Archäologisches Institut, Eurasien-Abteilung, Im Dol 2-6, D-14195 Berlin.

[3] The International Tree-Ring Data Bank (ITRDB) is located at http://www.ngdc.noaa.gov/paleo/treering.html.

[4] On the average means here averaged over the tested collections shown in the tables.

by up to 33% (on the average only by 15% though), and the number of mistakenly found edit operations increases by up to 0.9 (on the average only by 0.4).

To see how our algorithm performs on sequences that contain missing rings, take a look at Table 2 which shows test results for the *germ001* data with one randomly deleted element and for different values of α. Setting $\alpha = 0$ (again, this equals the simple sliding algorithm) the correct date is found in only 51% or 41% of the cases (for floating-average or log preprocessing, respectively). When allowing the algorithm to perform some editations by choosing α slightly greater than 1, the chance for a correct dating increases dramatically. But the farther α is away from the correct number of edit operations needed, the more false edit operations are performed and the more the chance for a correct date decreases. But since there are not many double or missing rings expected in a tree ring sequence, a small value of α (e.g. 2 or 3) should be sufficient most of the times.

Table 2. Test results concerning the *germ001* data with a random deletion of one sample element and different values for α.

α	date	date & edit	k	date	date & edit	k
0	51 %	0 %	0.0	41 %	0 %	0.0
1	94 %	91 %	1.0	95 %	91 %	1.0
2	93 %	89 %	1.0	92 %	87 %	1.0
3	81 %	77 %	1.4	84 %	76 %	1.4
4	82 %	78 %	1.5	84 %	76 %	1.4
5	85 %	78 %	2.2	85 %	76 %	1.9

Float-ave preprocessing   Log preprocessing

Tables 3, 4 and 5 show test results for different data collections with one random deletion, one random splitting and two random consecutive deletions. Although the percentages for a correct dating with a correct editation vary from 38% to 95%, the percentages are usually extremely higher than those for a dating with $\alpha = 0$. However, as can be seen in the column *date* for $\alpha > 0$, the program finds the correct date more often than the correct date plus the correct editation, because it proposes some wrong, additional or not enough edit operations. In fact, if the program finds the correct date, it usually proposes most of the editations at an almost correct position and skips necessary editations only if they are too close to the beginning or the end of the sequence. In any case the results of the program give more information to the user about possible missing or double rings than the standard crossdating methods do. Concerning two consecutive deletions, Table 5 shows that if two missing rings have been found, they lie only about 2 or 3 years apart.

Table 3. Test results with a random deletion of one sample element.

	date	date	date & edit	k	date	date	date & edit	k
	$\alpha = 0$	$\alpha = 2$			$\alpha = 0$	$\alpha = 2$		
kieftest	24 %	73 %	67 %	1.2	28 %	76 %	69 %	1.2
germ001	51 %	92 %	89 %	1.0	41 %	92 %	87 %	1.0
germ003	58 %	94 %	83 %	1.0	53 %	92 %	81 %	1.0
germ004	47 %	84 %	76 %	1.2	50 %	87 %	78 %	1.2
germ006	21 %	91 %	85 %	1.0	26 %	88 %	81 %	1.0
cana030	27 %	72 %	69 %	1.0	30 %	90 %	80 %	1.0
az052	44 %	97 %	94 %	1.0	42 %	95 %	95 %	1.0
az526	38 %	91 %	79 %	1.0	35 %	96 %	83 %	1.1
SET01	34 %	60 %	54 %	1.0	25 %	59 %	49 %	1.1
SET02	37 %	54 %	48 %	1.2	33 %	59 %	53 %	1.2
oh004	47 %	85 %	83 %	1.0	42 %	81 %	80 %	1.0
swed302	43 %	83 %	72 %	1.0	39 %	89 %	79 %	1.0

Float-ave preprocessing   Log preprocessing

Table 4. Test results with a random splitting of one sample element.

	date	date	date & edit	k	date	date	date & edit	k
	$\alpha = 0$	$\alpha = 2$			$\alpha = 0$	$\alpha = 2$		
kieftest	22 %	75 %	63 %	1.2	28 %	75 %	58 %	1.2
germ001	52 %	92 %	88 %	1.0	42 %	93 %	88 %	1.0
germ003	56 %	94 %	83 %	1.0	52 %	92 %	79 %	1.0
germ004	48 %	88 %	78 %	1.1	50 %	88 %	74 %	1.1
germ006	19 %	90 %	81 %	1.0	26 %	88 %	85 %	1.0
cana030	29 %	72 %	66 %	1.0	34 %	82 %	71 %	1.0
az052	47 %	98 %	88 %	1.0	42 %	98 %	80 %	1.0
az526	36 %	83 %	74 %	1.0	34 %	85 %	66 %	1.2
SET01	18 %	60 %	53 %	1.0	16 %	59 %	56 %	1.1
SET02	28 %	57 %	48 %	1.1	32 %	59 %	52 %	1.1
oh004	47 %	87 %	85 %	1.0	43 %	85 %	83 %	1.0
swed302	34 %	74 %	67 %	1.1	35 %	79 %	69 %	1.0

Float-ave preprocessing   Log preprocessing

4.3   Real Missing Rings

In this paragraph we present results from tests performed on data containing real missing rings. Test data like this is widely available because many dendrochronologists mark a missing ring as a ring with width 0. Test data for double rings is rather hard to find because dendrochronologists usually do not record the occurence of double rings. The reason for that is that there is a chance to visually identify a double ring on the wood (e.g. after some more preparation of the

Table 5. Test results with a random deletion of two consecutive sample elements.

	date	date	date & edit	k	distance betw. edits	date	date	date & edit	k	distance betw. edits
	$\alpha = 0$		$\alpha = 3$			$\alpha = 0$		$\alpha = 3$		
kieftest	13 %	67 %	55 %	2.2	2.7	21 %	73 %	56 %	2.1	2.5
germ001	56 %	91 %	85 %	2.0	2.2	54 %	89 %	78 %	2.0	2.2
germ003	54 %	91 %	74 %	2.0	3.5	58 %	89 %	68 %	2.0	3.5
germ004	38 %	80 %	79 %	2.1	3.1	49 %	85 %	69 %	2.1	2.7
germ006	16 %	85 %	75 %	2.0	2.5	21 %	84 %	66 %	2.0	2.5
cana030	20 %	71 %	66 %	2.0	2.8	34 %	86 %	73 %	2.0	3.0
az052	44 %	94 %	91 %	2.0	1.8	42 %	91 %	88 %	2.0	1.7
az526	44 %	89 %	79 %	2.0	2.2	48 %	92 %	78 %	2.1	2.3
SET01	24 %	55 %	49 %	2.0	2.8	34 %	55 %	39 %	2.2	3.1
SET02	30 %	51 %	39 %	2.1	3.1	38 %	52 %	38 %	2.1	2.4
oh004	45 %	79 %	76 %	2.0	2.3	49 %	76 %	72 %	2.0	2.2
swed302	34 %	77 %	65 %	2.0	2.7	48 %	81 %	65 %	2.0	2.6

Float-ave preprocessing   Log preprocessing

wood or using a better microscope), whereas for a missing ring there is not. We therefore restricted the tests on data with real inconsistencies to data containing missing rings.

Table 6. Test results for data with real missing rings; $\alpha = 4$.

	# of samples	ave. # of miss. rings	date	date & edit	date	date & edit
wa067	19	2.42	16 ≅ 84%	12 ≅ 63%	17 ≅ 89%	14 ≅ 74%
wa069	13	1.08	12 ≅ 92%	9 ≅ 69%	13 ≅ 100%	10 ≅ 77%
wa072	19	2.32	16 ≅ 84%	13 ≅ 68%	16 ≅ 84%	13 ≅ 68%
wa079	23	2.22	17 ≅ 74%	15 ≅ 65%	18 ≅ 78%	17 ≅ 74%
breclav	7	1.0	7 ≅ 100%	5 ≅ 71%	7 ≅ 100%	6 ≅ 86%
chin04	13	2.92	12 ≅ 92%	9 ≅ 69%	13 ≅ 100%	11 ≅ 85%
az052	6	3.33	5 ≅ 83%	4 ≅ 67%	5 ≅ 83%	4 ≅ 67%
az526	14	3.71	13 ≅ 93%	6 ≅ 43%	13 ≅ 93%	7 ≅ 50%

Float-ave preprocessing Log preprocessing

Table 6 shows test results for collections of samples where some samples contain missing rings. During the tests α has been chosen to be 4. The *breclav* data was supplied by Deutsches Archäologisches Institut, the others are available at the ITRDB. The *wa* data is from a subalpine larch from Washington State, the *chin* data is Armand's pine from China, and the *az* data is from Arizona, as mentioned above. The column *# of samples* contains the number of samples in the collection that contain missing rings. The *ave. # of miss. rings* column shows the average number of missing rings contained in the samples. The *data & edit* column contains the number of samples (and also the percentage relative to the *# of samples*) that the algorithm dates correctly with the correct number and position (with a tolerance of 10) of insertions. The master sequences to date against are built out of those samples in each collection that do not contain missing rings.

4.4   Runtime Tests

The crossdating algorithm has been tested on a Sparc Ultra 1 machine. The runtime the program needs for fixed $\alpha = 2$ is illustrated in Fig. 8, and for a fixed sample length $n = 300$ in Fig. 9. For a typical input consisting of a sample of length $n = 300$, a master of length $m = 1000$ and $\alpha = 2$ the program needs about 13 seconds.

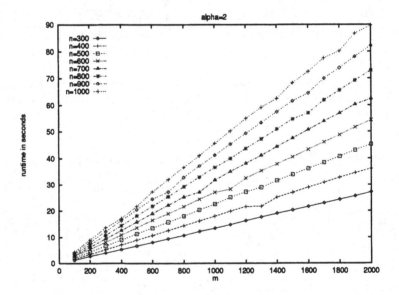

Fig. 8. Runtime tests with $\alpha = 2$ and n and m the lengths of the sample and master sequence, respectively.

5   Implementation and Conclusions

We investigated the problem of matching tree ring width sequences (crossdating) which is stated in dendrochronology. Assuming that a tree forms exactly one ring each year the matching can be performed by an easy $\theta(mn)$ algorithm. We presented a $\mathcal{O}(\alpha^2 mn + \alpha^4(m + n))$ crossdating algorithm which takes the

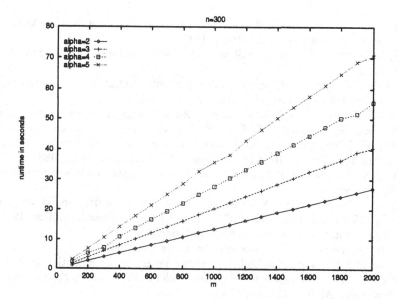

Fig. 9. Runtime tests with $n = 300$ fixed, where n and m are the lengths of the sample and master sequence, respectively.

possibility of missing and double rings into account by employing an edit distance as a distance measure.

The algorithm has been implemented and tested. The tests show that the dating quality of the algorithm varies depending strongly on the input data. It is best when α equals the number of inconsistencies to be found. It is therefore not possible to date a tree ring sequence according solely to the first dating proposition of the algorithm. However, in the usual dating process results of an automatic matching are taken as dating propositions only and are always visually verified by a dendrochronologist. Since our program allows the evaluation of the computed results for some different parameter settings, e.g. for a lower value for α, in rather fast $\mathcal{O}(\alpha^2(m + n))$ time, and it usually offers some good datings, the program should be eligible to serve as an additional dating tool searching for missing and double rings.

For further research it would be interesting to investigate whether it is possible to compare several tree ring sequences at once in order to produce a mean sequence (*master sequence, chronology*). The question could also be raised as to whether similar matching techniques based on the edit distance can be applied to other environmental archives like sea or glacier sediments, where certain environmental events produce different distortions of the underlying sequences.

References

[1] Paul C. Van Deusen. A dynamic program for cross-dating tree rings. *Canadian Journal of Forest Research*, 20:200–205, 1989.

[2] Douglas F. Elliott and K. Ramamohan Rao. *Fast Transforms - Algorithms, Analyses, Applications*. Academic Press, 1982.

[3] Dan Gusfield. *Algorithms on Strings, Trees and Sequences*. Cambridge University Press, 1997.

[4] Richard L. Holmes. Computer-assisted quality control in tree-ring dating and measurement. *Tree-Ring Bulletin*, 43:69–75, 1983.

[5] Joseph B. Kruskal and David Sankoff. An anthology of algorithms and concepts for sequence comparison. In David Sankoff and Joseph B. Kruskal, editors, *Time Warps, String Edits, and Mocromolecules: The Theory and Practice of Sequence Comparison*, chapter 10. Addison-Wesley Publishing Company, 1983.

[6] H.J. Nussbaumer. *Fast Fourier Transform and Convolution Algorithms*. Springer Verlag, 1981.

[7] William H. Press, Saul A. Teukolsky, William T. Vetterling, and Brian P. Flannery. *Numerical Recipes in C*. Cambridge University Press, 2. edition, 1992.

[8] Frank Rinn. *TSAP Reference Manual*. Heidelberg. http://ourworld.compuserve.com/homepages/frankrinn/.

[9] Hiroaki Sakoe and Seibi Chiba. Dynamic programming algorithm optimization for spoken word recognition. *IEEE Transactions on Acoustics, Speech, and Signal Processing*, ASSP-26(1):43–49, 1978.

[10] David Sankoff and Joseph B. Kruskal, editors. *Time Warps, String Edits, and Mocromolecules: The Theory and Practice of Sequence Comparison*. Addison-Wesley Publishing Company, 1983.

[11] F.H. Schweingruber. *Trees and Wood in Dendrochronology*. Springer-Verlag, 1993.

[12] Graham A. Stephen. *String Searching Algorithms*. World Scientific, 1994.

[13] Carola Wenk. Algorithmen für das Crossdating in der Dendrochronologie. Master's thesis, Freie Universität Berlin, Institut für Informatik, 1997.

Fast Multi-dimensional
Approximate Pattern Matching*

Gonzalo Navarro and Ricardo Baeza-Yates

Dept. of Computer Science, University of Chile
Blanco Encalada 2120 - Santiago - Chile
{gnavarro,rbaeza}@dcc.uchile.cl

Abstract. We address the problem of approximate string matching in
d dimensions, that is, to find a pattern of size m^d in a text of size n^d
with at most $k < m^d$ errors (substitutions, insertions and deletions
along any dimension). We use a novel and very flexible error model,
for which there exists only an algorithm to evaluate the similarity between two elements in two dimensions at $O(m^4)$ time. We extend the
algorithm to d dimensions, at $O(d!m^{2d})$ time and $O(d!m^{2d-1})$ space. We
also give the first search algorithm for such model, which is $O(d!m^d n^d)$
time and $O(d!m^d n^{d-1})$ space. We show how to reduce the space cost
to $O(d!3^d m^{2d-1})$ with little time penalty. Finally, we present the first
sublinear-time (on average) searching algorithm (i.e. not all text cells are
inspected), which is $O(kn^d/m^{d-1})$ for $k < (m/(d(\log_\sigma m - \log_\sigma d)))^{d-1}$,
where σ is the alphabet size. After that error level the filter still re-
mains better than dynamic programming for $k \le m^{d-1}/(d(\log_\sigma m - \log_\sigma d))^{(d-1)/d}$. These are the first search algorithms for the problem. As
side-effects we extend to d dimensions an already proposed algorithm for
two-dimensional exact string matching, and we obtain a sublinear-time
filter to search in d dimensions allowing k mismatches.

1 Introduction

Approximate pattern matching is the problem of finding a pattern in a text
allowing errors (insertions, deletions, substitutions) of characters. A number of
important problems related to string processing lead to algorithms for approximate string matching: text searching, pattern recognition, computational biology, audio processing, etc. Two dimensional pattern matching with errors has
applications, for instance, in computer vision (i.e. searching a subimage inside
a large image). In three dimensions, our algorithms may be useful for searching
allowing errors in video data (where the time would be the third dimension) or
in some types of medical data (e.g. MRI brain scans).

For one dimension this problem is well-known, and is modeled using the
edit distance. The *edit distance* between two strings a and b, $ed(a, b)$, is defined
as the minimum number of *edit operations* that must be carried out to make

---

* Supported in part by Fondecyt grant 1-990627.

M. Crochemore, M. Paterson (Eds.): CPM'99, LNCS 1645, pp. 243–257, 1999.

them equal. The allowed operations are insertion, deletion and substitution of characters in a or b. The problem of *approximate string matching* is defined as follows: given a *text* of length n, and a *pattern* of length m, both being sequences over an alphabet Σ of size σ, find all segments (or "occurrences") in *text* whose edit distance to *pattern* is at most k, where $0 < k < m$. The classical solution is $O(mn)$ time and involves dynamic programming [20].

Krithivasan and Sitalakshmi (KS) [17] proposed a simple extension to two dimensions. Given two images of the same size, the edit distance is the sum of the edit distance of the corresponding row images. This definition is justified when the images are transmitted row by row and there are not too many communication errors (e.g. photocopy images, where most errors come from the mechanical traction mechanism along one dimension only, or images transmitted by fax), but it is not appropriate otherwise. Using this model they define an approximate search problem where a subimage of size $m \times m$ is searched into a large image of size $n \times n$, which they solve in $O(m^2n^2)$ time using a generalization of the classical one-dimensional algorithm.

In [5], Baeza-Yates (BY) defined a more general extension (there called RC), where the errors can occur along rows or columns at any time. This model is much more robust and useful for more applications. We are interested in this general model in this work. Figure 1 gives an example.

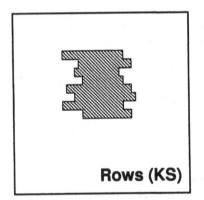
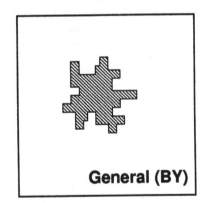

Rows (KS)   **General (BY)**

Fig. 1. Alternative error models.

Although in [5] they give an $O(m^4)$ time algorithm to compute the edit distance among two images of size $m \times m$, they do not give any algorithm to search a subimage inside a larger image allowing errors.

In this work, we first generalize the edit distance algorithm to d dimensions with complexity $O(d!m^{2d})$. We then give an $O(d!m^dn^d)$ time algorithm for the search problem, matching the same complexity of the simpler KS model in two dimensions, and show how to reduce the space requirements so that they depend only on the pattern size. We also give a new filtering algorithm that allows to quickly discard large parts of the text that cannot contain a match. This algorithm searches the pattern in average time $O(kn^d/m^{d-1})$ for $k < (m/(d(\log_\sigma m - \log_\sigma d)))^{d-1}$, where σ is the alphabet size. After that error level the filter changes its cost but remains better than dynamic programming for $k \leq m^{d-1}/(d(\log_\sigma m - \log_\sigma d))^{(d-1)/d}$. These are the first searching algorithms for this problem.

Two side-effects are obtained as well. First, we generalize to d dimensions and analyze a previously proposed algorithm to search in two dimensions not allowing errors. Second, we obtain a filter to search a pattern in d dimensions allowing up to k character substitutions.

2   Previous Work

The classical dynamic programming algorithm [20] to search a pattern in a text allowing errors uses dynamic programming and is $O(mn)$ time and $O(m)$ space.

This solution was later improved by a number of algorithms, which we do not cover here. The only one of interest to this work is a filtering algorithm [21, 8, 7]. It states that if a pattern is cut in $k + 1$ pieces, then any occurrence with up to k errors must contain one of the pieces unchanged. This is obvious since k errors cannot alter the $k + 1$ pieces given the edit operations that we consider (which cannot alter two pieces at the same time). The algorithm simply scans the text using a multipattern exact search algorithm for all the pieces. Each time a piece is found, it uses dynamic programming over an area of length $m + 2k$ where the approximate occurrence can be found.

The multipattern search can be carried out in $O(n)$ worst-case search time by using an Aho-Corasick machine [1], or in $O(n/m)$ best-case time using Commentz-Walter [12] or another Boyer-Moore type algorithm adapted to multipattern search. The total cost of verifications keeps below $O(n)$ if $k/m \leq 1/(3\log_\sigma m)$.

Two dimensional string matching was first considered by Bird and Baker [11, 10], who obtain $O(n^2)$ worst-case time. Good average results are presented by Zhu and Takaoka in [22]. The best average case result is due to Baeza-Yates and Régnier [9], who obtain $O(n^2/m)$ time on average and $O(n^2)$ in the worst case.

The case of two dimensional approximate string matching usually considers only substitutions for rectangular patterns, which is much simpler than the general case with insertions and deletions. For substitutions, the pattern shape matches the same shape in the text (e.g. if the pattern is a rectangle, it matches a rectangle of the same size in the text). For insertions and deletions, instead, rows and/or columns of the pattern can match pieces of the text of different length. Under the substitutions model, one of the best results on the worst case is due to Amir and Landau [4], which achieves $O((k + \log \sigma)n^2)$ time but uses $O(n^2)$ space. A similar algorithm is presented in [13]. Ranka and Heywood solve the same problem in $O((k+m)n^2)$ time and $O(kn)$ space. Amir and Landau also present a different algorithm running in $O(n^2 \log n \log \log n \log m)$ time. On average, the best algorithm is due to Kärkkäinen and Ukkonen [15], with its analysis and space usage improved by Park [19]. The expected time is $O(n^2k/m^2 \log_\sigma m)$

for $k < m^2/(4\log_\sigma m)$ using $O(m^2)$ space ($O(k)$ space on average). This time result is optimal for the expected case.

Krithivasan and Sitalakshmi (KS) [17] defined the edit distance in two dimensions as the sum of the edit distance of the corresponding row images. Using this model they search a subimage of size $m \times m$ into a large image of size $n \times n$, in $O(m^2 n^2)$ time using a generalization of the classical one-dimensional algorithm. Krithivasan [16] presents for the same model an $O(m(k + \log m)n^2)$ algorithm that uses $O(mn)$ space. Amir and Landau [4] give an $O(k^2 n^2)$ worst case time algorithm using $O(n^2)$ space. Amir and Farach [3] also considered non-rectangular patterns achieving $O(k(k + \sqrt{m \log m}\sqrt{k \log k})n^2)$ time.

In [6] we use the same model and improve the expected case to $O(n^2 k \log_\sigma m/ m^2)$ on average for $k < m(m + 1)/(5 \log_\sigma m)$, using $O(m^2)$ space. This time matches the optimal result allowing only substitutions, and is also optimal [15], being the restriction on k only a bit stricter. For higher error levels, [6] presents an algorithm with time complexity $O(n^2 k/(\sqrt{\sigma}\log n))$, which works for $k < m(m + 1)(1 - e/\sqrt{\sigma})$. It is also shown that this limit on k cannot be improved.

In [5], Baeza-Yates defined more general models, where the errors can occur along rows or columns. Three distances R, C and L are defined, and for the first two it is shown that the filters of [6] can be applied to obtain the same complexity and slightly reduced tolerance to errors, i.e. $k < m(m + 1)/(7 \log_\sigma m)$. A fourth model defined in [5] is called RC, which generalizes R and C since the errors can occur along rows or columns at any time. This model is much more robust and useful for more applications, and is the one we use in this work. We cover this model in detail in the next section.

3   Multidimensional Approximate Searching

The classical dynamic programming algorithm [18] to compute the edit distance between two one-dimensional strings A and B of length m_1 and m_2 computes a matrix $C_{0..m_1, 0..m_2}$. The value $C_{i,j}$ holds the edit distance between $A_{1..i}$ and $B_{1..j}$. The construction algorithm is as follows

$$C_{i,0} \leftarrow i \ , \quad C_{0,j} \leftarrow j$$
$$C_{i,j} \leftarrow \text{if } A_i = B_j \text{ then } C_{i-1,j-1} \quad \text{else } 1 + \min(C_{i-1,j-1}, C_{i-1,j}, C_{i,j-1})$$

and the distance $ed(A, B)$ is the final value of C_{m_1, m_2}. The rationale of the formula is that if $A_i = B_j$ then the cost to convert $A_{1..i}$ into $B_{1..j}$ is that of converting $A_{1..i-1}$ into $B_{1..j-1}$. Otherwise we have to make one error and select among three choices: (a) convert $A_{1..i-1}$ into $B_{1..j-1}$ and replace A_i by B_j, (b) convert $A_{1..i-1}$ into $B_{1..j}$ and delete A_i, and (c) convert $A_{1..i}$ into $B_{1..j-1}$ and insert B_j.

This algorithm takes $O(m_1 m_2)$ space and time. It is easily adapted to search a pattern P in a text T allowing up to k errors [20]. In this case we want to report all the text positions j such that a suffix of $T_{1..j}$ matches P with at most k errors. This time the matrix is $C_{0..n, 0..m}$ and the construction formula is

$$C_{i,0} \leftarrow 0 \;,\;\; C_{0,j} \leftarrow j$$
$$C_{i,j} \leftarrow \text{if } P_i = T_j \text{ then } C_{i-1,j-1} \text{ else } 1 + \min(C_{i-1,j-1}, C_{i-1,j}, C_{i,j-1})$$

where the only change is that a pattern of length zero matches with no errors at any text position. All the positions i such that $C_{i,m} \leq k$ are reported. This takes $O(mn)$ time. The space can be reduced to $O(m)$ by noticing that only the old and new column of the matrix need to be stored. We define $led(T, P)$ as the smallest edit distance among the pattern P and a suffix of T, and therefore $led(T_{1..i}, P) = C_{i,m}$.

In [5], a natural extension to the edit distance notion for two dimensional strings (or "images") A and B was defined (called RC in that paper, and ed_2 in this work). It allows the errors to occur along any dimension. An algorithm to compute the edit distance among two images is defined. For simplicity we assume that they are square and of the same size $m \times m$, although it is easy to remove that limitation. The algorithm computes a four-dimensional matrix $C_{0..m,0..m,0..m,0..m}$, so that $C_{i,j,p,q} = ed(A_{1..i,1..j}, B_{1..p,1..q})$. C is built using the formulas

$$C_{i,0,0,0} \leftarrow i \;,\;\; C_{0,j,0,0} \leftarrow j$$
$$C_{0,0,p,0} \leftarrow p \;,\;\; C_{0,0,0,q} \leftarrow q$$
$$C_{i,j,p,q} \leftarrow \min(\; C_{i-1,j,p-1,q} + ed(A_{i,1..j}, B_{p,1..q}),\; C_{i-1,j,p,q} + j,\; C_{i,j,p-1,q} + q,$$
$$C_{i,j-1,p,q-1} + ed(A_{1..i,j}, B_{1..p,q})\; C_{i,j-1,p,q} + i,\; C_{i,j,p,q-1} + p \;)$$

which has a very similar rationale of the one-dimensional case: at each point we can solve the last row (first line of the min() formula) or the last column (second line of the min() formula). In each case, we either insert the whole row, delete the whole row, or replace the row of A by the row of B (and $ed()$ gives the best way to do it). This algorithm is $O(m^6)$ time and $O(m^4)$ space. However, by precomputing all the values

$$Horiz_{i,j,p,q} = ed(A_{i,1..j}, B_{p,1..q}) \qquad Vert_{i,j,p,q} = ed(A_{1..i,j}, B_{1..p,q})$$

(i.e. all the row-wise and column-wise alignments), the search time drops to $O(m^4)$ and the space does not change. This is because the $ed()$ of the C formula are obtained in constant time, and $Horiz$ consists of m^2 one-dimensional edit distance computations, among $A_{i,*}$ and $B_{p,*}$. The same holds for $Vert$.

The space can also be reduced to $O(m^3)$, as shown in [5]. We select, say, i as the most external variable of the iteration to fill the matrix. Therefore, we need only the values at iteration $i - 1$ to compute the values at iteration i. Hence, we do not need to store all the cells of all the i-th iterations, just the last one. The same can be done with $Horiz$ and $Vert$, by using i as the most external iteration variable.

In [5] they mention that this algorithm extends to d dimensions in time $O(m^{2d})$ but they do not give the details. We give a detailed algorithm in the next section and show that the exact complexity is $O(d! m^{2d})$. Also, no algorithm was given in [5] to search a subimage in a larger image using the above distance function. We do so in the following sections. We finally extend the one-dimensional filtering algorithm to more dimensions.

4   Edit Distance in More Dimensions

The idea of the previous section can be extended to compute $ed_d()$, i.e. the edit distance generalized to d dimensions. The algorithm is $O(d!m^{2d})$ time and $O(m^{2d-1})$ space.

A $(2d)$-dimensional matrix C is computed (d dimensions for A and d dimensions for B), and the $ed()$ of the above formula is replaced by ed_{d-1}. If the values of ed_{d-1} are not precomputed then we have $O(m^{2d-1})$ space (by using the trick of selecting one variable as the most external in the iteration) plus the space needed to compute ed_{d-1} (only one at a time is computed). This gives the recurrence

$$S_1 = m, \qquad S_d = m^{2d-1} + S_{d-1}$$

which yields $O(m^{2d-1})$ space. The time, on the other hand, involves to fill m^{2d} cells, where each cell performs a minimum over $3d$ elements (i.e. insertion, deletion and ed_{d-1} in d dimensions). This makes it necessary to compute d times the function $ed_{d-1}()$. That is

$$T_1 = m^2, \qquad T_d = m^{2d}\,3d + m^{2d}\,d\,T_{d-1}$$

which yields $O(d!m^{d(d+1)})$. This matches the $O(m^6)$ result for two dimensions mentioned in [5].

However, we may precompute all the necessary values of $ed_{d-1}()$. Along each one of the d dimensions, we take all the m^2 (i,p) possible combinations of values of the selected dimension in A and B, and compute $ed_{d-1}()$ between the $(d-1)$-dimensional objects which result from restricting the selected dimension to i in A and to j in B. Once this is done, the ed_{d-1} computations can be taken as constants in the formula of $ed_d()$. The time cost is now

$$T_1 = m^2, \qquad T_d = m^{2d}\,3d + dm^2 T_{d-1}$$

which yields $O(d!m^{2d})$ time (which matches the improved $O(m^4)$ algorithm of [5] for two dimensions). This is a big improvement over the naive algorithm. The space requirements are, however, higher. We have to store, for the d-dimensional object, m^{2d} cells plus the precomputed values, along each dimension, of all the m^2 combinations of (i,p) values for that dimension, and all the space for the lower dimensions resulting for each pair (i,p). That is

$$S_1 = m, \qquad S_d = m^{2d} + dm^2 S_{d-1}$$

which yields

$$S_d = d!m^{2d}\left(\frac{1}{1!} + \frac{1}{2!} + \ldots + \frac{1}{d!}\right) \le d!m^{2d}e = O(d!m^{2d})$$

and we can use the trick of the external variable to reduce this to $O(d!m^{2d-1})$.

5   A Dynamic Programming Search Algorithm

We modify the edit distance algorithm so that instead of computing the edit distance between two elements, it searches a small pattern P of size m^d inside a large text T of size n^d. The idea is a simple modification of the edit distance algorithm. For two dimensions the formula is as follows

$$C_{i,0,0,0} \leftarrow 0 \;,\; C_{0,j,0,0} \leftarrow 0$$
$$C_{0,0,p,0} \leftarrow p \;,\; C_{0,0,0,q} \leftarrow q$$
$$C_{i,j,p,q} \leftarrow \min(\; C_{i-1,j,p-1,q} + led(A_{i,1..j}, B_{p,1..q}), \; C_{i-1,j,p,q} + q, \; C_{i,j,p-1,q} + q,$$
$$C_{i,j-1,p,q-1} + led(A_{1..i,j}, B_{1..p,q}) \; C_{i,j-1,p,q} + p, \; C_{i,j,p,q-1} + p \;)$$

where the only differences are that the basic values are zero when the pattern is of size zero, that we penalize insertions and deletions according to the pattern size, and that instead of $ed()$ we use $led()$, so that we select the best suffix of the text along each dimension. If we are searching allowing up to k errors, then we report all text (i, j) positions such that $C_{i,j,m,m} \leq k$.

The form to extend this to more dimensions is immediate. By repeating the analysis of the above section, we see that the naive algorithm is $O(d!(mn)^{\frac{d(d+1)}{2}})$ time and $O(m^d n^{d-1})$ space (since n is much larger than m, we select one of the text coordinates as the most external variable). By precomputing the distances in lower dimensions, the search algorithm is $O(d!m^d n^d)$ time and $O(d!m^d n^{d-1})$ space.

5.1   Correctness

We now prove that the above algorithm is correct (in two dimensions). This extends easily to more dimensions.

Lemma: For each text position (i, j), it is possible to perform $C_{i,j,m,m}$ edit operations in the pattern P (converting it into P') so that the pattern P' matches the text suffix $T..i, ..j$, and this is not possible with less operations.

Proof: We prove the Lemma for any $C_{i,j,p,q}$. The Lemma is obviously true for the base case of the formula. For the recursive case, we inductively assume that the Lemma is true for the subproblems. We consider the first line of the update formula, which corresponds to the rows (the other cases are equivalent).

If the value for $C_{i,j,p,q}$ is obtained using a row insertion in the pattern, then we can inductively align $P_{1..p,1..q}$ at $T_{..i-1,j}$ with cost $C_{i-1,j,p,q}$, and then insert the text segment $T_{i,j-p+1..j}$ in P at the cost of p more errors so as to align $P_{1..p,1..q}$ at $T_{..i,j}$.

If the value for $C_{i,j,p,q}$ is obtained using a row deletion in the pattern, then we can inductively align $P_{1..p-1,1..q}$ at $T_{..i,j}$ with cost $C_{i,j,p-1,q}$, and then delete the pattern row $P_{p,1..q}$ from P at the cost of p more errors so as to align $P_{1..p,1..q}$ at $T_{..i,j}$.

Finally, if we obtain $C_{i,j,p,q}$ by replacing $P_{p,1..q}$ with a row suffix of $T_{i,..j}$, then the $led()$ of the formula gives the optimal way to do it, so that we align

$P_{1..p-1,1..q}$ at $T_{..i-1,j}$ with cost $C_{i-1,j,p-1,q}$, and then convert the pattern row $P_{p,1..q}$ to some text row suffix of $T_{i,..j}$, at $led(T_{i,1..j}, P_{p,1..q})$ cost.

Alternatively, we can use the recursion on the column values. It is also clear that this cannot be done better. On the other hand, we can use induction over the number of dimensions to show that the Lemma is correct for any d-dimensional problem.

5.2 Reducing the Space Requirements

The space requirement of the algorithm is $O(d!m^d n^{d-1})$, which is too high. This is awkward since the problem exhibits high locality. That is, the fact that a text position matches or not depends only on the last $(m+k)^d$-size text "suffix" that ends at that point. In fact, if $k > m$ we just need to start $2m$ positions behind the subtext at each dimension, since if more than m errors are made along a given line, it is better to just perform m replacements.

Therefore, if we cut the text in $(n/s)^d$ subtexts (of d dimensions) of size s^d, we can work separately at each subtext provided we start, at each dimension, $m + \min(m, k)$ positions behind the cube so as to have the context properly initialized when we reach the cube. The total time is $(n/s)^d d! m^d (m + \min(m, k) + s)^d$, and the total space is $d! m^d (m + \min(m, k) + s)^{d-1}$.

For instance, we may select $s = m$, and then we obtain an algorithm which is at most $O(d! 3^d m^d n^d)$ time and $O(d! 3^d m^{2d-1})$ space (and less if $k < m$), which is much more reasonable. The minimum possible space requirement is $O(d! 2^d m^{2d-1})$, at time cost $O(d! 2^d m^{2d} n^d)$ (that is, $s = 1$).

6 Multidimensional Exact String Matching

In [9], they allow to search, in two dimensions, a pattern in a text in $O(n^2/m)$ average time. They traverse only the text rows of the form $i \times m$ searching for all the pattern rows at the same time (using Aho-Corasick [1]), and verify all potential matches. Clearly, no match can be missed with the filter.

In [9], the authors briefly mention that their technique can be extended to more dimensions by selecting one dimension and recursively using an algorithm for $(d-1)$ dimensions on the m-th "rows" of such text. However no more details are given, nor any analysis.

We give now a more detailed version of the algorithm and analyze it. We select one dimension (say, coordinate i) and obtain n/m different $(d-1)$ dimensional objects of the form $T_{m,1..n,1..n,...}, T_{2m,1..n,1..n,...}, ..., T_{im,1..n,1..n,...}$, and so on. On the other hand, we obtain m patterns of $(d-1)$ dimensions, namely $P_{1,1..m,1..m,...}, P_{2,1..m,1..m,...}, ..., P_{p,1..m,1..m,...}$ and so on. All the m subpatterns are searched in each one of the $(d-1)$ dimensional subtexts. See Figure 2. Each time one of the $(d-1)$ dimensional subpatterns is found in a text position, the complete d-dimensional pattern is checked.

An important part of the analysis of [9] for two dimensions is that the total cost to verify potential matches is not too large. It is not immediate that this

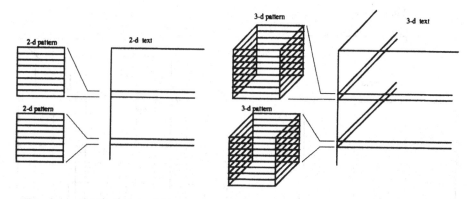

Fig. 2. Algorithm for exact searching. All the pattern "rows" are searched in n/m text "rows" at the same time.

is still valid for more dimensions, since a very large number of verifications are finally triggered.

The cost to verify a potential match in d dimensions is always $O(1)$ on average, since we have to check if m^d letters of the pattern are equal to the text at a given position. Since we stop the checking as soon as we find a mismatch, we verify more than c characters with probability $1/\sigma^c$. Hence, the average number of characters checked is $\sum 1/\sigma^c = O(1)$ (even for patterns of unbounded size).

We denote by $E_{d,r}$ the average search cost for r patterns in d dimensions. The existence of the Aho-Corasick [1] algorithm implies that $E_{1,r} = n$. Now, for d dimensions, we perform n/m searches for rm patterns on $d-1$ dimensions, and check all the candidates that occur. The probability of a pattern of size m^{d-1} occurring in a text position is $1/\sigma^{m^{d-1}}$, but we multiply that by rm because we search for rm different patterns. As the average cost to verify each potential match is $O(1)$, and the $(d-1)$ dimensional texts are of size n^{d-1}, we have that

$$E_{d,r} = \frac{n}{m}\left(E_{d-1,rm} + n^{d-1}\frac{rm}{\sigma^{m^{d-1}}}\right) = \frac{n}{m}E_{d-1,rm} + \frac{n^d r}{\sigma^{m^{d-1}}}$$

which gives

$$E_{d,r} = \frac{n^d}{m^{d-1}} + \sum_{w=1}^{d-1}\frac{n^d r}{\sigma^{m^w}} = O\left(n^d\left(\frac{1}{m^{d-1}} + \frac{r}{\sigma^m}\right)\right)$$

(where the first term corresponds to the actual searches which are all done in one dimension).

To search for one pattern we replace r by 1 in this final formula (although the algorithm internally uses multipattern search). This formula matches the result for two dimensions, since $1/\sigma^m = o(1/m)$. In general, if d is considered fixed, the above result for $r = 1$ can be bounded by $O(n^d/m^{d-1})$.

The space complexity of the algorithm corresponds to the Aho-Corasick machine, whose space requirements are proportional to the total size of all the patterns, i.e. $O(rm^d)$. We use now this algorithm as a building block.

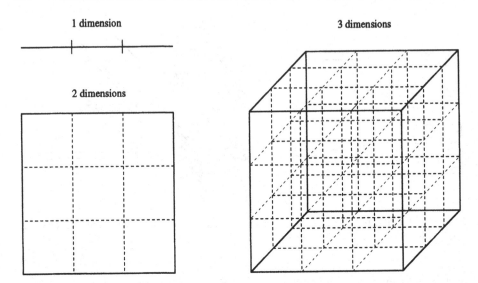

Fig. 3. Filtering algorithm for $j = 3$. The maximum possible k so that some block appears unchanged is 2, 2, and 8 as the dimension grows.

7   A Fast Filter for Multidimensional Approximate Search

We present now an effective filter to quickly discard large parts of the text which cannot contain a match, so that we use the dynamic programming algorithm to verify only the text areas which could contain an occurrence of the pattern.

The filter is based on a generalization of the one-dimensional filter explained in Section 2. In that case, we cut the pattern in $(k + 1)$ pieces, and since each error can destroy at most one piece, we have always one piece left untouched inside each occurrence.

In two and more dimensions, we cut the pattern in j pieces along each dimension, for some $1 \leq j \leq m$ (see Figure 3). Since each error occurs along one dimension only, at most kj pieces are destroyed. Therefore, since there are j^d pieces in total, it is enough that $j^d > kj$ to ensure that at least one of the pieces is left untouched (although we do not know which one). Hence, we search for all the j^d pieces at the same time in the text without allowing errors. Those pieces are of size $(m/j)^d$, and can be searched with the algorithm of the previous section in $O(m^d)$ space and an average time of

$$n^d \left(\frac{1}{(m/j)^{d-1}} + \frac{j^d}{\sigma^{m/j}} \right) = j^d n^d \left(\frac{1}{jm^{d-1}} + \frac{1}{\sigma^{m/j}} \right)$$

Each time one such piece is found, we have to verify a surrounding text area to check for a possible match. This area extends $(m + 2\min(m, k))$ positions along each dimension (since the match could start at most $\min(m, k)$ positions backward or finish up to $\min(m, k)$ positions forward). Hence, the cost of a verification is the same as that of searching the pattern in a text of size

$(m + 2 \min(m, k))^d$ allowing errors, which is $O(d!m^d(m + 2\min(m, k))^d)$. The total number of verifications is obtained by multiplying the number of pattern pieces j^d by the probability of a piece matching, i.e. $1/\sigma^{(m/j)^d}$. Hence, the total expected cost for verifications is $j^d d!m^d(m + 2\min(m, k))^d n^d/\sigma^{(m/j)^d}$.

Notice that, since we only verify pieces of the text of size $(m + 2\min(m, k))^d$, the space requirement of this algorithm is $O(d!m^d(m + 2\min(m, k))^{d-1})$ (this corresponds to the verification phase, since the search of the pieces needs much less, i.e. $O(m^d)$). This is a form of our previous technique to reduce space requirements (recall Section 5.2) equivalent to using $s = \min(m, k)$. However, in this case we only check a few portions of the text.

Both the search and the verification cost worsen as j grows, so we are interested in the minimum j that works. As said, we need that $j^d > kj$, hence

$$ j = \left\lfloor k^{\frac{1}{d-1}} \right\rfloor + 1 $$

is the best choice. The formula does not work for one dimension (because it is not true that kj pieces are destroyed), and for 2 dimensions it sets $j = k + 1$ as in the traditional one-dimensional case. Notice that we need that $j \le m$, and therefore the mechanism works for $k < k_3 = m^{d-1}$. Using this optimum (and minimum) j, the total cost of searching plus verifying is

$$ n^d k^{\frac{d}{d-1}} \left(\frac{1}{m^{d-1}k^{\frac{1}{d-1}}} + \frac{1}{\sigma^{m/k^{1/(d-1)}}} + \frac{d!m^d(m + 2\min(m, k))^d}{\sigma^{m^d/k^{d/(d-1)}}} \right) $$

which worsens as k grows. This search complexity has three terms, each of which dominates for a different range of k values. The first one dominates for

$$ k \le k_0 = \frac{m^{d-1}}{(d\log_\sigma m)^{d-1}}(1 + o(1)) $$

while the second dominates from $k > k_0$ until

$$ k \le k_1 = \frac{m^{d-1}}{(d(\log_\sigma d + 2\log_\sigma m))^{\frac{d-1}{d}}}(1 + o(1)) $$

In the maximum acceptable value $k = m^{d-1} - 1$, the search complexity becomes $O(d!3^d m^{3d} n^d)$, which is worse than using dynamic programming. We want to know which is the k value for which the filter is better than dynamic programming. We can compare against the version that uses the same amount of space (which corresponds to $s = \min(m, k)$), whose time complexity is $O(d!2^d m^{2d} n^d)$; or we can compare it against the fastest version of dynamic programming, which needs much more space and whose time cost is $O(d!m^d n^d)$. In either case we have that the k range for which the filter is better than dynamic programming is

$$ k \le k_2 = \frac{m^{d-1}}{(2d\log_\sigma m)^{\frac{d-1}{d}}}(1 + o(1)) $$

where the difference in the version of dynamic programming used affects lower order terms only.

Finally, the most stringent condition we can ask to the filter is to be sublinear, i.e. faster than $O(n^d)$. If we try to consider the third term of the search complexity as dominant, we arrive to a k value which is smaller than k_1, which means that the solution is in a stricter k range. By considering the second term of the search complexity, we arrive to the condition $k \leq k_0$. That is, the search time is sublinear precisely when the first term of the summation dominates.

To summarize, the search algorithm is sublinear (i.e. $O(kn^d/m^{d-1})$) for $k < (m/(d\log_\sigma m))^{d-1}$, and it improves over dynamic programming for $k \leq m^{d-1}/(2d\log_\sigma m)^{(d-1)/d}$. Figure 4 illustrates the result of the analysis.

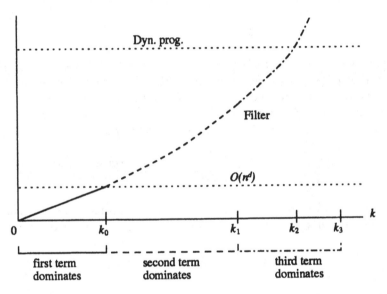

Fig. 4. The complexity of the proposed filter, depending on k.

7.1 A Stricter Filter

We have assumed up to now that we verify the presence of the pattern allowing errors as soon as any of the j^d pieces appears. However, we can do better. We know that $j^d - jk$ pieces must appear, at their correct positions, for a match to be possible. Therefore, whenever a piece appears, we can check the neighborhood for the exact occurrences of other pieces. On average, the verification of each piece will fail in $O(1)$ character comparisons, and we will check $O(jk)$ pieces until jk of them fail the test (this is because both are geometric processes). Therefore, we have a preverification test which occurs with probability $j^d/\sigma^{(m/j)^d}$, costs $O(jk)$ and is able to discard more text positions before actually verifying the candidate area. The probability that a text position passes the preverification test and undergoes the dynamic programming verification can be computed by considering that $j^d - jk$ cells need to match, which means that $m^d - km^d/j^{d-1}$ characters match. On the other hand, we can select as we want which jk cells match out of j^d, which multiplies the probability by $\binom{j^d}{jk}$. Finally, if the text area passes this filter, we verify it at the same cost as before (i.e. $d!m^d(m + 2\min(m, k))^d$). The new search cost is therefore

$$
n^d \left(\frac{j^{d-1}}{m^{d-1}} + \frac{j^d}{\sigma^{m/j}} + \frac{j^d jk}{\sigma^{(m/j)^d}} + \frac{\binom{j^d}{jk} d! m^d (m + 2\min(m, k))^d}{\sigma^{m^d - km^d/j^{d-1}}} \right)
$$

where the first term dominates for $j \leq m/(d\log_\sigma m)$, the second one up to $j \leq m/(\log_\sigma m + \log_\sigma k)^{1/d}$, and the third one for larger j. The fourth term decreases with j, and therefore it is not immediate that the minimum j is the optimum (in fact it is not). We have not been able to determine the optimum j, but we can still obtain the maximum k value up to where the filter is better than dynamic programming. The first two terms are never worse than dynamic programming, and the third improves over dynamic programming for

$$
j \leq \frac{m}{(\log_\sigma m + \log_\sigma k - d\log_\sigma d)^{1/d}} (1 + o(1))
$$

which gives a condition on k since $j^{d-1} > k$:

$$
k \leq k_2' = \frac{m^{d-1}}{(d(\log_\sigma m - \log_\sigma d))^{\frac{d-1}{d}}} (1 + o(1))
$$

Now, we introduce this maximum j value into the fourth term to determine whether it is also better than dynamic programming at that point. The result is that, using that j value, the fourth term is dominated by the third precisely for $k \leq k_2'$. Therefore we improve over dynamic programming for $k \leq k_2'$ (which is better than our previous k_2 limit). The proposed j is the best for high k values, but smaller values are better for lower k values. In particular, we may be interested in obtaining the sublinearity limit for this filter. The first three terms put an upper bound on j, the strictest one being

$$
j \leq \frac{m}{d(\log_\sigma m - \log_\sigma d)} (1 + o(1))
$$

and using this maximum j value the fourth term gives us the maximum k that allows sublinear search time:

$$
k \leq k_0' = \frac{m^{d-1}}{(d(\log_\sigma m - \log_\sigma d))^{d-1}} (1 + o(1))
$$

which is slightly better than our previous k_0 limit.

7.2   Adapting the Filter to Substitutions

The problem of searching a pattern allowing k substitutions is much simpler, and we can apply our machinery to that case as well. A brute force search algorithm checks any possible text position until it finds k mismatches. Being a geometric process, this occurs after $O(k)$ character comparisons, which makes the total search cost $O(kn^d)$ on average.

The same filter proposed in this section works for the case of k substitutions, the only difference being that in this case the cost to verify a candidate text position is $O(k)$, i.e. much cheaper. The search cost still has three terms, the first one being dominant for $k \leq k_0$. The second component is now dominant for

$$k \ \leq \ k_1' \ = \ \frac{m^{d-1}}{(d \log_\sigma m)^{\frac{d-1}{d}}} (1 + o(1))$$

and the last one dominates for $k > k_1'$. This filter is sublinear (i.e. does not inspect all the text characters) on average for $k < k_0$ as before. On the other hand, it turns out to be better than brute force (i.e. $O(kn^d)$) for $k \leq k_1'$, i.e. before the verification step dominates the search cost.

8   Conclusions

We have presented the first algorithms to search a multidimensional pattern in multidimensional text allowing editing errors along any dimension. This is a new model recently proposed in [5]. We have generalized to d dimensions their algorithm to compute edit distance, where we obtained $O(d!m^{2d})$ time and $O(d!m^{2d-1})$ space (where the compared elements are of size m^d).

We have obtained and proved the correctness of the first search algorithm for this model, where a pattern of size m^d is searched in a text of size n^d at $O(d!m^dn^d)$ time and $O(d!m^dn^{d-1})$ space. We have shown how to trade time for space, for instance with $O(d!3^dm^{2d-1})$ space we have $O(d!3^dm^dn^d)$ time.

Finally, we have proposed a filter which obtains roughly $O(kn^d/m^{d-1})$ (i.e. sublinear) average search time for $k < (m/(d(\log_\sigma m - \log_\sigma d)))^{d-1}$, where σ is the alphabet size. After that error level the filter changes its cost but remains better than dynamic programming for $k \leq m^{d-1}/(d(\log_\sigma m - \log_\sigma d))^{(d-1)/d}$. For instance, in two dimensions the filter is sublinear for $k < m/(2 \log_\sigma m)$ and better than dynamic programming for $k \leq m/\sqrt{2 \log_\sigma m}$.

These are the first search algorithms and fast filters for the first model which extends successfully the concept of approximate string matching to more than one dimension. Although we present the algorithms for square d-dimensional pattern and text, they also work for hyper-rectangular elements.

Our work is a (very preliminary) step towards presenting a combinatorial alternative to the current image processing technology. However, for this to be successful, we must allow not only errors but also rotations, scalings and deformations in the images. There are some works addressing those issues separately [2, 14], but they have not been merged. We are currently working on this integration.

References

[1] A. Aho and M. Corasick. Efficient string matching: an aid to bibliographic search. *CACM*, 18(6):333–340, June 1975.

[2] A. Amir and G. Calinescu. Alphabet independent and dictionary scaled matching. In *Proc. CPM'96*, number 1075 in LNCS, pages 320–334, 1996.

[3] A. Amir and M. Farach. Efficient 2-dimensional approximate matching of non-rectangular figures. In *Proc. SODA'91*, pages 212–223, 1991.

[4] A. Amir and G. Landau. Fast parallel and serial multidimensional approximate array matching. *Theoretical Computer Science*, 81:97–115, 1991.

[5] R. Baeza-Yates. Similarity in two-dimensional strings. In *Proc. COCOON'98*, number 1449 in LNCS, pages 319–328, Taipei, Taiwan, August 1998.

[6] R. Baeza-Yates and G. Navarro. Fast two-dimensional approximate pattern matching. In *Proc. LATIN'98*, number 1380 in LNCS, pages 341–351. Springer-Verlag, 1998.

[7] R. Baeza-Yates and G. Navarro. Faster approximate string matching. *Algorithmica*, 23(2):127–158, 1999. To appear. Preliminary version in *Proc. CPM'96*.

[8] R. Baeza-Yates and C. Perleberg. Fast and practical approximate pattern matching. In *Proc. CPM'92*, LNCS 644, pages 185–192, 1992.

[9] R. Baeza-Yates and M. Régnier. Fast two dimensional pattern matching. *Information Processing Letters*, 45:51–57, 1993.

[10] T. Baker. A technique for extending rapid exact string matching to arrays of more than one dimension. *SIAM Journal on Computing*, 7:533–541, 1978.

[11] R. Bird. Two dimensional pattern matching. *Inf. Proc. Letters*, 6:168–170, 1977.

[12] B. Commentz-Walter. A string matching algorithm fast on the average. In *Proc. ICALP'79*, number 6 in LNCS, pages 118–132. Springer-Verlag, 1979.

[13] M. Crochemore and W. Rytter. *Text Algorithms*. Oxford University Press, Oxford, UK, 1994.

[14] K. Fredriksson and E. Ukkonen. A rotation invariant filter for two-dimensional string matching. In *Proc. CPM'98*, number 1448 in LNCS, pages 118–125, 1998.

[15] J. Karkkäinen and E. Ukkonen. Two and higher dimensional pattern matching in optimal expected time. In *Proc. SODA'94*, pages 715–723. SIAM, 1994.

[16] K. Krithivasan. Efficient two-dimensional parallel and serial approximate pattern matching. Technical Report CAR-TR-259, University of Maryland, 1987.

[17] K. Krithivasan and R. Sitalakshmi. Efficient two-dimensional pattern matching in the presence of errors. *Information Sciences*, 43:169–184, 1987.

[18] S. Needleman and C. Wunsch. A general method applicable to the search for similarities in the amino acid sequences of two proteins. *J. of Molecular Biology*, 48:444–453, 1970.

[19] K. Park. Analysis of two dimensional approximate pattern matching algorithms. In *Proc. CPM'96*, LNCS 1075, pages 335–347, 1996.

[20] P. Sellers. The theory and computation of evolutionary distances: pattern recognition. *J. of Algorithms*, 1:359–373, 1980.

[21] S. Wu and U. Manber. Fast text searching allowing errors. *CACM*, 35(10):83–91, October 1992.

[22] R. Zhu and T. Takaoka. A technique for two-dimensional pattern matching. *Comm. ACM*, 32(9):1110–1120, 1989.

Finding Common RNA Secondary Structures
from RNA Sequences*

Zhuozhi Wang and Kaizhong Zhang

Dept. of Computer Science, University of Western Ontario,
London, Ont. N6A 5B7, Canada
kzhang@csd.uwo.ca

1 Introduction

RNAs (Ribonucleic Acids) play an important role when organisms reproduce
themselves. RNAs are single-stranded, however they tend to form higher order
structures such as secondary or tertiary structures by folding onto themselves.
It is the RNA structures that determine the functions of RNA sequences. Since
it is very difficult to crystallize and/or get nuclear magnetic resonance spectrum
data for large RNA molecules, reliable methods to determine RNA structures
from the primary sequences is important. An important step toward the determination of RNA structure is the prediction of RNA secondary structures. Based
on a reliable RNA secondary structure, possible tertiary interactions that occur
between secondary structural elements and between these elements and single-
stranded region can be characterized. Thermodynamic stability methods have
been developed [5] to fold a single RNA into secondary structures with minimum
or near minimum energy with some success. Phylogenetic comparative methods
are more successful which try to determine the common secondary structures
from a set of RNA sequences by checking a large number of possible base pair-
ings for their possible conservation. However this method is very tedious since it
is basically performed manually. In this abstract, we propose an algorithm using
dynamic programming trying to automate the phylogenetic comparative process. Given three RNA sequences, we first apply the folding algorithms for each
sequence to determine the frequently recurring stems which are considered to be
thermodynamically favourable. We then apply our algorithm to the three stem
lists generated from the folding algorithm to determine the common secondary
structures. We have applied our method to three viruses: cocksackievirus, human
rhinovirus (type 14), and poliovirus (type 3). Our method successfully produced
the main components of the common secondary structures of these viruses.

2 Notations

An RNA molecule is made up of a long chain of subunits (ribonucleotides) linking
together. Each ribonucleotide contains one of four possible bases, **A** (adenine),

* Research supported partially by the Natural Sciences and Engineering Research
Council of Canada under Grant No. OGP0046373.

C (cytosine), **G** (guanine), and **U** (uracil). Thus an RNA molecule is uniquely determined by its sequence of bases. RNAs fold by intramolecular base pairing. RNA secondary structures are stabilized by the hydrogen bonds that results from these base pairing. In addition, the stacking of base pairs in a helix stabilizes the molecule and decreases the free energy of the folded structure. However the appearance of loops destabilizes the RNA structure. In an RNA structure, base pairs will usually be formed as one of the three kinds: G-C, A-U and G-U. There are three hydrogen bonds between G-C, two between A-U, and one between G-U.

It is clear that the RNA secondary structure is much more complicated than the RNA primary structure. Given an RNA $R = a_1a_2...a_n$, we use $a_i \cdot a_j$ to denote a base pair between a_i and a_j where $1 \leq i < j \leq n$. Following the tradition, we will refer to the first base a_i as the 5' end of the pair and the second base a_j as the 3' end of the pair. Formally we say S is a secondary structure if if satisfies the following conditions.

1. If S contains $a_i \cdot a_j$, then a_i and a_j are either **A** and **U**, **U** and **A**, **C** and **G**, **G** and **C**, **G** and **U**, or **U** and **G**.
2. If S contains $a_i \cdot a_k$, then it cannot contain $a_i \cdot a_j$ (with j ≠ k) or $a_j \cdot a_k$ (with i ≠ j). (one-to-one)
3. If $h < i < j < k$, then S cannot contain both $a_h \cdot a_j$ and $a_i \cdot a_k$. (non-crossing)
4. If S contains $a_i \cdot a_j$, then $|j - i| \geq 4$

RNA secondary structure can be decomposed into five kinds of substructures, namely *stems, hairpin loops, bulge loops, interior loops* and *multiple loops*. If S contains $a_i \cdot a_j$, $a_{i+1} \cdot a_{j-1}$, ..., $a_{i+h-1} \cdot a_{j-h+1}$, we say each of these pairs (except the last) is stack on the following pair. We refer to these consecutive pairs as a *stacked pair* or as a *stem* and denote it as (i, j, h). RNA secondary structure is determined by its set of stems since this set is just another representation of its set of base pairs.

We now consider the relationship between two stems $I = (i_1, j_1, h_1)$ and $II = (i_2, j_2, h_2)$ from a stem list (not necessarily a secondary structure), see Figure 1 and 2. In Figure 1 and 2, we use the so-called *Domes Representation*.

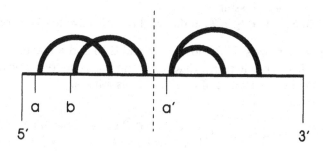

Fig. 1. Crossing stems

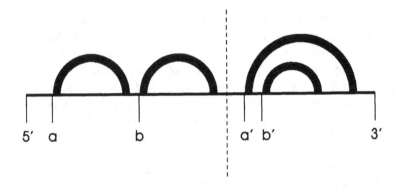

Fig. 2. Non-crossing stems

We say that stem I and stem II are *crossing* if and only if $i_1 \le i_2 \le j_1 \le j_2$. In Figure 1 are crossing stems. We say stem I is *before* stem II or stem II is *after* stem I if and only if $j_1 < i_2$, see Figure 2. We say stem I is *outside* stem II or stem II is *inside* stem I if and only if $i_1 + h \le i_2 < j_2 \le j_1 - h$, see Figure 2.

3   Algorithm

3.1   Definitions

Since our algorithm with deal with three stem lists generated from folding algorithm, we now consider three stem lists S, T, and U.

Given a triple (s, t, u), where $s = (i_1, j_1, h_1)$ is from S, $t = (i_2, j_2, h_2)$ is from T, and $u = (i_3, j_3, h_3)$ is from U, we define $score(s, t, u) = \min\{h_1, h_2, h_3\}$.

Given two triples (s_1, t_1, u_1) and (s_2, t_2, u_2), where s_1 and s_2 are from stem list S, t_1 and t_2 are from stem list T, and u_1 and u_2 are from stem list U, we say (s_1, t_1, u_1) and (s_2, t_2, u_2) are *compatible* if and only if

1. s_1 and s_2 are not crossing;
2. t_1 and t_2 are not crossing;
3. u_1 and u_2 are not crossing;
4. s_1 and s_2 in S, t_1 and t_2 in T, and u_1 and u_2 in U have the same relationship.

Given three stem lists S, T and U, our goal is to find a maximal set of non-crossing stems from each list such that they form the same topological structure. Formally we define the *weight* of S, T and U as follows:

$weight(S, T, U) =$

$$\max_{k_1, k_2, \ldots, k_n} \left\{ \sum_{i=1}^{n} score(s_{k_i}, t_{k_i}, u_{k_i}) \,\middle|\, \begin{array}{l} s_{k_i} \in S, t_{k_i} \in T, u_{k_i} \in U; \text{ for any } i \text{ and } j, \\ (s_{k_i}, t_{k_i}, u_{k_i}) \text{ and } (s_{k_j}, t_{k_j}, u_{k_j}) \text{ are compatible} \end{array} \right\}$$

Suppose that S, T, and U are sorted by the $3'$ end of the stems. Let s_i be the ith stem in S, t_j be the jth stem in T, and u_k be the kth stem in U. Let

S_i' be the stem list containing the stems below s_i, T_j' be the stem list containing the stems below t_j, and U_k' be the stem list containing the stems below u_k. We define $forest\_weight(i, j, k)$ and $tree\_weight(i, j, k)$ as follows.

$$forest\_weight(i, j, k) = weight(\{s_1, ..s_i\}, \{t_1, ..t_j\}, \{u_1, ..u_k\})$$

$$tree\_weight(i, j, k) = score(s_i, t_j, u_k) + weight(S_i', T_j', U_k')$$

3.2 Properties

Lemma 1. *Let i' be the largest index such $3'$ end of $s_{i'}$ is less than $5'$ end of s_i, j' be the largest index such $3'$ end of $t_{j'}$ is less than $5'$ end of t_j, and k' be the largest index such $3'$ end of $u_{k'}$ is less than $5'$ end of u_k, then*

$$forest\_weight(i, j, k) = \min \begin{cases} forest\_weight(i-1, j, k) \\ forest\_weight(i, j-1, k) \\ forest\_weight(i, j, k-1) \\ forest\_weight(i', j', k') + tree\_weight(i, j, k) \end{cases}$$

Proof. Consider the best way to match $s_1, ..., s_i$, $t_1, ..., t_j$, and $u_1, ..., u_k$, there are four possibilities. First, s_i does not match to any stem in T or U, therefore $forest\_weight(i, j, k) = forest\_weight(i-1, j, k)$. Second, t_j does not match to any stem in S or U, therefore $forest\_weight(i, j, k) = forest\_weight(i, j-1, k)$. Third, u_k does not match to any stem in S or T, therefore $forest\_weight(i, j, k) = forest\_weight(i, j, k-1)$. Fourth, s_i, t_j, and u_k are matched up with each other, we have $forest\_weight(i, j, k) = forest\_weight(i', j', k') + tree\_weight(i, j, k)$. □

Based on this lemma, we can implement the algorithm by using six nested loops. The resulting algorithm is reasonable for two stem lists. However for three stem lists, it is extremely slow. Note that for the practical application we always consider RNAs with some sequence similarity. This means that we do not need to consider all the triples. Instead, we only need to consider triples which are close. This will speed up the algorithm.

We now refine our definition of $weight(S, T, U)$. We introduce one parameter $end\_control$ to reduce the number of stems that we need to consider. Given two stems $s = (i_1, j_1, h_1)$ and $t = (i_2, j_2, h_2)$, we say s and t are *semi-matchable* if $|j_1 - j_2| \leq end\_control$. We say they are *matchable* if they are semi-matchable and in addition $|i_1 - i_2| \leq end\_control$. We use $s <> t$ to denote this relation. $weight(S, T, U) =$

$$\max_{k_1, k_2, ..., k_n} \left\{ \sum_{i=1}^{n} score(s_{k_i}, t_{k_i}, u_{k_i}) \middle| \begin{array}{l} s_{k_i} \in S, t_{k_i} \in T, u_{k_i} \in U; \text{ for any } i \text{ and } j, \\ (s_{k_i}, t_{k_i}, u_{k_i}) \text{ and } (s_{k_j}, t_{k_j}, u_{k_j}) \text{ are compatible} \\ \text{for any } i, \ s_{k_i} <> t_{k_i}, t_{k_i} <> u_{k_i}, u_{k_i} <> s_{k_i}. \end{array} \right\}$$

With this refined definition, the definitions of $forest\_weight(i, j, k)$ and $tree\_weight(i, j, k)$ remain the same as before.

Lemma 2. *Let i' be the largest index such 3' end of $s_{i'}$ is less than 5' end of s_i, j' be the largest index such 3' end of $t_{j'}$ is less than 5' end of t_j, and k' be the largest index such 3' end of $u_{k'}$ is less than 5' end of u_k, then*

$$forest\_weight(i,j,k) = \min \begin{cases} forest\_weight(i-1,j,k) \\ forest\_weight(i,j-1,k) \\ forest\_weight(i,j,k-1) \\ forest\_weight(i',j',k') + tree\_weight(i,j,k) \\ \quad if \ s_i <> t_j, \ t_j <> u_k, \ and \ u_k <> s_i \end{cases}$$

Proof. The proof is exactly the same as that of lemma 1 except that in order for s_i, t_j, and u_k to match they have to satisfy the condition that $s_i <> t_j$, $t_j <> u_k$, and $u_k <> s_i$. □

From this lemma we know that we only need to compute $tree\_weight(i,j,k)$ in case $s_i <> t_j$, $t_j <> u_k$, and $u_k <> s_i$.

For stem s_i in stem list S, consider its semi-matchable stems in stem list T. Since T is sorted, there is an interval $[s,e]$ such that t_j is semi-matchable with s_i if and only if $s \le j \le e$.

For any stem s_i in S, let $s_T^S(i)$ and $e_T^S(i)$ be the starting and ending indices of s_i's semi-matchable stems in stem list T. Similarly we can define $s_U^S(i)$, $e_U^S(i)$, $e_S^T(j)$, and $e_S^U(k)$.

Lemma 3.

$forest\_weight(i,j,k) = forest\_weight(i,j-1,k) \ \ if \ j > e_T^S(i)$
$forest\_weight(i,j,k) = forest\_weight(i,j,k-1) \ \ if \ k > e_U^S(i)$
$forest\_weight(i,j,k) = forest\_weight(i-1,j,k) \ \ if \ j < s_T^S(i) \ or \ k < s_U^S(i)$

Proof. When $k > e_U^S(i)$, u_k is useless since it cannot match any s_l where $1 \le l \le i$. Therefore $forest\_weight(i,j,k) = forest\_weight(i,j,k-1)$. Similarly, we can prove that if $j > e_T^S(i)$, then $forest\_weight(i,j,k) = forest\_weight(i,j-1,k)$. If $j < s_T^S(i)$ or $k < s_U^S(i)$, then s_i is useless since either it cannot match to t_l where $1 \le l \le j$ or it cannot match to u_l where $1 \le l \le k$. Therefore $forest\_weight(i,j,k) = forest\_weight(i-1,j,k)$. □

From this lemma, we know that for each s_i in stem list S, we only need to compute $forest\_weight(i,j,k)$ such that $s_T^S(i) \le j \le e_T^S(i)$ and $s_U^S(i) \le k \le e_U^S(i)$.

Lemma 4.
If $j > e_T^S(i)$ and $s_U^S(i) \le k \le e_U^S(i)$, then
$$forest\_weight(i,j,k) = forest\_weight(i,e_T^S(i),k).$$

If $s_T^S(i) \le j \le e_T^S(i)$ and $k > e_U^S(i)$, then
$$forest\_weight(i,j,k) = forest\_weight(i,j,e_U^S(i)).$$

If $j > e_T^S(i)$ and $k > e_U^S(i)$, then
$$forest\_weight(i, j, k) = forest\_weight(i, e_T^S(i), e_U^S(i)).$$

Proof. We can prove this lemma by applying lemma 3 repeatedly.   □

Lemma 5.
If $j < s_T^S(i)$ and $k \geq s_U^S(i)$, then
$$forest\_weight(i, j, k) = forest\_weight(e_S^T(j), j, k).$$
If $j \geq s_T^S(i)$ and $k < s_U^S(i)$, then
$$forest\_weight(i, j, k) = forest\_weight(e_S^U(k), j, k).$$
If $j < s_T^S(i)$ and $k < s_U^S(i)$, then
$$forest\_weight(i, j, k) = forest\_weight(\min\{e_S^T(j), e_S^U(k)\}, j, k).$$

Proof. We can prove this lemma by applying lemma 3 repeatedly.   □

Lemma 6. *If $s_T^S(i) \leq j$ and $s_U^S(i) \leq k$, then*
$$forest\_weight(i, j, k) = forest\_weight(i, \min\{j, e_T^S(i)\}, \min\{k, e_U^S(i)\}))$$

Proof. Immediately from lemma 4.   □

Lemma 7. *If $j < s_T^S(i)$ or $k < s_U^S(i)$, let $i_1 = \min\{e_S^T(j), e_S^U(k)\}$, then*
$$forest\_weight(i, j, k) = forest\_weight(i_1, \min\{j, e_T^S(i_1)\}, \min\{k, e_U^S(i_1)\})$$

Proof. If $k \geq s_U^S(i)$, then $e_S^U(k) \geq i$ and If $j \geq s_T^S(i)$, then $e_S^T(j) \geq i$. Therefore by lemma 5, $forest\_weight(i, j, k) = forest\_weight(\min\{e_S^T(j), e_S^U(k)\}, j, k)$. Let $i_1 = \min\{e_S^T(j), e_S^U(k)\}$, by lemma 6 $forest\_weight(i, j, k) = forest\_weight(i_1, \min\{j, e_T^S(i_1)\}, \min\{k, e_U^S(i_1)\})$.   □

3.3 Algorithm

The algorithm works as follows:

- For each triple of stems (s_i, t_j, u_k) that are matchable, calculate $tree\_weight(i, j, k)$;
- Compute $weight(S, T, U)$.
- Trace back to collect the matchable stems in each stem list that contribute to the $weight(S, T, U)$.

The algorithm to calculate $tree\_weight(i, j, k)$ and $weight(S, T, U)$ is given in figure 3.

Input: Three stem lists S, T, and U.
Output: $weight(S, T, U)$.

compute $s_T^S(i)$ and $e_T^S(i)$; $1 \leq i \leq |S|$
compute $s_U^S(i)$ and $e_U^S(i)$; $1 \leq i \leq |S|$
compute $e_S^T(j)$; $1 \leq j \leq |T|$
compute $e_S^U(k)$; $1 \leq k \leq |U|$

compute $before\_stem(S, i)$, $1 \leq i \leq |S|$;
compute $before\_stem(T, j)$, $1 \leq j \leq |T|$;
compute $before\_stem(U, k)$, $1 \leq k \leq |U|$;

```
for i := 1 to |S| do
    ii = before_stem(S, i);
    for j := s_T^S(i) to e_T^S(i) do
        jj = before_stem(T, j);
        for k := s_U^S(i) to e_U^S(i) do
            kk = before_stem(U, k);
            (i', j', k') = adjust(i - 1, j, k);
            m[0] = forest_weight(i', j', k');
            (i', j', k') = adjust(i, j - 1, k);
            m[1] = forest_weight(i', j', k');
            (i', j', k') = adjust(i, j, k - 1);
            m[2] = forest_weight(i', j', k');
            m[3] = 0;
            if s_i <> t_j and t_j <> u_k and u_k <> s_i then
                (i', j', k') = adjust ( ii, jj, kk );
                m[3] = forest_weight(i', j', k');
                    + tree_weight[i][j][k];
            forest_weight[i][j][k] = max(m[0], m[1], m[2], m[3])

return forest_weight(|S|, |T|, |U|);
```

Fig. 3. Procedure: Computing $weight(S, T, U)$

We first compute $s_T^S(i)$, $e_T^S(i)$, $s_U^S(i)$, $e_U^S(i)$, $e_S^T(j)$, and $e_S^U(k)$. This step can be done in linear time.

We then compute $before\_stem(S, i)$ which denotes the largest indexed stem in S whose 3' end is less than s_i's 5' end. We also compute $before\_stem(T, j)$ and $before\_stem(U, k)$. After a sorting, this step can be done in linear time.

The main part of the algorithm is three nested loops. For any given i, where $1 \leq i \leq |S|$, we only calculate $forest\_weight(i, j, k)$ for $s_T^S(i) \leq j \leq e_T^S(i)$ and $s_U^S(i) \leq k \leq e_U^S(i)$. When calculating $forest\_weight(i, j, k)$, we may refer to locations which are outside our calculating range. In this situation, we need to adjust the indices by using lemma 6 or lemma 7.

Let n_T be the maximum number of stems in T that are semi-matchable with a single stem in S. Let n_U be the maximum number of stems in U that are semi-matchable with a single stem in S. The time complexity of the algorithm in figure 3 bounded by $O(|S| \times n_T \times n_U)$. The toltal time complexity of our algorithm is is $O(|S|^2 \times n_T^2 \times n_U^2)$. With the *end_control* parameter, we can control n_T and n_U.

4 Experiment Results

In our experiments, we used three viruses — *cocksackievirus*, *human rhinovirus* (type 14) and *poliovirus* (type 3) [2]. We call them CVB, HRV, and POL. These viruses are believed to belong to the same family, and therefore they should have common secondary structures.

We first apply the folding algorithms [5] for each sequence to determine the frequently recurring stems which are considered to be thermodynamically favourable. We then apply our algorithm to the stem lists generated to determine the common secondary structures.

The results shown in figure 4, 5, and 6 are obtained by setting parameters *end3_ control* to 20. RNA secondary structure display is provided by Structure-Lab from U.S. National Cancer Institute.

Compared with the published results in [2] using phylogenetic comparative methods, our algorithm produced main components of the common secondary structures.

From base *1* to base *89*, our results are almost the same as those in [2].

From base *241* to base *441*, the shape is also almost the same except for missing some short stems, which are missing from the input stem lists as well. These short stems can be generated by running the folding algorithm again with the stems we have found fixed. This will make the large internal loops disappear.

From base *445* to base *560*, we also got the same shape.

The substructure from base *90* to base *240* differs considerably. Again, this is caused by the lack of appropriate input stems.

Note that using folding algorithm alone cannot produce the correct secondary structure model. In our experiment, for each sequence, the folding algorithm generated thousands of secondary structures none of which is close to the correct model.

In conclusion, together with the folding algorithms, our algorithm can produce main components of common RNA secondary structures from RNA sequences. Base on these components, a more accurate model can be developed.

5 Conclusion and Future Work

We present an algorithm which produced reasonable models for common secondary structures from three RNA sequences.

We are currently improving our algorithms. The score between three stems is too simple to be realistic. We plan to change it into more meaningful measure.

We would also like to do a preprocessing to check the situation where a stem has a matchable stem in the second list but not in the third list. In this case we may want check output from folding algorithm to see if some corresponding stem exist with lower frequence. These stems can be added to the third stem list.

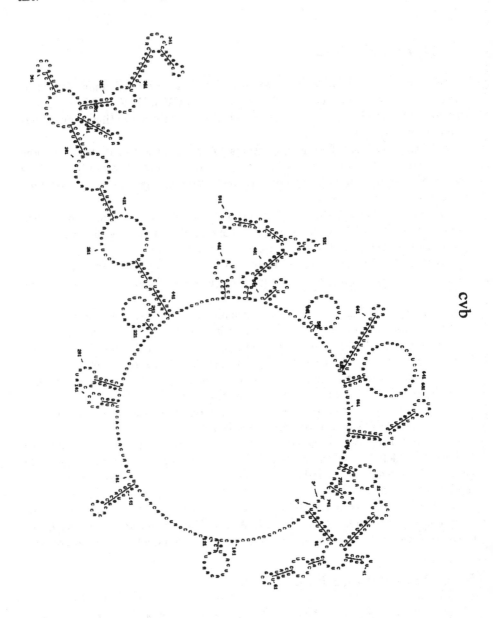

Fig. 4. Figure of CVB

6 Acknowledgements

We would like to thank Dr. Shapiro (U.S. National Cancer Institute) for his help on providing viruses RNA data and displaying RNA secondary structures using StructrueLab.

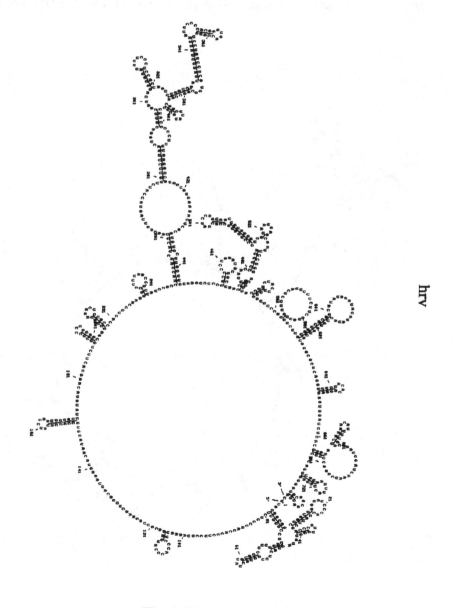

Fig. 5. Figure of HRV

References

1. S. Y. Le, K. Zhang, and J.V. Maizel, Jr., 'A method for predicting common structures of homologous RNAs', *Computers and Biomedical Research*, 128, pp.53-66, 1995.
2. S. Y. Le and M. Zuker, 'Common structure of the 5' non-coding RNA in enteroviruses and Rhinovirusws – Thermodynamical stability and statistical significance', *J. Mol. Biol.*, 216, pp.729-741, 1990.
3. Waterman, M.S. Eds., 'Mathematical methods for DNA sequence', *CRC Press*, Boca Raton, FL 1989.

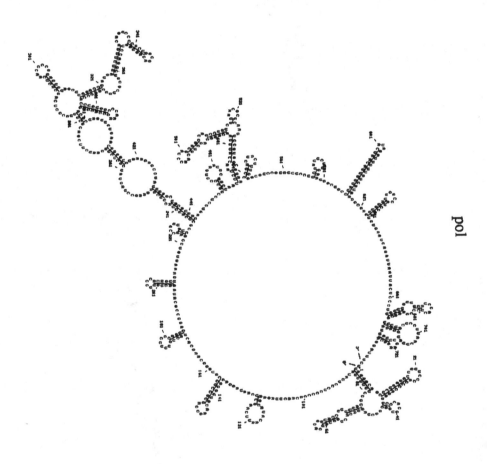

pol

Fig. 6. Figure of POL

4. A.L. Wiliams and I. Tinoco, Jr., 'A dynamic programming algorithm for finding alternate RNA secondary structures', *Nucleic Acids Research*, 14, pp.199-315, 1986.
5. M. Zuker, 'On finding all suboptimal foldings from of an RNA molecule', *Science* 244, pp.48-52, 1989.
6. M. Zuker and D. Sankoff, 'RNA secondary structure and their prediction', *Bull. Math. Biol.* 46, pp.591-621, 1984.
7. M. Zuker and P. Stiegler, 'Optimal computer folding of large RNA sequences using thermodynamics and auxiliary information', *Nucleic Acid Res.* 9, pp.133-148, 1981.

Finding Common Subsequences with Arcs and Pseudoknots

Patricia A. Evans

Computer Science
University of New Brunswick
Fredericton, NB, Canada

Abstract. This paper examines the complexity of comparing sequences that have arcs linking symbol pairs. Such arc-annotated sequences can represent molecular sequences with bonds between bases, such as RNA sequences. Crossing arcs that can represent sequence pseudoknots are included. The problem of finding the longest common subsequence, on which pairwise sequence comparison algorithms are frequently based, is modified to require common subsequences to preserve the arcs induced by the selected symbol positions. This problem is then analyzed using classical and parameterized complexity. It is shown to be NP-complete, and also W[1]-complete when parameterized by desired length of common subsequence. If it is parameterized instead by arc cutwidth k, however, it becomes fixed-parameter tractable, and usable for sequences with arc structures of limited cutwidth. An algorithm is given that runs in time $\in O(9^k nm)$.

1 Introduction

Genetic and protein sequence similarity can indicate evolutionary similarity and some functional similarity between the sequences. One common way to measure the similarity of two sequences is pairwise sequence alignment, a method of comparison based on the longest common subsequence algorithm (see [5] and [7]). Arcs that link bases within a sequence can be used to indicate secondary structure of molecular sequences by representing molecular bonds and links between pieces of the molecule's structure. These arcs can be incorporated into sequence comparison to produce an overall measure of similarity between sequences annotated with arcs.

Previous work on aligning arc-annotated sequences has focused on RNA sequences, where the arcs represent bonds. Matched arcs are used to enhance or guide the sequence alignment and improve its similarity score. Structures with pseudoknots, which require the representative arcs to cross, are usually excluded. Early work on RNA alignment involved predicting common secondary structure while aligning the sequences [6]. Corpet and Minchot [2] produced an algorithm that aligns a new sequence with a bank of aligned sequences, matching the new sequence to the bank while also preserving as much as possible of the common secondary structure of the sequence bank. This algorithm runs in time $\in O(n^5)$

M. Crochemore, M. Paterson (Eds.): CPM'99, LNCS 1645, pp. 270–280, 1999.

for sequence length n. The algorithm of Bafna et al. [1] aligns two sequences with nested arcs only, and uses weights for both sequence and arc matching. However, it does not detect mismatches for the arcs, and ignores the arc information if it does not improve the score. It also has worst-case time complexity $\Theta(n^2m^2)$, where n and m are the sequence lengths. For long sequences, this time complexity can be too high. This time complexity is also independent of the complexity and depth of the arc structure, and thus could not exploit this structure to reduce the time required. Lenhof et al. do include pseudoknots in their graph-based work on RNA sequence alignment [4]. Their algorithm, however, aligns sequences where only one sequence has an associated structure. Like the work before it, the links between base pairs are used to enhance the alignment.

This paper examines the problem of finding the longest common subsequence of a pair of arc-annotated sequences. The common subsequence must not only match both sequences, but also preserve all arcs that link subsequence symbols. This analysis specifically looks into the complexity of solving this problem when the arcs can cross; pseudoknots in sequences can thus be represented. Different parameters of the problem are examined, determining the conditions for usable or effective computation of the arc-preserving longest common subsequence. This problem is proved to be NP-complete, and it is also W[1]-complete when parameterized by the desired length of common subsequence. If a bound on the arc structure's cutwidth, the number of arcs that cross any position, is used as the parameter instead, the problem is fixed-parameter tractable. An algorithm is given for this variant that finds the length of the longest common subsequence in $O(9^k nm)$ for cutwidth k and input sequence lengths n and m. For $k < \log_9(\max(n, m))$, this algorithm's time complexity is less than that of earlier work.

2 Problem Definition

The Arc-Preserving Longest Common Subsequence problem for sequences with crossing arcs is defined as:

Input: target length l and the pair of annotated sequences (S_1, P_1) and (S_2, P_2).

These annotated pairs consist of the sequences S_1 and S_2 over some fixed alphabet Σ, with arc annotations $P_1 \subset \{1, \ldots, |S_1|\}^2$ and $P_2 \subset \{1, \ldots, |S_2|\}^2$. Each set of arcs P_A is further restricted so that $\forall (i_1, i_2)$ and $(i_3, i_4) \in P_A$, $i_1 = i_3$ if and only if $i_2 = i_4$, and $i_2 \neq i_3$. These conditions require that each sequence position can be an endpoint for at most one arc; it is linked to at most one other position. The length of S_1 is n and the length of S_2 is m.

Output: returns *true* if and only if there was some mapping $MS \subset \{1, \ldots, |S_1|\} \times \{1, \ldots, |S_2|\}$ between the positions of S_1 and S_2 such that $|MS| = l$ and

1. the mapping is one-to-one and preserves the order of the subsequence :

$$\forall (i_1, j_1) \in MS \text{ and } (i_2, j_2) \in MS$$
$$i_1 = i_2 \text{ if and only if } j_1 = j_2$$
$$i_1 < i_2 \text{ if and only if } j_1 < j_2$$

2. the arcs induced by the mapping are preserved :
   $\forall (i_1, j_1) \in MS$ and $(i_2, j_2) \in MS$, $(i_1, i_2) \in P_1$ if and only if $(j_1, j_2) \in P_2$
3. the mapping produces a common subsequence :
   $\forall (i, j) \in MS$, $S_1[i] = S_2[j]$.

If any of these conditions is not met, *false* is returned.

This problem is slightly different from those addressed by others. The preservation of the arcs induced by the subsequence is enforced. The algorithm of Bafna et al. [1] instead allows the arcs to be ignored if they do not contribute positively to the alignment score. The arcs in that case are used to enhance the alignment and its value, but do not control it; mismatches between arcs are not detected, and are thus disregarded. Extensions of the algorithm in this paper can include weights for symbols and arcs, and the arc weights can be both positive and negative. Negative arc weights are only possible because the information represented by the arcs is not ignored by the algorithm. This additional feature enables the alignment parameters to be adjusted so that the weight of matched pairs does not overwhelm the entire alignment. The only restriction on this use of reducing weights is that the reduction needs to be no greater than the smaller of the two endpoint weights; otherwise, the endpoint will not be matched at all. Checking for arc mismatches, as this algorithm does, also shows the difference between matching unbonded bases, and matching a bonded with an unbonded base. Previous work does not allow for this distinction. Furthermore, previous work does not allow crossing arcs (or limits them to one sequence only), and therefore exclude sequences with pseudoknots from these comparisons.

3   Hardness Results

This problem can be analyzed using the techniques from both classical and parameterized complexity. For classical complexity, it is proved to be NP-complete. For parameterized complexity, which can show the effect of specific parameters on the problem's complexity, it is W[1]-complete when the desired subsequence length l is used as the parameter.

The parameterized complexity hierarchy [3] is composed of the classes

$$FPT \subseteq W[1] \subseteq W[2] \subseteq \cdots \subseteq W[t] \subseteq \cdots \subseteq W[SAT] \subseteq \cdots \subseteq W[P]$$

and uses parametric reductions from problem A to problem B that require the parameter of B to be a function only in the parameter of A, independent of problem size. Clique is one problem that is complete for the class W[1] (shown in [3]).

Lemma.   Clique is polynomially reducible and strongly uniformly parametrically reducible to Arc-Preserving LCS.
Reduction.   From k-Clique of a graph $G = (V, E)$, where $V = \{1, 2, \ldots, n\}$.

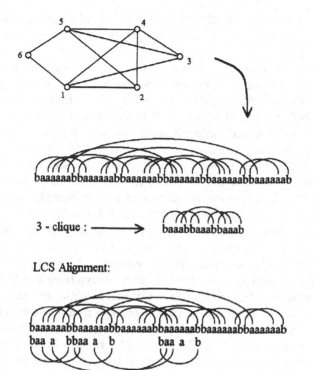

Fig. 1. Example of transformation from Clique to Arc-Preserving LCS.

Construct $S_1[1..(n^2 + 2n)]$ and P_1 as follows:

$$S_1 = (ba^n b)^n$$

$$P_1 = \{((u-1)(n+2) + 1, u(n+2)) | u \in V\} \cup$$
$$\{((u-1)(n+2) + v + 1, (v-1)(n+2) + u + 1) | (u,v) \in E\}$$

Construct $S_2[1..(k^2 + k)]$ and P_2 as follows:

$$S_2 = (ba^k b)^k$$

$$P_2 = \{((u-1)(k+2) + 1, u(k+2)) | u \in V\} \cup$$
$$\{((u-1)(k+2) + v + 1, (v-1)(k+2) + u + 1) | u, v \in \{1, \ldots, k\}, u \neq v\}$$

The parameter is $l = k^2 + 2k$, while the maximum sequence length is $n' = n^2 + 2n$. Since the length of the sequences is bounded by a polynomial in n, this is a polynomial reduction. The sequence parameter l is a polynomial only in k and is independent of n, so this reduction is also a parameterized reduction.

The target subsequence size is the same as the length of the second sequence, so we are really asking if the arc sequence S_2 is a subsequence of the arc sequence S_1. Figure 1 illustrates an example of this reduction.

Proof Sketch. We need to show that the graph $G = (V, E)$ has a clique of size k if and only if (S_1, P_1) and (S_2, P_2) have an arc-preserving common subsequence of length $l = k^2 + 2k$.

$\Rightarrow$:   Let $V' \subseteq V$ be a clique in G of size k. Each vertex $u \in V'$ corresponds to a segment of S_1, $ba^n b$, and a segment of S_2, $ba^k b$. If this pair of segments are matched, they will contribute $k + 2$ to the length of a common subsequence. The pair of b symbols are joined by an arc in both segments, so those arcs are preserved. The a symbols in each segment are linked, in order, to the a symbols in the other segments that correspond to vertices from the clique V'. Thus k of the a symbols of S_1, those linked to the other selected segments (plus one for the segment itself), can also be matched to the corresponding segment of S_2 while preserving the arcs. The common subsequence is the concatenation of the segment pairs' common subsequences, and has length $k(k + 2)$.

$\Leftarrow$:   Let the annotated sequences (S_1, P_1) and (S_2, P_2) have an arc-preserving subsequence of length $l = k^2 + 2k$. The linked pairs of b symbols in both sequences enforce the matching of symbols from only k segments of S_1, with exactly $k + 2$ symbols matched from each segment. From these $k + 2$ symbols, 2 of them are b symbols. Of the remaining k symbols, one is not linked, while the others are all linked to symbols from different segments that are also from the selected set of k segments. Since these arcs were defined to link segments that corresponded to vertices in G that were joined by an edge, a pair of vertices is linked if and only if their corresponding segments are linked. The subsequence must preserve the arcs from (S_2, P_2) that link all of its k segments. Therefore all selected segments from S_1 are also linked pairwise by arcs, and the set of k corresponding vertices of the graph G are a clique in G.   □

Theorem 1.   Arc-Preserving LCS is NP-complete.

Proof: All the requirements for a solution to the Arc-Preserving LCS problem can be checked in polynomial time, so it is in NP. It is polynomially reducible from Clique by the above Lemma, so it is also NP-hard. Thus Arc-Preserving LCS is NP-complete.   □

Theorem 2.   Arc-Preserving LCS is W[1]-complete when parameterized by desired subsequence length l.

Proof: Arc-Preserving LCS, parameterized by l, is in W[1] since any instance of the problem can be converted into a decision circuit with weft 1 whose accepted input assignments of weight l correspond to the arc-preserving common subsequences of length l. This circuit can be constructed with one input bit for each possible sequence position match (i, j). Each condition for acceptable input can be checked with a 2-input gate, and the circuit output can be set to 0 if any condition is violated, or 1 if all conditions hold.

Clique, which is hard for W[1], is parametrically reducible to Arc-Preserving LCS (parameterized by l) by the reduction in the Lemma given above. Thus Arc-Preserving LCS is W[1]-complete for the parameter l.   □

4   Sequences with Bounded Cutwidth

4.1   Necessary Data Structures

While the problem is NP-complete, and is also W[1]-complete when parameterized by desired subsequence length l, the problem becomes fixed-parameter tractable if it is instead parameterized by the arc cutwidth. This cutwidth is defined as the maximum number of arcs that cross or end at any arbitrary position of the sequence. If both sequences have their cutwidth bounded by some k, the problem can be solved in time $O(f(k)nm)$ where $f(k)$ is a function in k independent of both n and m. Sufficient bounds on k will make the problem tractable.

On initial examination, this problem should be able to be solved by splitting the tables whenever an initial endpoint is encountered. Restricting the cutwidth of the sequences, then, should make this problem fixed-parameter tractable. These tables can be used to store the initial endpoint matches that are made on the various computation paths, and these paths can then be searched when a final endpoint is encountered. This searching can determine if any maximum computation path has matched the initial endpoints that correspond to the final endpoints encountered. However, this network of paths can, potentially, cover all $4^k n^2$ positions in the tables, and may need to be searched completely.

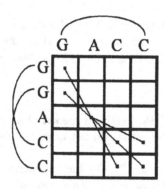

Fig. 2. Computation Path Network Example.

The network of computation paths cannot be searched simply using a breadth-first search technique, nor can we only maintain a list of all arc assignments on the maximum computation paths that lead to each table entry. Since matching the final endpoint depends on whether (and to what) the initial endpoint was matched, the network of paths also includes path combinations that are not allowed. Symbol matches inside an arc would merge lists that need to be kept separate. In Figure 2, for example, the entry that matches the two a symbols brings together two computation paths that match initial endpoint g symbols;

however, the subsequent matching of the final endpoint symbols can only be done one way for each initial match, not two as the network of paths would indicate. Looking for an initial match in this network, then, would require searching each possible path through the $4^k n^2$ different table positions in order to correctly compute the new table value.

Since searching through the network of paths is very costly, the different valid computation paths can instead be kept in a tree. Each of the table positions should have its own tree of all valid computation paths. In order to be able to minimize the length of the paths in the tree, each position will need to have its own copy, instead of referencing trees at previous positions. These paths are kept short by removing initial endpoint matches of arcs that are no longer active (whose final endpoints have been encountered). This editing reduces the length of each path to at most k, and the storing of the endpoint matches in a tree means that all paths to be searched are valid ones. Thus the time to search for an initial endpoint match is reduced while both the space used and the complexity of the data structure are increased.

4.2   Bounded Cutwidth Algorithm

The tree structure outlined above is used in the following algorithm that finds the arc-preserving longest common subsequence for arcs with bounded cutwidth.

Theorem 3.   The Arc-Preserving LCS problem, parameterized by cutwidth k, is Fixed-Parameter Tractable, and can be solved in time $O(9^k nm)$.
Proof.   A solution for the Arc-Preserving LCS problem is found by the following algorithm.
Algorithm:

Step 1. For each of P_1 and P_2, partition the set of arcs into k sets, where each set contains a chain of arcs that do not cross or nest. Number these chains from 0 through $k - 1$.

Step 2. For each of (S_1, P_1) and (S_2, P_2), look at each subset of the set of chains. For each subset of chains of P_1, create a copy of the sequence S_1 with the initial endpoints of all arcs in those chains removed and replaced by some $x \notin \Sigma$. The set of sequences thus created is $\mathcal{S}_1$, and is generally indexed by h_1, where $h_1 = \sum_{i \in subset} 2^i$. For (S_2, P_2), create the set of sequences $\mathcal{S}_2$ in the same way, indexing it using h_2.

Step 3. For each combination of h_1 and h_2, create a two-dimensional table, $n \times m$, that uses strings $\mathcal{S}_1[h_1]$ and $\mathcal{S}_2[h_2]$. These tables will be used to calculate the length of the longest common subsequence. Each table position includes both a value $T^{(h_1, h_2)}[i, j]$, the length of the longest common subsequence so far, and a tree $M^{(h_1, h_2)}[i, j]$ of the initial arc endpoints matched along the computation paths that produce that value. These matches between initial endpoints are tentative assignments that will be checked when the final endpoints of the arcs are encountered.

Step 4. Calculate the longest common arc-preserving subsequence of (S_1, P_1) and (S_2, P_2) by traversing the tables. A table is considered *active* if one arc from each of the chains in its subset is active.

At each step, the values for that position in all active tables are calculated. The trees are also merged and manipulated. This tree data structure needs to support the following operations.

merge: is applied to a finite list of trees and their corresponding subsequence length values, and returns the merge of those trees that have the maximum corresponding values. When the trees are merged, they are copied and also simplified from the root down by uniting identical children of the same parent node.

test: looks for a given arc assignment pair in the tree; returns *true* if it is found, *false* otherwise.

prune: given a tree and an arc assignment pair, removes all paths that do not contain the pair, and then removes the pair itself.

trim: given a tree, an arc number k', and a flag value, removes all nodes in the tree that contain an arc assignment that involves an arc with that number k'. This operation checks either the i or j values, depending on the value of the flag.

extend: given a tree and an arc assignment pair, add the pair to the tree as its new root.

The complexity of each of these operations, except for *extend*, is proportional to the size of the tree, so $\in O(|M^{(h_1, h_2)}[i, j]|)$. The *extend* operation runs in constant time.

These operations are used to keep the trees up to date as the table values are being calculated. The basic longest common subsequence formula

$$T[i, j] = \max(T[i-1, j], T[i, j-1], T[i-1, j-1] + w(S_1[i], S_2[j]))$$

where

$$w(x, y) = \begin{cases} 1 \text{ if } x = y \\ 0 \text{ otherwise} \end{cases}$$

is used to calculate the table values, although it can be changed to any LCS-based alignment weighting scheme. This calculation is varied when arc endpoints are encountered, as follows:

1. When an initial arc endpoint is encountered, all tables that include that arc's chain in its subset are activated and initialized by copying over needed values into the preceding row or column.
2. When a final arc endpoint is encountered, the table without the initial endpoint is calculated normally. The other table, where the initial endpoint was allowed to match, is calculated without matching the final endpoint. These two tables are then *merged* to find the maximum. The trees are *trimmed* to remove all assignments that use that arc.

3. If a pair of initial endpoints is encountered, one from each sequence, the algorithm attempts to match their arcs. The tables are activated and initialized as in 1, but one table – that has both initial endpoints – requires the tree at that position to be *extended* by adding that arc assignment pair.
4. If a pair of final endpoints is encountered, the algorithm must determine (using *test*) if the corresponding pair of initial endpoints are in the tree. If they are, and the maximum value is produced by matching the final endpoints, those endpoints are matched and the tree is *pruned*. Otherwise, the trees and tables are *merged* as in 2.

After the table computation, the decision algorithm returns *true* if and only if the length of the longest common arc-preserving subsequence, stored in $T^{(0,0)}[n, m]$, is at least l, and returns *false* otherwise. This algorithm computes table entries for up to 4^k tables, each table having nm entries.

4.3   Time Complexity Analysis:

The computation of each table entry $M^{h_1, h_2}[i, j]$ takes $O(|M^{h_1, h_2}(i, j)|)$ time. In this algorithm, the trees are kept minimal so that they only store matches of starting endpoints of currently active arcs.
To find the size of $M^{(h_1, h_2)}[i, j]$:

Each path in the tree is a sequence both on matched i values (i') and on matched j values (j'). Let $p_1(i')$ be the position of i' among the starting endpoints of active arcs on S_1, so $p_1(i') = |\{(i_1, i_2) : (i_1, i_2) \in P_1, i_1 \le i \le i_2,$ and $i_1 \le i' \}|$. Similarly, let $p_2(j')$ be the position of j' among the starting endpoints of active arcs on S_2.

At each position in the tree, replace the label (i', j') with $(p_1(i'), p_2(j'))$. The sequences of $p_1(i')$ values and $p_2(j')$ values along any path from root to leaf are strictly decreasing. The maximum number of nodes in such a tree with (x, y) as its root is thus given by the recurrence relation

$$S(1, y) = 1 \quad \forall y \quad \text{and} \quad S(x, 1) = 1 \quad \forall x$$

$$S(x, y) = 1 + \sum_{t=1}^{x-1} \sum_{r=1}^{y-1} S(t, r) \quad \forall y > 1, \forall x > 1$$

which converts to

$$S(x, y) = S(x - 1, y) + S(x, y - 1)$$

and is equivalent to the closed form

$$S(x, y) = \binom{x + y - 2}{x - 1} .$$

Since the root of the entire tree $M^{(h_1, h_2)}[i, j]$ may be blank with all possible children, the number of nodes in $M^{(h_1, h_2)}[i, j]$ is at most $\binom{r+s}{r}$ where r is the number of active arcs from (S_1, P_1) and s is the number of active arcs from (S_2, P_2).

To determine the total size of the trees over all the tables, consider that both sequences have bounded cutwidth k, and that the algorithm is currently computing any specific position $[i, j]$ in each of the tables. Each time the tables are split, half of the tables are allowed to include the new assignment, while the other half cannot include it. So for each possible r and s, the number of tables that can have r active arcs from S_1 and s active arcs from S_2 is $\binom{k}{r} \cdot \binom{k}{s}$. Thus the total number of tree nodes at position $[i, j]$ over all the 4^k tables is at most

$$S(k) = \sum_{r=0}^{k} \sum_{s=0}^{k} \binom{k}{r} \binom{k}{s} \binom{r+s}{s} .$$

This expression can be convoluted to get

$$S(k) = \sum_{t=0}^{k} \binom{k}{t}^2 2^{2t} = \sum_{t=0}^{k} \left(\binom{k}{t} 2^t \right)^2 \leq \left(\sum_{t=0}^{k} \binom{k}{t} 2^t \right)^2 = 9^k ,$$

so the sum over all the tables of the number of tree entries that must be copied is no more than 9^k. An asymptotic estimate for a lower bound on $S(k)$ can be found by looking at one specific term of the sum, where $t = \frac{2k}{3}$. From this term,

$$S(k) \geq \left(\binom{k}{\frac{k}{3}} 2^{\frac{2k}{3}} \right)^2 .$$

This term can be expanded using factorials

$$\left(\frac{k!}{\frac{2k}{3}! \frac{k}{3}!} \cdot 2^{\frac{2k}{3}} \right)^2$$

and estimated using Stirling's approximation $(n! \sim \sqrt{2\pi n}(\frac{n}{e})^n)$ to get

$$\left(\binom{k}{\frac{k}{3}} 2^{\frac{2k}{3}} \right)^2 \sim \frac{9^k}{k} .$$

This approximation reveals that the upper bound of 9^k is very close to $S(k)$; it can be off by at most a factor of k.

Since $S(k)$ is the upper bound on the total size of the trees for each position (r, s) over all tables, and there are nm such positions to be computed, the algorithm runs in time $O(9^k nm)$.   □

5   Conclusion

The examination of this problem using classical complexity shows that it is NP-complete. A parameterized investigation, however, reveals that it is W[1]-complete for desired subsequence length l, but fixed-parameter tractable for bounded cutwidth k. Both hardness results come from a single dual-purpose reduction. The algorithm presented to show fixed-parameter tractability for k runs in time $\in O(9^k nm)$. This time complexity means that if the complexity of the arc structure is bounded by a logarithm of the maximum sequence length n, the longest arc-preserving common subsequence can be found in time $\in O(n^2 m)$. This time complexity is an improvement over earlier results, and shows conditions under which the problem becomes tractable. The algorithm given also handles pseudoknots on both sequences, while previous work does not.

The parameterized analysis indicates that the problem is tractable for sequences with arc structure of bounded cutwidth. Different kinds of structures can be looked at to determine if they meet this restriction, or if they can be manipulated to meet it. Many RNA structures contain highly repetitive arcs. This repetition could be exploited to compress the arc structure into something that has bounded cutwidth. This algorithm can also be extended to work with weights for both match and mismatch for symbols and arcs. The algorithm detects arc mismatch, so it can apply weight penalties for this, and can also use negative weights for arcs. Using negative weights can allow for a small reduction in the score if both symbols of a linked pair are matched; this can alleviate the sometimes overpowering effect of the weights of matched pairs, allowing other matches to have more relative effect on the sequence similarity score. The use of negative arc weights is only possible with an algorithm that preserves induced arcs, such as the one given in this paper.

References

1. V. Bafna, S. Muthukrishnan, and R. Ravi. Computing similarity between RNA strings. *DIMACS Technical Report* **96-30**, 1996.
2. F. Corpet and B. Minchot. RNAlign program: alignment of RNA sequences using both primary and secondary structures. *Computer Applications in the Biosciences* **10** (1994), 389-399.
3. R. Downey and M. Fellows. Fixed-parameter intractability. *Proceedings of the Seventh Annual Conference on Structure in Complexity Theory* (1992), 36-49.
4. H. Lenhof, K. Reinert, and M. Vingron. A polyhedral approach to RNA sequence structure alignment. *Proceedings of the Second Annual International Conference on Computational Molecular Biology (RECOMB 98)* (1998), 153-159.
5. S. Needleman and C. Wunsch. A general method applicable to the search for similarities in the amino-acid sequence of two proteins. *Journal of Molecular Biology* **48** (1970), 443-453.
6. D. Sankoff. Simultaneous solution of the RNA folding, alignment, and protosequence problems. *SIAM Journal of Applied Mathematics* **45** (1985), 810-825.
7. T. Smith and M. Waterman. Identification of common molecular subsequences. *Journal of Molecular Biology* **147** (1981), 195-197.

Computing Similarity between RNA Structures*

Kaizhong Zhang[1], Lusheng Wang[2], and Bin Ma[3]

[1] Dept. of Computer Science, University of Western Ontario,
London, Ont. N6A 5B7, Canada
kzhang@csd.uwo.ca

[2] Dept. of Computer Science, City University of Hong Kong,
83 Tat Chee Avenue, Kowloon, Hong Kong
lwang@cs.cityu.edu.hk

[3] Dept. of Mathematics, Peking University,
Beijing 100871, P.R. China
bma@sxx0.math.pku.edu.cn

Abstract. The primary structure of a ribonucleic acid (RNA) molecule is a sequence of nucleotides (bases) over the alphabet $\{A, C, G, U\}$. The secondary or tertiary structure of an RNA is a set of base-pairs (nucleotide pairs) which forms bonds between $A - U$ and $C - G$. For secondary structures, these bonds have been traditionally assumed to be one-to-one and non-crossing.

This paper considers a notion of similarity between two RNA molecule structures taking into account the primary, the secondary and the tertiary structures. We show that in general this problem is NP-hard for tertiary structures. We present algorithms for the case where at least one of the RNA involved is of secondary structures. We then show that this algorithm might be used to deal with the practical application. We also show an approximation algorithm.

1 Introduction

Ribonucleic Acid (RNA) is an important molecule which performs a wide range of functions in biological system. In particular it is RNA (not DNA) that contains genetic information of virus such as HIV and therefore regulates the functions of such virus. RNA has recently become the center of much attention because of its catalytic properties, leading to an increased interest in obtaining structural information.

It is well known that secondary and tertiary structural features of RNAs are important in the molecular mechanism involving their functions. The presumption, of course, is that to a preserved function there corresponds a preserved molecular confirmation and, therefore, a preserved secondary and tertiary structure. Therefore the ability to compare RNA structures is useful.

In RNA secondary or tertiary structure, a bonded pair of bases (base-pair) is usually represented as an edge between the two complementary bases involved

---

* Research supported partially by the Natural Sciences and Engineering Research Council of Canada under Grant No. OGP0046373.

M. Crochemore, M. Paterson (Eds.): CPM'99, LNCS 1645, pp. 281–293, 1999.

in the bond. It is assumed that any base participates in at most one such pair. For the secondary structure, the edges of the bonded pairs are non-crossing.

Following the notion of similarity in comparing sequences, we define a similarity between two RNA molecule structures taking into account the primary, the secondary and the tertiary structures.

Results

We show that computing this similarity between RNA tertiary structures is NP-complete. We present an algorithm for the case where at least one of the RNA involved is of secondary structure. We then show this algorithm could be used to compare tertiary structures in practical application Finally we will give an approximation algorithm.

Related work

Since the secondary structure appears as tree-like structure, there are works considering comparison using tree comparison [7,4,5,8,3]. However these methods do not directly use base-paired nucleotides and unpaired nucleotides. Instead loops and stems (stacked pairs) are used as the basic unit making it difficult to define the semantic meaning in the process of converting one RNA into another. To overcome this difficulty, the method we propose in this paper directly use base-paired and unpaired nucleotides in the representation and apply some basic operations on them.

Another line of works are primary structure based where the comparison is basically done on the primary structure while trying to incorporate secondary structure data [1,2]. The weakness of this approach is that it does not treat a base-pair as a whole entity. For example, in the comparison of two RNAs, a base-pair from one RNA can have one nucleotide deleted while the other nucleotide matched to nucleotide (unpaired or even paired) in the other RNA. Our method treat base-pair as a unit, it can be matched to another base-pair, it can be deleted, or it can be inserted. This is closer to the spirit of the comparative analysis method currently being used in the analysis of RNA secondary structures either manually or automatically.

2   Comparing Two RNA Structures

2.1   RNA Structures and Basic Operations

The primary structure of a ribonucleic acid (RNA) molecule is a sequence of nucleotides (bases) over the four-letter alphabet $\sum = \{A, C, G, U\}$. The secondary or tertiary structure of an RNA is a set of base-pairs (nucleotide pairs) which formed bonds between $A - U$ and $C - G$. Following Zuker [14,15,16], we assume a model where there is no knots in the secondary structure. This means that for the secondary structure, the bonds are non-crossing. For tertiary structure, there is no restriction of non-crossing.

Given an RNA structure R, we use $R[i]$ to represent the ith nucleotide of R. We use $R[i..j]$ to represent the sequence of nucleotides from $R[i]$ to $R[j]$.

We use $S(R)$ to represent the set of structural elements consisting of both its set of base-pairs and the remaining unpaired nucleotides.

$$S(R) = \begin{array}{l} \{(i,j)|i < j \text{ and } (R[i], R[j]) \text{ is a base pair in } R\} \\ \cup\{(i,i)|R[i] \text{ is not involved in any base pair in } R\} \end{array}$$

We use $S(R)[i..j]$ to represent the set of structural elements in sequence $R[i..j]$.

$$S(R)[i..j] = \{r|r = (k,l) \in S(R), \ i \le k, \ l \le j\}$$

For $r = (i,j) \in S(R)$, we use $label_R(r)$ to represent label of r in R. If $i = j$, then $label_R(r) = R[i] = R[j]$, otherwise $label_R(r) = R[i]R[j]$. For $r = (i,j) \in S(R)$, i and j are often called the $5'$ end and $3'$ end of r. We define $left(r) = i$ and $right(r) = j$.

Following the tradition in sequence comparison [6,9,10], we define three operations, relabel, delete, and insert, on RNA structures. For a given RNA structure R, each operation can be applied to either a base-pair in $S(R)$ or an unpaired base. Relabelling a base-pair is to replace one base-pair in $S(R)$ with another. This means that at the sequence level, two bases may be changed at the same time. Deleting a base-pair is to delete the pair from $S(R)$. At the sequence level, this means to delete two bases at the same time. Inserting a base-pair is to insert a new base-pair into $S(R)$. At the sequence level, this means to insert two bases at the same time. Relabelling an unpaired base is to replace it with another base. Deleting an unpaired base is to delete the base from the sequence. Inserting a base is to insert a new base into the sequence as an unpaired base. Note that there is no relabel operation that can change a base-pair to an unpaired base or vice versa.

Following [11,13], we represent an edit operation as $a \to b$, where a and b are either λ or labels of base-pair from $\{A, C, G, U\} \times \{A, C, G, U\}$, or unpaired base from $\{A, C, G, U\}$.

We call $a \to b$ a change operation if $a \ne \lambda$ and $b \ne \lambda$; a delete operation if $b = \lambda$; and an insert operation if $a = \lambda$.

Let S be a sequence $s_1, ..., s_k$ of edit operations. An S-derivation from RNA structure A to RNA structure B is a sequence of RNA structures $A_0, ..., A_k$ such that $A = A_0$, $B = A_k$, and $A_{i-1} \to A_i$ via s_i for $1 \le i \le k$.

Let γ be a cost function which assigns to each edit operation $a \to b$ a nonnegative real number $\gamma(a \to b)$. We constrain γ to be a distance metric. That is, i) $\gamma(a \to b) \ge 0$, $\gamma(a \to a) = 0$; ii) $\gamma(a \to b) = \gamma(b \to a)$; and iii) $\gamma(a \to c) \le \gamma(a \to b) + \gamma(b \to c)$.

We extend γ to a sequence of edit operations S by letting $\gamma(S) = \sum_{i=1}^{|S|} \gamma(s_i)$.

The *edit distance* between two RNA structures is defined by considering the minimum cost edit operation sequence that transforms one structure to the other. Formally the edit distance between R_1 and R_2 is defined as:

$$D(R_1, R_2) = \min_S \ \{\gamma(T) \mid T \text{ is an edit operation sequence taking } S(R_1) \text{ to } S(R_2)\}$$

2.2 Mapping between RNA Structures

Let $r = (r_l, r_r)$ and $s = (s_l, s_r)$ be two elements in $S(R)$ of an RNA R, we define the relation between r and s as follows. We say r is *before* s if $r_r < s_l$. We say r is *inside* s if $s_l < r_l$ and $r_r < s_r$. We say r is *cross-before* s if $r_l < s_l$ and $r_r < s_r$.

Let R_1 and R_2 be two RNA structures. We define a triple (M, R_1, R_2) to be a mapping from R_1 to R_2, where M is a binary relation on $S(R_1) \times S(R_2)$ such that

(1) For any (r, s) in M,
     r is a base-pair in R_1 if and only if s is a base-pair in R_2.
(2) For any pair of (r_1, s_1) and (r_2, s_2) in M,
     (a) $r_1 = r_2$ if and only if $s_1 = s_2$ (one-to-one)
     (b) r_1 is *before* r_2 if and only if s_1 is *before* s_2.
     (c) r_1 is *inside* r_2 if and only if s_1 is *inside* s_2.
     (d) r_1 is *cross_before* r_2 if and only if s_1 is *cross_before* s_2.

We will use M instead of (M, R_1, R_2) if there is no confusion. Let M be a mapping from R_1 to R_2. Then we can similarly define the cost of M:

$$\gamma(M) = \sum_{(r,s) \in M} \gamma(label_{R_1}(r) \to label_{R_2}(s)) \\ + \sum_{r \notin M} \gamma(label_{R_1}(r) \to \lambda) + \sum_{s \notin M} \gamma(\lambda \to label_{R_2}(s))$$

Mappings can be composed. Let M_1 be a mapping from R_1 to R_2 and M_2 be a mapping from R_2 to R_3. Define

$$M_1 \circ M_2 = \{(r, t) \mid \exists s \text{ s.t. } (r, s) \in M_1 \text{ and } (s, t) \in M_2\}.$$

Lemma 1. *1)* $M_1 \circ M_2$ *is a mapping between* R_1 *and* R_3. *2)* $\gamma(M_1 \circ M_2) \leq \gamma(M_1) + \gamma(M_2)$.

Proof. 1) follows from the definition of mapping. Let us check condition (2) only. Suppose that (r_1, t_1) and (r_2, t_2) are in $M_1 \circ M_2$, by definition of mapping, there exist s_1 and s_2 such that (r_1, s_1) and (r_2, s_2) are in M_1 and (s_1, t_1) and (s_2, t_2) are in M_2. If r_1 is before r_2, then by the definition of mapping, s_1 is before s_2. Therefore t_1 is before t_2, again by the definition of mapping. Similarly if r_1 is inside r_2 or r_1 is cross-before r_2, then if t_1 is inside t_2 or t_1 is cross-before t_2.

2) Let M_1 be the mapping from R_1 to R_2, M_2 be the mapping from R_2 to R_3, and $M_1 \circ M_2$ be the composed mapping from R_1 to R_3. Three general situations occur. $(r, s) \in M_1 \circ M_2$, $r \notin M_1$, or $s \notin M_2$. In each case this corresponds to an edit operation $\gamma(x \to y)$ where x and y may be labels or may be λ. In all such cases, the triangle inequality on the distance metric γ ensures that $\gamma(x \to y) \leq \gamma(x \to z) + \gamma(z \to y)$.   □

The relation between a mapping and a sequence of edit operations is as follows:

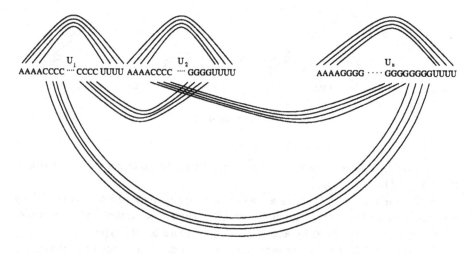

Fig. 1. RNA structure 1

Lemma 2. *Given S, a sequence $s_1, \ldots, s_k$ of edit operations from R_1 to R_2, there exists a mapping M from R_1 to R_2 such that $\gamma(M) \leq \gamma(S)$. Conversely, for any mapping M_e, there exists a sequence of edit operations such that $\gamma(S) = \gamma(M)$.*

Proof. The first part can be proved by induction on k. The base case is $k = 1$. This case holds because any single edit operation preserves the mapping conditions. In general case, let S_1 be the sequence $s_1, \ldots, s_{k-1}$ of edit operations. There exist a mapping M_1 such that $\gamma(M_1) \leq \gamma(S_1)$. Let M_2 be the mapping for s_k. From lemma 1, we have $\gamma(M_1 \circ M_2) \leq \gamma(M_1) + \gamma(M_2) \leq \gamma(S)$.   □

Based on the lemma, the following theorem states the relation between the distance and the mappings.

Theorem 1. $D(R_1, R_2) = \min\limits_{M} \{\gamma(M) \mid M$ *is a mapping from R_1 to $R_2\}$*

Proof. Immediately from lemma 2.   □

3 NP-Hard Result

We now consider the problem of comparing RNA structures where both structures are tertiary structures. We show that this is in general NP-hard.

We will reduce the 3-SAT problem to this problem.

Problem of 3-SAT

Let $S = C_1 \cdot C_2 \ldots C_n$, where $C_i = (v_{i_1} \cup v_{i_2} \cup v_{i_3})$, be an instance of 3-SAT problem. We will construct two RNA structures R_1 and R_2 as in Figure 1 and Figure 2.

In R_1, there are n segments each of which is enclosed by four base pairs. These base pairs are all AU pairs. And each segment is connected to every other

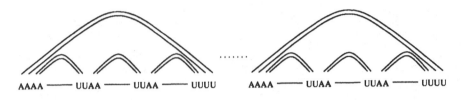

Fig. 2. RNA structure 2

segment by four base pairs of CG type. Note that the number of base pairs in R_1 is $4 \cdot n \cdot (1 + (n-1)/2)$.

In R_2, for each v_{i_k}, there is a corresponding segment which is enclosed by two base pairs of AU type. Each clause C_i is then represented by segments of $v_{i_1}, v_{i_2}, v_{i_3}$ and is enclosed by another two base pairs of AU type.

We now consider base pairs between segments in R_2. For each v_{i_k}, define s_{i_k} as follows.

$$s_{i_k} = \{v_{j_l} | i \neq j \text{ and } v_{j_l} \text{ is not complement of } v_{i_k}\}$$

For each v_{j_l} in s_{i_k} there are four bases in the segment for v_{i_k}. If $j < i$ then these bases are G's, otherwise these are C's. Suppose that v_{j_l} and v_{g_h} are in s_{i_k}, then the bases for v_{j_l} is before the bases for v_{g_h} if either $j > g$ or $j = g$ and $l > h$. Note that if v_{j_l} is in s_{i_k} then v_{i_k} is also in s_{j_l}. Now suppose that $i < j$, then the bases in segment v_{i_k} for v_{j_l} are C's and the bases in segment v_{j_l} for v_{i_k} are G's. In the RNA structure R_2, they form base pairs. Figure 3 shows an example involving two clauses. Let N be the number of base pairs in R_2, then $N = 8 \cdot n + 2 \cdot \sum_{i=1}^{n} \sum_{k=1}^{3} |s_{i_k}|$.

It is clear that R_1 and R_2 can be constructed in polynomial time from an instance of 3-SAT problem S. In the following, we assume that each operation has unit cost. We will show that S can be satisfied if and only if $D(R_1, R_2) = N - 4 \cdot n \cdot (1 + (n-1)/2)$.

Let S be an instance of 3-SAT problem and R_1 and R_2 be as in Figure 1 and Figure 2. The following lemmas give the relationship between S and $D(R_1, R_2)$.

Lemma 3. *If S can be satisfied, then $D(R_1, R_2) = N - 4 \cdot n \cdot (1 + (n-1)/2)$.*

Proof. If S can be satisfied, then for each clause C_i there is at least one v_{i_k} whose value is true. Consider R_2, for each clause, we can first delete any segment which does not correspond to v_{i_k} and its enclosing base pairs. For the segment of v_{i_k}, we can delete the bases which base paired with these segments that have already been deleted. The resulting structure after these deletions is exactly the same as R_1. Therefore $D(R_1, R_2) = N - 4 \cdot n \cdot (1 + (n-1)/2)$ since the number of base pairs in R_2 is N and in R_1 is $4 \cdot n \cdot (1 + (n-1)/2)$. □

Lemma 4. *If $D(R_1, R_2) = N - 4 \cdot n \cdot (1 + (n-1)/2)$, then S can be satisfied.*

$$S = (X+Y+\bar{Z})(\bar{X}+\bar{Y}+Z)$$

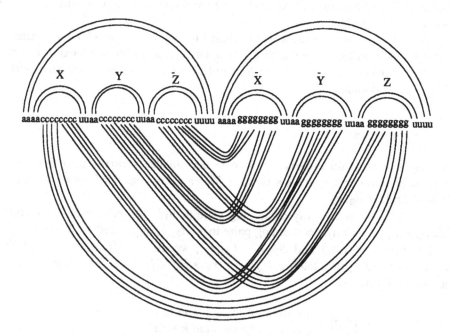

Fig. 3. An example

Proof. In this case, every base pair in R_1 is in the optimal mapping M. In addition, each base pair in R_1 is matched to an identical base pair in R_2. This means that the four base pairs enclosing each segment must map to four base pairs for each clause in R_2. The only possibility for this to happen is that for each clause two (out of three) segments have to be deleted. Therefore mapping M in R_2 keeps one variable in each clause. And this variable is connected to all the variables left in other clauses by means of base pairing. So for any two variables left there is no conflict. Hence we can assign the value true to all the variables left and S is satisfied.   □

Theorem 2. *The problem of determining if $D(R_1, R_2) \leq k$ is NP-complete.*

Proof. This problem is clearly in NP since one can guess a mapping in R_1 and R_2 and check to see if the cost is less or equal to k or not.

By lemma 3 and lemma 4, S is satisfied if and only if $D(R_1, R_2) = N - 4 \cdot n \cdot (1 + (n - 1)/2)$. Therefore this problem is NP-hard.

Hence this problem is NP-complete.   □

4   Algorithms

When both RNAs are secondary structures, since there is no crossing, we can represent RNA structures as ordered forests and then use the tree edit distance algorithm to solve this problem [12,13].

We now consider the case where at most one of the RNA involved is tertiary structure. We present an algorithm which solves this problem An extension of our algorithm can handle the case where both RNAs are tertiary structures with H-type pseudo-knots.

4.1   Properties

We use a bottom-up approach. We consider smaller substructures first and eventually consider the whole structure. We can now consider how to compute $D(R_1[l_1..r_1], R_2[l_2..r_2])$.

Let $S_1[1..m]$ be an array containing pairs in $S(R_1)[l_1..r_1]$ sorted by $3'$ end. Let $S_2[1..n]$ be an array containing pairs in $S(R_2)[l_2..r_2]$ sorted by $3'$ end.

Let $S_1[i] = (s_1, t_1)$ and $S_2[j] = (s_2, t_2)$, we define $left_1[i]$, $cross\_left_1[i]$ and $cross\_weight_1[i]$ as follows. $left_2[j]$, $cross\_left_2[j]$ and $cross\_weight_2[j]$ are defined similarly.

$$left_1[i] = \begin{cases} \max\{k\} \ S_1[k]\text{'s } 3' \text{ end is less than } s_1 \\ 0 \qquad\qquad if \ no \ such \ k \ exist \end{cases}$$

$$cross\_left_1[i] = \begin{cases} 1 \ if \ there \ exist \ a \ k < i, \ such \ that \ S_1[k] \ cross\_before \ S_1[i] \\ 0 \ if \ no \ such \ k \ exist \end{cases}$$

$$cross\_weight_1[i] = \sum_{1 \le k < i, S_1[k] cross\_before S_1[i]} \gamma(label_{R_1}(S_1[i]) \to \lambda)$$

Again let $S_1[i] = (s_1, t_1)$ and $S_2[j] = (s_2, t_2)$, we now define $D_1(i, j)$ and $D_2(i, j)$ as follows.

$$D_1(i, j) = D(R_1[l_1..t_1], R_2[l_2..t_2])$$

$$D_2(i, j) = D(R_1[s_1..t_1], R_2[s_2..t_2])$$

Lemma 5. If $left_1[i] \ne 0$, $left_2[j] \ne 0$, $cross\_left[i] \ne 0$, or $cross\_left[j] \ne 0$, then

$$D_1(i, j) = \min \begin{cases} D_1(i-1, j) & + \gamma(label_{R_1}(S_1[i]) \to \lambda) \\ D_1(i, j-1) & + \gamma(\lambda \to label_{R_2}(S_2[j])) \\ D_1(left_1[i], left_2[j]) + & D_2(i, j) \\ & +cross\_weight_1[i] + cross\_weight_2[j] \end{cases}$$

Proof. Let $S_1[i] = (s_1, t_1)$ and $S_2[j] = (s_2, t_2)$. Consider the best mapping between $R_1[l_1..t_1]$ and $R_2[l_2..t_2]$. If $S_1[i] = (s_1, t_1)$ is not in the mapping, then $D_1(i,j) = D_1(i-1, j) + \gamma(label_{R_1}(S_1[i]) \rightarrow \lambda)$. If $S_2[j] = (s_2, t_2)$ is not in the mapping, then $D(i,j) = D_1(i, j-1) + \gamma(\lambda \rightarrow label_{R_2}(S_2[j]))$. If both $S_1[i] = (s_1, t_1)$ and $S_2[j] = (s_2, t_2)$ are in the mapping, then they should map to each other by the definition of mapping. In this case, since one of the structures is a secondary structure, any base pair cross_before $S_1[i]$ or $S_2[j]$ will not be in the mapping and should be deleted. Therefore, if $left_1[i] \neq 0$, or $left_2[j] \neq 0$, $D(i,j) = D_1(left_1[i], left_2[j]) + D_2(i,j) + cross\_weight_1[i] + cross\_weight_2[j]$. If $left_1[i] = 0$ and $left_2[j] = 0$, and $cross\_left[i] \neq 0$, or $cross\_left[j] \neq 0$, then $D(i,j) = D_2(i,j) + cross\_weight_1[i] + cross\_weight_2[j]$. If we define $D(0,0)=0$, then we can combine the above two cases. Note that one of the cross_weights is zero since in secondary structure, there is no crossing. Also if $S_1[i]$ and $S_2[j]$ are both single bases, both cross_weights are zero. $\qquad\square$

Lemma 6. *If $left_1[i] = 0$, $left_2[j] = 0$, $cross\_left[i] = 0$, and $cross\_left[j] = 0$, then*

$$D_1(i,j) = \min \begin{cases} D_1(i-1,j) & + \gamma(label_{R_1}(S_1[i]) \rightarrow \lambda) \\ D_1(i,j-1) & + \gamma(\lambda \rightarrow label_{R_2}(S_2[j])) \\ D_1(i-1,j-1) & + \gamma(label_{R_1}(S_1[i]) \rightarrow label_{R_2}(S_2[j])) \end{cases}$$

Proof. Let $S_1[i] = (s_1, t_1)$ and $S_2[j] = (s_2, t_2)$. Consider the best mapping between $R_1[l_1..t_1]$ and $R_2[l_2..t_2]$. The first two cases are similar to lemma 5. For the last case, since there is no pair before or cross_before $S_1[i]$ or $S_2[j]$, $S_1[k]$, $1 \leq k < i$, is inside $S_1[i]$ and $S_2[k]$, $1 \leq k < j$, is inside $S_2[j]$. Therefore $D_1(i,j) = D_1(i-1, j-1) + \gamma(label_{R_1}(S_1[i]) \rightarrow label_{R_2}(S_2[j])) + cross\_weight_1[i] + cross\_weight_2[j]$. $\qquad\square$

From the above lemmas, we can compute $D(R_1, R_2)$ using bottom up approach. Moreover, it is clear that we do not need to compute all $D(R_1[l_1..r_1], R_2[l_2..r_2])$. Since we only use $D_2(i,j)$ in lemma 5 and 6, we only need to compute these $D(R_1[l_1..r_1], R_2[l_2..r_2])$ such that (l_1, r_1) is a base pair in R_1 and (l_2, r_2) is a base pair in R_2. Furthermore, by lemma 6, if (l_1, r_1) and (l_1+1, r_1-1) are both base pairs in R_1 and (l_2, r_2) and (l_2+1, r_2-1) are both base pairs in R_2, then we only need to compute $D(R_1[l_1..r_1], R_2[l_2..r_2])$. $D(R_1[l_1..r_1], R_2[l_2+1..r_2-1])$, $D(R_1[l_1+1..r_1-1], R_2[l_2..r_2])$, and $D(R_1[l_1+1..r_1-1], R_2[l_2+1..r_2-1])$ will be by-product of the computation of $D(R_1[l_1..r_1], R_2[l_2..r_2])$.

These base pairs are called stacked pairs. A stem in an RNA R is a set of stack pairs of maximum size. More formally, we say $s = (i, j, k)$ is a stem in $R(S)$ if $(i, j), (i+1, j-1), ...(i+k-1, j-k+1)$ are all base pairs in $R(S)$ and $(i-1, j+1)$ and $(i+k, j-k)$ are not base pairs in $R(S)$.

4.2   Algorithm

Given R_1 and R_2, we can first compute sorted stem lists L_1 for R_1 and L_2 for R_2. It follows from the above discussion that, for each pair of stems $L_1[i] = (i_1, j_1, k_1)$

To compute $D(R_1[i_1, j_1], R_2[i_2, j_2])$

compute a sorted list S_1 of base pairs inside (i_1, j_1);
compute a sorted list S_2 of base pairs inside (i_2, j_2);

compute $left_1[]$ and $left_2[]$;
compute $cross\_left_1[]$ and $cross\_left_2[]$;
compute $cross\_weight_1[]$ and $cross\_weight_2[]$;

$D_1(0,0) = 0$
for $i := 1$ to $|S_1|$
    for $j := 1$ to $|S_2|$
        if $left_1[i] \neq 0$ or $cross\_left_1[i] \neq 0$ or
        $left_2[j] \neq 0$ or $cross\_left_2[j] \neq 0$ then
            Compute $D_1(i, j)$ as in Lemma 5
        else
            Compute $D_1(i, j)$ as in Lemma 6

Fig. 4. Procedure: Computing $D(R_1[i_1, j_1], R_2[i_2, j_2])$

and $L_2[j] = (i_2, j_2, k_2)$, we have to compute $D(R_1[i_1, j_1], R_2[i_2, j_2])$. Figure 5 shows the algorithm. We use lemma 5 and 6 to compute $D(R_1[i_1, j_1], R_2[i_2, j_2])$. Figure 4 shows this computation.

Let $R_1[1..m]$ and $R_2[1..n]$ be the two given RNA structures. Let $stem(R_1)$ and $stem(R_2)$ be the number of stems in R_1 and R_2 respectively. The time compute $D(R_1[i_1, j_1], R_2[i_2, j_2])$ is bounded by $O(|S(R_1)| \times |S(R_2)|)$. Since $|S(R_1)| < m$ and $|S(R_2)| < n$, the time complexity of the algorithm is $O(stem(R_1) \times stem(R_2) \times m \times n)$. The space complexity of the algorithm is $O(|S(R_1)| \times |S(R_2)|) = O(m \times n)$.

Note that when one of the RNA is secondary structure, this algorithm compute the optimal solution of the problem. This algorithm can be modified to handle the case where the input RNAs are tertiary structures with H-type pseudoknots (a stem crosses with at most one other stem).

If we represent the secondary structure by a forest, then by using the technique of Klein [3] we can compute similarity between a secondary structure and a tertiary structure in $O(m^2 n \log n)$ time where $m \leq n$.

Note also that since the number of tertiary interactions is relatively small compared with the number of secondary interactions, we can also use this algorithm to compute the similarity when both structures are tertiary structures. Essentially the algorithm tries to find the best secondary structures to match and delete tertiary interactions. Although this is not an optimal solution, in practice it would produce a reasonable result by matching most of the base pairs. A

Input: $R_1[1..m]$ and $R_2[1..n]$.

Compute a sorted (by $3'$ end) stem list L_1 for R_1.
Compute a sorted (by $3'$ end) stem list L_2 for R_2.

for $i := 1$ **to** $|L_1|$
    for $j := 1$ **to** $|L_2|$
        let $L_1[i] = (i_1, j_1, k_1)$
        let $L_1[j] = (i_2, j_2, k_2)$
        compute $D(R_1[i_1, j_1], R_2[i_2, j_2])$

compute $D(R_1[1, m], R_2[2, n])$

Fig. 5. An algorithm: Computing $D(R_1, R_2)$

post-processing step can be applied to add some matching tertiary interactions which satisfy the mapping constraints.

5 Approximation Algorithms

In this section, we consider a maximization version of the problem. Let M be a mapping from R_1 to R_2. The value $\delta(M)$ of M is defined to be the number of identical pairs of base-pairs in M. Suppose that we define $\gamma(a, b)$ to be 0 if a and b are identical; 2 if a and b are non-identical base-pairs; and 1 if one of them is λ. Then $\delta(M) + \gamma(M) = n_1 + n_2$, where n_1 and n_2 are the number of base-pairs in $S(R_1)$ and $S(R_2)$. Instead of finding a M with the smallest cost $\gamma(M)$, we want to find a M with the largest value $\delta(M)$. Obviously, the maximization version is also NP-complete.

We give an ratio-$(b-1) + \frac{2}{b+1}$ approximation algorithm for the case where each base-pair *crosses* with at most b other base-pairs. Due to space limitation, we only present the basic idea here.

Our *basic idea* is as follows: We start with an arbitrary base-pair (i, j) in $S(R_1)$ and consider (i, j) and the other at most b base-pairs (i_1, j_1), (i_2, j_2), ..., and (i_b, j_b) crossing (i, j) in $S(R_1)$. Call the $b + 1$ base-pairs (i, j), (i_1, j_1), (i_2, j_2), ..., and (i_b, j_b) a *b-component* for $S(R_1)$. We use (i', j'), (i'_1, j'_1), (i'_2, j'_2), ..., and (i'_b, j'_b) to denote a b-component for $S(R_2)$. For each pair of subsequences $R_1[p..q]$ and $R_2[p'..q']$, we consider all pairs of b-components for them. A *match* between the two b-components contains $k + 1$ matched pairs of base-pairs such that (i, j) matches (i', j') and the $k + 1$ matched pairs of base-pairs satisfy (a)-(d) in the definition of a mapping. (i, j) and (i', j') form an *imposed* pair of base-pairs. The $k + 1$ base-pairs form $2(k + 1)$ positions in both $R_1[p..q]$ and $R_2[p'..q']$ that decompose both $R_1[p..q]$ and $R_2[p'..q']$ into $2k + 3$ segments, called *matched* segments. For each pair of b-components for $R_1[p..q]$ and $R_2[p'..q']$, we try all possible matches between the two b-components. For each match, we forbid any other base-pairs not in the b-components to cross any base-pair in the b-components. The match between the corresponding matched segments are computed recursively. (See Figure 6.)

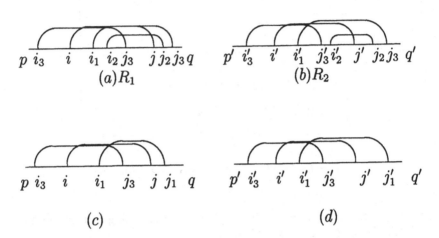

Fig. 6. (a) the set of specified links for R_1. (b) the set of specified links for R_2. (c) the preserved links for R_1 in a match. (d) the preserved links for R_2 in a match. (i, j) matches (i', j') and (i_l, j_l) matches (i'_l, j'_l) for $l = 1$ and 3. Such a match form 7 matched segments for both R_1 and R_2.

References

1. V. Bafna, S. Muthukrishnan, and R. Ravi, 'Comparing similarity between RNA strings', Proc. Combinatorial Pattern Matching Conf. 95, LNCS 937, pp.1-14, 1995
2. F. Corpet and B. Michot, 'RNAlign program: alignment of RNA sequences using both primary and secondary structures', *Comput. Appl. Biosci* vol. 10, no. 4, pp.389-399, 1995
3. P.N. Klein, 'Computing the edit-distance between unrooted ordered trees', Proc. Annual European Symposium on Algorithms 98 LNCS 1461, pp.91-102, 1998.
4. S.Y. Le, R. Nussinov and J.V. Mazel, 'Tree graphs of RNA secondary structures and their comparisons' *Comput. Biomed. Res.* vol. 22, pp.461-473, 1989
5. S.Y. Le, J. Owens, R. Nussinov, J.H. chen, B. Shapiro, and J.V. Mazel, 'RNA secondary structures: comparisons and determination of frequently recurring substructures by consensus', *Comput. Appl. Biosci* vol. 5, pp.205-210, 1989
6. S.E. Needleman and C.D. Wunsch, 'A general method applicable to the search for similarities in the amino-acid sequences of two proteins', *J. Mol. Bio.*, 48, pp.443-453, 1970
7. B. Shapiro, 'An algorithm for comparing multiple RNA secondary structures', *Comput. Appl. Biosci* vol. 4, no. 3, pp.387-393, 1988
8. B. Shapiro and K. Zhang, 'Comparing multiple RNA secondary structures using tree comparisons', *Comput. Appl. Biosci* vol. 6, no.4, pp.309-318, 1990

9. T.F. Smith and M.S. Waterman, 'The identification of common molecular subsequences', *J. Mol. Bio.* 147, pp.195-197, 1981

10. T.F. Smith and M.S. Waterman, 'Comparison of biosequences', *Adv. in Appl. Math.* 2, pp.482-489, 1981

11. K.C. Tai, 'The tree to tree correction problem', *JACM* vol.26, no.3, pp.422-433, 1979

12. Kaizhong Zhang, 'Computing similarity between RNA secondary structures', *Proceedings of IEEE International Joint Symposia on Intelligence and Systems*, Rockville, Maryland, May 1998, pp. 126-132.

13. K. Zhang and D. Shasha, 'Simple fast algorithms for the editing distance between trees and related problems', *SIAM J. Computing* vol. 18, no. 6, pp.1245-1262, 1989

14. M. Zuker, 'On finding all suboptimal foldings from of an RNA molecule', *Science* 244, pp.48-52, 1989

15. M. Zuker and D. Sankoff, 'RNA secondary structure and their prediction', *Bull. Math. Biol.* 46, pp.591-621, 1984

16. M. Zuker and P. Stiegler, 'Optimal computer folding of large RNA sequences using thermodynamics and auxiliary information', *Nucleic Acid Res.* 9, pp.133-148, 1981

Author Index

Springer
and the
environment

At Springer we firmly believe that an
international science publisher has a
special obligation to the environment,
and our corporate policies consistently
reflect this conviction.

We also expect our business partners –
paper mills, printers, packaging
manufacturers, etc. – to commit
themselves to using materials and
production processes that do not harm
the environment. The paper in this
book is made from low- or no-chlorine
pulp and is acid free, in conformance
with international standards for paper
permanency.

Springer

Lecture Notes in Computer Science

For information about Vols. 1–1562
please contact your bookseller or Springer-Verlag

Vol. 1606: J. Mira, J.V. Sánchez-Andrés (Eds.), Foundations and Tools for Neural Modeling. Proceedings, Vol. I, 1999. XXIII, 865 pages. 1999.

Vol. 1607: J. Mira, J.V. Sánchez-Andrés (Eds.), Engineering Applications of Bio-Inspired Artificial Neural Networks. Proceedings, Vol. II, 1999. XXIII, 907 pages. 1999.

Vol. 1608: S. Doaitse Swierstra, P.R. Henriques, J.N. Oliveira (Eds.), Advanced Functional Programming. Proceedings, 1998. XII, 289 pages. 1999.

Vol. 1609: Z. W. Raś, A. Skowron (Eds.), Foundations of Intelligent Systems. Proceedings, 1999. XII, 676 pages. 1999. (Subseries LNAI).

Vol. 1610: G. Cornuéjols, R.E. Burkard, G.J. Woeginger (Eds.), Integer Programming and Combinatorial Optimization. Proceedings, 1999. IX, 453 pages. 1999.

Vol. 1611: I. Imam, Y. Kodratoff, A. El-Dessouki, M. Ali (Eds.), Multiple Approaches to Intelligent Systems. Proceedings, 1999. XIX, 899 pages. 1999. (Subseries LNAI).

Vol. 1612: R. Bergmann, S. Breen, M. Göker, M. Manago, S. Wess, Developing Industrial Case-Based Reasoning Applications. XX, 188 pages. 1999. (Subseries LNAI).

Vol. 1613: A. Kuba, M. Šámal, A. Todd-Pokropek (Eds.), Information Processing in Medical Imaging. Proceedings, 1999. XVII, 508 pages. 1999.

Vol. 1614: D.P. Huijsmans, A.W.M. Smeulders (Eds.), Visual Information and Information Systems. Proceedings, 1999. XVII, 827 pages. 1999.

Vol. 1615: C. Polychronopoulos, K. Joe, A. Fukuda, S. Tomita (Eds.), High Performance Computing. Proceedings, 1999. XIV, 408 pages. 1999.

Vol. 1616: P. Cointe (Ed.), Meta-Level Architectures and Reflection. Proceedings, 1999. XI, 273 pages. 1999.

Vol. 1617: N.V. Murray (Ed.), Automated Reasoning with Analytic Tableaux and Related Methods. Proceedings, 1999. X, 325 pages. 1999. (Subseries LNAI).

Vol. 1618: J. Bézivin, P.-A. Muller (Eds.), The Unified Modeling Language. Proceedings, 1998. IX, 443 pages. 1999.

Vol. 1619: M.T. Goodrich, C.C. McGeoch (Eds.), Algorithm Engineering and Experimentation. Proceedings, 1999. VIII, 349 pages. 1999.

Vol. 1620: W. Horn, Y. Shahar, G. Lindberg, S. Andreassen, J. Wyatt (Eds.), Artificial Intelligence in Medicine. Proceedings, 1999. XIII, 454 pages. 1999. (Subseries LNAI).

Vol. 1621: D. Fensel, R. Studer (Eds.), Knowledge Acquisition Modeling and Management. Proceedings, 1999. XI, 404 pages. 1999. (Subseries LNAI).

Vol. 1622: M. González Harbour, J.A. de la Puente (Eds.), Reliable Software Technologies – Ada-Europe'99. Proceedings, 1999. XIII, 451 pages. 1999.

Vol. 1625: B. Reusch (Ed.), Computational Intelligence. Proceedings, 1999. XIV, 710 pages. 1999.

Vol. 1626: M. Jarke, A. Oberweis (Eds.), Advanced Information Systems Engineering. Proceedings, 1999. XIV, 478 pages. 1999.

Vol. 1627: T. Asano, H. Imai, D.T. Lee, S.-i. Nakano, T. Tokuyama (Eds.), Computing and Combinatorics. Proceedings, 1999. XIV, 494 pages. 1999.

Col. 1628: R. Guerraoui (Ed.), ECOOP'99 - Object-Oriented Programming. Proceedings, 1999. XIII, 529 pages. 1999.

Vol. 1629: H. Leopold, N. García (Eds.), Multimedia Applications, Services and Techniques - ECMAST'99. Proceedings, 1999. XV, 574 pages. 1999.

Vol. 1631: P. Narendran, M. Rusinowitch (Eds.), Rewriting Techniques and Applications. Proceedings, 1999. XI, 397 pages. 1999.

Vol. 1632: H. Ganzinger (Ed.), Automated Deduction – Cade-16. Proceedings, 1999. XIV, 429 pages. 1999. (Subseries LNAI).

Vol. 1633: N. Halbwachs, D. Peled (Eds.), Computer Aided Verification. Proceedings, 1999. XII, 506 pages. 1999.

Vol. 1634: S. Džeroski, P. Flach (Eds.), Inductive Logic Programming. Proceedings, 1999. VIII, 303 pages. 1999. (Subseries LNAI).

Vol. 1636: L. Knudsen (Ed.), Fast Software Encryption. Proceedings, 1999. VIII, 317 pages. 1999.

Vol. 1638: A. Hunter, S. Parsons (Eds.), Symbolic and Quantitative Approaches to Reasoning and Uncertainty. Proceedings, 1999. IX, 397 pages. 1999. (Subseries LNAI).

Vol. 1639: S. Donatelli, J. Kleijn (Eds.), Application and Theory of Petri Nets 1999. Proceedings, 1999. VIII, 425 pages. 1999.

Vol. 1640: W. Tepfenhart, W. Cyre (Eds.), Conceptual Structures: Standards and Practices. Proceedings, 1999. XII, 515 pages. 1999. (Subseries LNAI).

Vol. 1643: J. Nešetřil (Ed.), Algorithms – ESA '99. Proceedings, 1999. XII, 552 pages. 1999.

Vol. 1644: J. Wiedermann, P. van Emde Boas, M. Nielsen (Eds.), Automata, Languages, and Programming. Proceedings, 1999. XIV, 720 pages. 1999.

Vol. 1645: M. Crochemore, M. Paterson (Eds.), Combinatorial Pattern Matching. Proceedings, 1999. VIII, 295 pages. 1999.

Vol. 1647: F.J. Garijo, M. Boman (Eds.), Multi-Agent System Engineering. Proceedings, 1999. X, 233 pages. 1999. (Subseries LNAI).

Vol. 1649: R.Y. Pinter, S. Tsur (Eds.), Next Generation Information Technologies and Systems. Proceedings, 1999. IX, 327 pages. 1999.

Vol. 1650: K.-D. Althoff, R. Bergmann, L.K. Branting (Eds.), Case-Based Reasoning Research and Development. Proceedings, 1999. XII, 598 pages. 1999. (Subseries LNAI).

Vol. 1651: R.H. Güting, D. Papadias, F. Lochovsky (Eds.), Advances in Spatial Databases. Proceedings, 1999. XI, 371 pages. 1999.

Vol. 1653: S. Covaci (Ed.), Active Networks. Proceedings, 1999. XIII, 346 pages. 1999.

Vol. 1654: E.R. Hancock, M. Pelillo (Eds.), Energy Minimization Methods in Computer Vision and Pattern Recognition. Proceedings, 1999. IX, 331 pages. 1999.

Vol. 1663: F. Dehne, A. Gupta. J.-R. Sack, R. Tamassia (Eds.), Algorithms and Data Structures. Proceedings, 1999. X, 367 pages. 1999.